Ordinary National Certificate Mathematics

Volume II

by the same author

ORDINARY NATIONAL CERTIFICATE MATHEMATICS
Volume I

Ordinary National Certificate Mathematics
Volume II

H. A. HORNER
M.A. (Cantab), F.I.M.A.

Head of Department of Mathematics
Luton College of Technology

HEINEMANN EDUCATIONAL BOOKS LTD · LONDON

Heinemann Educational Books Ltd
LONDON   MELBOURNE   TORONTO
SINGAPORE   JOHANNESBURG
HONG KONG   NAIROBI
AUCKLAND   IBADAN

SBN   435 71081 8

Published by
Heinemann Educational Books Ltd
48 Charles Street, London W.1
Printed in Great Britain at the
Pitman Press, Bath

Preface

This is the second volume of a two-volume series, intended to cover the syllabuses of the new Ordinary National Certificate schemes.

The presentation of the text is similar in form to that adopted in Volume I. Each chapter presents a pattern such that the amount of theory can be dealt with in a period of $2\frac{1}{2}$ hours. Clearly, if class periods are longer it will be possible to absorb the material more comfortably.

Students are again reminded that, in order to acquire a proper understanding of the subject and its applications it is necessary to learn a number of sound basic rules and methods. If these are practised using the examples provided, the interpretation and break down of practical problems will be less difficult. The number of unworked examples provided in each chapter is not excessive so that it should be possible for students to do most of these examples.

It will be found that, provided all the unworked examples are attempted conscientiously, the necessary skills in analysis, manipulation and computation will be acquired.

Throughout the text, emphasis is on correct methods of computation, correct methods for the transposition of formulae, logical layout of solutions, neatness and clarity of arrangement of material and the systematic use of all the normal mathematical and other tables found in books of standard tables.

It cannot be stressed too highly that students who are seriously concerned to develop their mathematical approach and skills must work steadily at the subject. No scientific or engineering subject can be fully comprehended and satisfactorily studied without a sound mathematical background.

As the second book in the series is being written near the beginning of the new National Certificate schemes it should be appreciated that the greatest importance is laid on methods of analysis and computation. The questions used are of suitable standard, but the usual practice of quoting from public examination bodies and Technical College examinations is not followed.

The final chapter of the book provides a fair number of revision examples which cover the subject matter in sequence. Because the style of examination papers tends to vary between the different schemes, no specimen examination papers are provided.

Some of the material included in the text will not be found in other books for Ordinary National Certificate schemes. As technology has developed it has become necessary for an earlier appreciation of certain topics to be obtained.

The development of high speed computing facilities is revolutionizing the methods of calculation necessary in industrial, scientific, and commercial applications of mathematics. Extensive use is now made of slide rules, calculators, and high speed digital and analogue computers. Because of these developments in high speed calculation and analysis it becomes necessary to understand some of the methods of calculation and analysis which are particularly applicable to these rapid methods of calculation and analysis.

These developments explain why this volume puts more accent on numerical analysis, iterative methods, statistical computation, and differential equations.

It is sincerely hoped that this volume will help to give students a good grounding in old and new techniques. There is no doubt that these techniques will be developed even further in Higher National Certificate courses.

The majority of the material is based on teaching, by the author to ONC, HNC, and other courses over a long period, and it is hoped that the experience will be helpful to both students and lecturers.

I would like to thank Biometrika for permission to include the table of the Normal Integral in the Appendix.

I must also express sincere thanks to John Chambers and Hamish MacGibbon of Heinemann's for their excellent advice and help during the preparation of both Volume I and Volume II. Also I wish to thank G. Hartfield for his excellent draughtsmanship. The presentation of text is largely due to their advice and efforts.

A few errors discovered in the first printing have been corrected in the second impression, however I will be grateful to receive notification of any further corrections.

H. A. HORNER

Contents

1

Logarithms and Exponential Equations

1.1 Logarithms

If a number N is expressed in the form a^x, i.e. $N = a^x$ then x may be called the logarithm to base a of N, i.e. $x = \log_a N$.

The base a may be any positive number, but for practical purposes, for the tabulation of logarithms, a is usually 10 or e ($= 2 \cdot 7183$).

Common logarithms are logarithms to base 10, *Natural (Napierian or hyperbolic) logarithms* are logarithms to base e.

1.1.1 Rules of logarithms

These rules are applicable when any base is used. The rules have previously been derived (Volume I) using the *laws of indices* and are only recorded here for reference. (The base is omitted.)

(i) *Multiplication:* $\log(ABC) = \log A + \log B + \log C$

(ii) *Division:* $\log\left(\dfrac{A}{B}\right) = \log A - \log B$

(iii) *Powers:* $\log a^x = x \log a$

(iv) *Roots:* $\log \sqrt[n]{a} = \dfrac{1}{n} \log a$

Basically logarithms are only employed for the four operations shown and these rules, if correctly applied, are sufficient for the solution of most problems.

1.2 Antilogarithms

These are values of the appropriate base raised to the power of the logarithm. Separate tables are not really necessary to find such values, and indeed, when 6, 7 or, 9 figure tables are used it may be found that antilogarithms are not provided.

Examples:

(i) Antilog$_{10}$ $1.8723 = 10^{1.8723} = \underline{74.52}$

  (found from four figure logarithm tables).

(ii) Antilog$_e$ $3.8762 = e^{3.8762} = \exp(3.8762)$
  $$= 10^1 \times 4.824 = \underline{48.24}$$

  (found from four figure *hyperbolic* or *Napierian* logarithm tables—no antilogarithm tables provided).

1.2.1 Evaluation of a^x

Example: Find the value of $(0.387)^{1.42}$.

Let $\qquad N = (0.387)^{1.42}$, $\log_{10} N = 1.42 \log_{10} 0.387$

$\therefore \quad \log_{10} N = 1.42 \times \bar{1}.5877 = 1.42 \times (-0.4123)$
$$= -1.42 \times 0.4123 = -0.5854 = \bar{1}.4146$$

(*Note:* Logarithms are only used to evaluate 1.42×0.4123.)

$\therefore \qquad N = $ antilog $\bar{1}.4146 = \underline{0.2598}$.

1.3 Change of Base of Logarithms

Sometimes it is necessary to calculate logarithms to bases other than 10. In order to perform such a calculation the following *change of base* formula is derived.

Let $\qquad N = a^x$ and consider a second base b,

i.e. $\qquad N = a^x = b^y$, then $\log_a N = x$, $\log_b N = y$.

Taking logarithms to base a

$$\log_a N = x \log_a a = y \log_a b = (\log_b N) \times \log_a b$$
$$\therefore \quad \log_b N = \log_a N \div \log_a b$$

or $\log_b N = \dfrac{\log_a N}{\log_a b}$

LOGARITHMS AND EXPONENTIAL EQUATIONS

3

Examples:

(i) Calculate $\log_5 17 \cdot 28$.

Using the formula $a = 10$, $b = 5$, $N = 17 \cdot 28$,

$\therefore$ $\log_5 17 \cdot 28 = (\log_{10} 17 \cdot 28)/(\log_{10} 5) = 1 \cdot 2375/0 \cdot 6990 = \underline{1 \cdot 77}$

(*Note:* Interpolation used to obtain $\log_{10} 1 \cdot 2375$.)

(ii) Calculate $\log_e 53 \cdot 7$ (e = $2 \cdot 718$).

$\log_e 53 \cdot 7 = (\log_{10} 53 \cdot 7)/(\log_{10} 2 \cdot 718)$
$= 1 \cdot 7300/0 \cdot 4343 = 3 \cdot 983$

Napierian logarithms are dealt with in more detail in 1.5.

1.4 Exponential Equations

These equations usually involve the unknown quantity in the exponents (or indices). This type of equation can only be solved by the use of logarithms.

Examples:

(i) Solve the equation $3^{2x-1} = 5^{x+2}$.

Since the equation states the equality of two numbers, the logarithms of these numbers must also be equal.

$\therefore$ $(2x - 1) \log 3 = (x + 2) \log 5$ (base 10 used)
$(2x - 1) \times 0 \cdot 4771 = (x + 2) \times 0 \cdot 6990$
(*Note:* Use of brackets.)
$0 \cdot 9542x - 0 \cdot 4771 = 0 \cdot 6990x + 1 \cdot 3980$
$x(0 \cdot 9542 - 0 \cdot 6990) = 1 \cdot 3980 + 0 \cdot 4771$
$0 \cdot 2552x = 1 \cdot 8751$
$\therefore$ $x = (1 \cdot 8751)/(0 \cdot 2552) = \underline{7 \cdot 347}$

(ii) If the law connecting two variables x and y is of the form $y = ax^n$ and corresponding numerical values are found as follows: $x = 5$, $y = 30 \cdot 2$; $x = 7 \cdot 5$, $y = 9 \cdot 3$, estimate values for a and n.

If $y = ax^n$, then
$$\log y = \log a + n \log x \qquad (1)$$

Substituting the numerical values:

$$\log 30 \cdot 2 = \log a + n \log 5 \tag{2}$$
$$\log 9 \cdot 3 = \log a + n \log 7 \cdot 5 \tag{3}$$
$$1 \cdot 4800 = \log a + n \times 0 \cdot 6990 \tag{4}$$
$$0 \cdot 9685 = \log a + n \times 0 \cdot 8751 \tag{5}$$

Subtracting equation (4) from (5)

$$-0 \cdot 5115 = 0 \cdot 1761n; \quad n = (-0 \cdot 5115)/(0 \cdot 1761) = \underline{-2 \cdot 904}$$

from equation (4):

$$\log a = 1 \cdot 4800 - 0 \cdot 6990n$$
$$\therefore \qquad \log a = 1 \cdot 4800 - 0 \cdot 6990 \times (-2 \cdot 904)$$
$$= 1 \cdot 4800 + 0 \cdot 6990 \times 2 \cdot 904$$
$$= 1 \cdot 4800 + 2 \cdot 0300 = 3 \cdot 5100$$
$$a = \text{antilog } 3 \cdot 5100 = \underline{3236}$$

$\therefore$ law is of the form $\underline{y = 3236x^{-2 \cdot 904}}$

1.5 Napierian Logarithms

The base of these logarithms is e (natural base). In a number of practical problems this base appears and it is often necessary to calculate, or find, a Napierian logarithm. The value of e may be found from the natural series:

$$e = 1 + \frac{1}{1!} + \frac{1}{2!} + \frac{1}{3!} + \frac{1}{4!} + \ldots \text{ to infinity}$$

$$= 2 \cdot 000\ 0000 + 0 \cdot 500\ 0000 + 0 \cdot 166\ 6667 + 0 \cdot 041\ 6666$$
$$+ 0 \cdot 008\ 3333 + 0 \cdot 001\ 3888 + 0 \cdot 000\ 1984 + 0 \cdot 000\ 0248$$

$$= 2 \cdot 718\ 2786 = 2 \cdot 7183 \text{ to 5 s.f.} = \underline{2 \cdot 718} \text{ to 4 s.f.}$$

Logarithms to this base may be calculated using the following change of base formula:

$$\log_e N = (\log_{10} N)/(\log_{10} e) = \log_{10} N/0 \cdot 4343 = \underline{2 \cdot 3026 \log_{10} N}$$

Examples:

(i) $\log_e 59 \cdot 4 = 2 \cdot 3026 \log_{10} 59 \cdot 4 = 2 \cdot 3026 \times 1 \cdot 7738$
$\qquad = \underline{4 \cdot 086}$ (using logarithms to multiply)

(ii) $\log_e 0.3874 = 2.3026 \log_{10} 0.3874 = 2.3026 \times \bar{1}.5882$
$= -2.3026 \times 0.4118 = \underline{-0.9483}$

(This could be written as $\bar{1}.0517$ but the characteristic does **not** have the same significance as for base 10 logarithms.)

(iii) antilog$_e$ $1.8764 = e^{1.8764} = N$
$\log_{10} N = 1.8764 \times \log_{10} e = 1.8764 \times 0.4343 = 0.8149$
$\therefore \quad N = \text{ant}_{10}\ 0.8149 = \underline{6.530}$

1.5.1 Use of Napierian tables

Logarithms and antilogarithms to base e may be read from these tables. This is illustrated by the following examples:

(i) $\log_e 7.286 = 1.9859$ (read directly from the tables)
(ii) $\log_e 59.73 = \log_e (10 \times 5.973) = \log_e 10 + \log_e 5.973$
$= 2.3026 + 1.7872 = \underline{4.0898}$

(iii) $\log_e 0.00785 = \log_e (7.85 \div 10^3) = \log_e 7.85 - \log_e 10^3$
$= 2.0605 - 6.9078 = \underline{-4.8473}\,(\bar{5}.1527)$

or $\quad \log_e 0.00785 = \log_e (7.85 \times 10^{-3}) = \log_e 7.5 + \log_e 10^{-3}$
$= 2.0605 + \bar{7}.0922 = \underline{\bar{5}.1527}\,(-4.8473)$

(iv) antilog$_e$ 3.2876; subtract $\log_e 10 = 2.3026$
$3.2876 - 2.3026 = 0.9850$

The antilog$_e$ of this is found to be 2.678. Since $\log_e 10$ was subtracted, this is equivalent to dividing by 10. Hence

antilog$_e$ $3.2876 = 10 \times 2.678 = \underline{26.78}$

(v) antilog$_e$ (-4.6287); add 6.9078 ($\log_e 10^3$)
$-4.6287 + 6.9078 = 2.2791$
antilog$_e$ $2.2791 = 9.768$
$\therefore \quad$ antilog$_e$ $(-4.6287) = 9.768 \div 10^3 = \underline{0.009768}$

(vi) antilog$_e$ $(\bar{5}.3862)$; subtract $\bar{7}.0922$ ($\log_e 10^{-3}$)
$\bar{5}.3862 - \bar{7}.0922 = 2.2940$
antilog$_e$ $2.2940 = 9.915$
$\therefore \quad$ antilog$_e$ $(\bar{5}.3862) = 9.915 \times 10^{-3} = \underline{0.009915}$

It will be noticed that the 'characteristics' involved are logarithms to base e of 10, 10^2, 10^3, etc., and are integral multiples of 2·3026.

With sufficient practice it will be found that these tables are more convenient for obtaining:

(1) $\log_e$ (numbers),   (2) values of e^x (antiloge x),
(3) values of e^{-x} (x positive).

WORKED EXAMPLES

(a) If $2 \log x + \log (3y - 1) = 3 + 4 \log y - \frac{1}{2} \log z$ (base 10), express x in terms of y and z.

Using the rules of logarithms this equation may be written as:
$\log x^2 + \log (3y - 1) = \log 10^3 + \log y^4 - \log \sqrt{z}$.

$$\log [x^2(3y - 1)] = \log \left(\frac{10^3 \times y^4}{\sqrt{z}} \right)$$

Hence $x^2(3y - 1) = \dfrac{1{,}000y^4}{\sqrt{z}}$; $x^2 = \dfrac{1{,}000y^4}{(3y - 1)\sqrt{z}}$

$$\therefore \qquad x = \left(\frac{1{,}000y^4}{(3y - 1)\sqrt{z}} \right)^{\frac{1}{2}} = \frac{10\sqrt{10}y^2}{\sqrt{(3y - 1)z^{\frac{1}{2}}}}$$

(b) If $pv^n = c$, evaluate c given $p = 3{,}000$, $v = 1\cdot7$, $n = 1\cdot38$.
$\log c = \log p + n \log v = \log 3{,}000 + 1\cdot38 \log 1\cdot7$
$\qquad = 3\cdot4771 + 1\cdot38 \times 0\cdot2304 = 3\cdot4771 + 0\cdot3179 = 3\cdot7950$
$\therefore \quad c = \text{antilog}_{10}\, 3\cdot7950 = \underline{6{,}237}$

(c) Solve the equation $3\cdot1^{x^2-1} = 5^{0\cdot5x}$.

$$(x^2 - 1) \log 3\cdot1 = 0\cdot5x\,(\log 5)$$
$$(x^2 - 1) \times 0\cdot4914 = x(0\cdot5 \times 0\cdot6990) = 0\cdot3495x$$
$$\therefore \qquad (x^2 - 1) = \frac{0\cdot3495}{0\cdot4914}\, x \qquad = 0\cdot711x$$
$$\underline{x^2 - 0\cdot711x - 1 = 0}$$

$x = [0\cdot711 \pm \sqrt{(0\cdot711^2 + 4)}]/2 = [0\cdot711 \pm \sqrt{(0\cdot5056 + 4)}]/2$
$\quad = (0\cdot711 \pm \sqrt{4\cdot5056})/2 \qquad = (0\cdot711 \pm 2\cdot122)/2$
$\quad = 2\cdot833/2 \text{ or } -1\cdot411/2 \qquad = \underline{1\cdot4165 \text{ or } -0\cdot7055}$

(d) The voltage V volts in a circuit at time t sec from zero is given in terms of the initial voltage E by the formula $V = Ee^{-t/CR}$.

Transpose the formula to find t and hence calculate the time required for the voltage to fall from 12 to 8 volts given $C = 5 \times 10^{-6}$, $R = 2 \cdot 2 \times 10^6$.

$$V = E/e^{t/CR}, \qquad e^{t/CR} = E/V,$$

$$\frac{t}{CR} = \log_e\left(\frac{E}{V}\right), \qquad \therefore \quad t = CR \log_e\left(\frac{E}{V}\right)$$

Using the given values

$$t = 5 \times 10^{-6} \times 2 \cdot 2 \times 10^6 \log_e (12/8) = 11 \log_e 1 \cdot 5$$
$$= 11 \times 0 \cdot 4055 = 4 \cdot 4605 \text{ sec} = \underline{4 \cdot 46 \text{ sec to 3 s.f.}}$$

Examples 1

1. The displacement s in a damped vibration is given by the formula $s = a\,e^{-Kt}\sin(nt)$ where t is the time in seconds. Find the value of s when $a = 3 \cdot 5$, $K = 0 \cdot 4$, $n = 2$, $t = 0 \cdot 75$.
2. During a resisted motion, the initial velocity u ft/s, final velocity v ft/s and time t sec are connected by the formula $t = \dfrac{1}{K}\log_e\left(\dfrac{g - Ku}{g - Kv}\right)$.
   Calculate the time for the velocity to increase from $u = 20$ ft/s to $v = 120$ ft/s, given $K = 0 \cdot 15$, $g = 32 \cdot 2$.
3. The current i in an electrical circuit is given by the formula
   $$i = I\,e^{-Rt/L}\sin 200\,\pi t$$
   Calculate i when $I = 50$, $R = 5$, $L = 0 \cdot 05$, $t = 0 \cdot 004$ (angle in radians).
4. The law connecting the pressure and volume of a gas is $pv^\gamma = c$. Find the constant c given the observed values $p = 15$, $v = 6 \cdot 6$, $\gamma = 1 \cdot 5$.
5. The percentage efficiency η of a compressor is given by the formula
   $$\eta = \frac{100 \log_e r}{\dfrac{2n}{n-1}\left[r^{(n-1)/n} - 1\right]}.$$
   Calculate the efficiency when $r = 7 \cdot 4$, $n = 1 \cdot 4$.
6. Solve the following equations:
   (i) $3^{2x-1} = e^{x+1}$, (ii) $5^x \times 7^{2x} = 15$, (iii) $(0 \cdot 7)^x = 0 \cdot 4$.
7. Solve the following equations:
   (i) $9^{x^2} = 3^{2x+6}$, (ii) $5^x = 3^{x^2-3}$.
8. The formula $T_2 = T_1 e^{\mu\theta}$ is used in connection with belt drives, where T_1 and T_2 are tensions, μ the coefficient of friction and θ the angle of lap in radians. Determine the angle of lap in degrees when $T_1 = 8$ lbf, $T_2 = 14$ lbf and $\mu = 0 \cdot 4$.

9. Transpose the formula $I = \dfrac{V}{R}(1 - e^{-Rt/L})$ to find t and hence calculate the value of t when $R = 300$, $I = 70$, $L = 7\cdot5$, $V = 32{,}500$.

10. (i) Solve the equation $\log_e\left(\dfrac{x^2 + 1}{2x}\right) = 0\cdot787$.

    (ii) Given $W = nB^x$, find the value of x when $W = 4{,}300$, $B = 11{,}900$, $n = 0\cdot001\ 296$.

11. A gas expands adiabatically according to the law $pv^n = c$. The work W done by the gas when it expands from volume v_1 at pressure p_1, to volume v_2 at pressure p_2 is given by $W = \dfrac{c}{n-1}\left(\dfrac{1}{v_1^{\,n-1}} - \dfrac{1}{v_2^{\,n-1}}\right)$.

    Show that this is equivalent to $W = \dfrac{1}{n-1}(p_1v_1 - p_2v_2)$.

    Find the work done when $c = 147$, $n = 1\cdot37$, $v_1 = 0\cdot8$, $v_2 = 9$.

12. (i) Evaluate $p_1v_1 \log_e(v_2/v_1)$ when $p_1 = 50 \times 12^2$, $v_1 = 3$, $v_2 = 10$.

    (ii) Evaluate R if $R = t/[K\log_e(v_0/v_1)]$ when $t = 20$, $K = 0\cdot75 \times 10^{-6}$, $v_0 = 33\cdot6$, $v_1 = 20$.

13. (i) Express x in terms of p, q, and r if

    $\log x = 3\log p + \tfrac{1}{2}\log q - \log r + 3\log 10$ (base 10).

    (ii) If a is an integer and $\log_a 192\cdot8 = 3\cdot269$, find a.

14. If $2\log_e M = 2 + \log_e N - \tfrac{1}{3}\log_e P$ expresses M in terms of e, N and P.

15. If $f = \dfrac{P}{\pi t^2}\left[\dfrac{4}{3}\log_e\left(\dfrac{D}{d}\right) + 1\right]$, calculate the value of D when $P = 1{,}600$, $t = 0\cdot56$, $d = 1\cdot4$ and $f = 5{,}060$.

16. The tensions T_1 and T_2 at the ends of a rope passing round a circular post are related by the formula $T_2 = T_1\,e^{\mu\theta}$ where $\mu = 0\cdot22$ and θ is in radians. Find (i) θ given $T_2 = 3T_1$, (ii) T_1 and T_2 given $\theta = 3\cdot1$, $T_2 - T_1 = 745$.

17. Evaluate A and B from the formulae $A/B = e^{K\theta}$, $A - B = 550$ given $K = 0\cdot25$, $\theta = 2\cdot9$.

18. M and t are related by the formula $M = a\,e^{bt}$. Find the values of a and b given $M = 13\cdot0$, when $t = 3\cdot1$, $M = 28\cdot6$ when $t = 5\cdot9$.

19. The relation between pressure p and volume v of a gas is $pv^n = c$. Determine the values of n and c given $p = 200$ when $v = 0\cdot8$, $p = 12\cdot6$ when $v = 6$.

20. The values of current i amp required to fuse a wire of diameter d thousandths of an inch are given by the law $i = Ad^n$. Find the constants A and n given $i = 2\cdot1$ when $d = 22$, $i = 5\cdot8$ when $d = 36$.

2

Quadratic Equations—Quadratic Functions—Miscellaneous Equations

2.1 Quadratic Equations (One Variable)

The general equation may be written as $ax^2 + bx + c = 0$, where a is the coefficient of x^2, b the coefficient of x, and c the constant term. In general the equation will have two *roots* (or *solutions*), which may be (i) real and different, (ii) real and equal, (iii) complex (or imaginary).

The various methods of solution have been dealt with adequately in Volume I, but they are summarized here:

Methods of Solution:

(i) Using factors. (ii) Using completion of the square. (iii) Using the formula. (iv) Using graphical methods.

Generally it will be found that methods (ii) and (iii) are the most widely used. The *formula* is derived by applying method (ii) as follows:

$$ax^2 + bx + c = 0; \text{ divide through by } a$$

$$x^2 + \frac{b}{a}x + \frac{c}{a} = 0; \quad x^2 + \frac{b}{a}x = -\frac{c}{a}$$

Add $\left(\frac{1}{2}\frac{b}{a}\right)^2 = \frac{b^2}{4a^2}$ to each side, then

$$x^2 + \frac{b}{a}x + \frac{b^2}{4a^2} = \frac{-c}{a} + \frac{b^2}{4a^2} = \frac{b^2 - 4ac}{4a^2}$$

$$\left(x + \frac{b}{2a}\right)^2 = \frac{b^2 - 4ac}{4a^2}$$

take the square root of each side,

$$x + \frac{b}{2a} = \pm \sqrt{\left(\frac{b^2 - 4ac}{4a^2}\right)} = \frac{\pm \sqrt{(b^2 - 4ac)}}{2a}$$

$$\therefore x = \frac{-b \pm \sqrt{(b^2 - 4ac)}}{2a}$$

2.1.1 Discriminant of the quadratic equation

$\Delta = b^2 - 4ac$ (the *discriminant*) may be used to determine the *types* of roots of the equation.

       Real and unequal roots: $b^2 - 4ac$ *positive*, or $\Delta > 0$.
       Real and equal roots: $b^2 - 4ac = 0$, or $\Delta = 0$.
       Complex roots: $b^2 - 4ac$ *negative*, or $\Delta < 0$.

In addition, when Δ is a simple square the roots will be rational and the equation factorizes into two simple linear factors.

2.1.2 Relations between the roots

Let the roots of the equation be α and β, then the *function* may be *factorized* as

$$ax^2 + bx + c = a\left(x^2 + \frac{b}{a}x + \frac{c}{a}\right) = a(x - \alpha)(x - \beta)$$

$$\therefore \qquad (x - \alpha)(x - \beta) = x^2 + \frac{b}{a}x + \frac{c}{a} \text{ is an } identity$$

or $\qquad x^2 - (\alpha + \beta)x + \alpha\beta = x^2 + \frac{b}{a}x + \frac{c}{a} \text{ is an } identity$

Equating corresponding *coefficients*

$$-(\alpha + \beta) = \frac{b}{a}, \quad \text{or} \quad \alpha + \beta = -\frac{b}{a} \quad \text{and} \quad \alpha\beta = \frac{c}{a}$$

or, in general, for a quadratic equation:

$$\alpha + \beta = \text{sum of roots} = -\frac{b}{a} = -\frac{\text{Coefficient of } x}{\text{Coefficient of } x^2}$$

$$\alpha\beta = \text{product of roots} = \frac{c}{a} = \frac{\text{Constant term}}{\text{Coefficient of } x^2}$$

These expressions enable new equations to be formed with roots which are *symmetrical* functions of α and β. This may be done without the actual determination of the roots.

Examples:

(i) Solve the equation $3x^2 + 5x - 1 = 0$.

$$\Delta = b^2 - 4ac = 5^2 - 4(3)(-1) = 25 + 12 = 37$$

$$x = \frac{-5 \pm \sqrt{37}}{6} = \frac{-5 \pm 6 \cdot 083}{6} = \frac{1 \cdot 083}{6} \text{ or } \frac{-11 \cdot 083}{6}$$

$$\therefore \quad x = 0 \cdot 1805 \text{ or } -1 \cdot 8472$$

(ii) Solve the equation $5x^2 + 2x + 1 = 0$.

$$\Delta = b^2 - 4ac = 2^2 - 4(5)(1) = 4 - 20 = -16$$

∴ the roots will be *complex*.

$$x = \frac{-2 \pm \sqrt{(-16)}}{(2)(5)} = \frac{-2 \pm 4\sqrt{-1}}{10}; \text{ let } j = \sqrt{-1}$$

then

$$x = \frac{-2 \pm 4j}{10} = -0 \cdot 2 \pm 0 \cdot 4j$$

(iii) If the roots of the equation $2x^2 + 3x - 5 = 0$ are α and β, form the equation whose roots are α^2 and β^2.

The equation may be written as

$$(x - \alpha^2)(x - \beta^2) = 0 \quad \text{or} \quad x^2 - (\alpha^2 + \beta^2)x + \alpha^2\beta^2 = 0$$

Now $\alpha + \beta = -\frac{3}{2}, \alpha\beta = -\frac{5}{2}$,

$$\alpha^2 + \beta^2 = (\alpha + \beta)^2 - 2\alpha\beta = \left(-\frac{3}{2}\right)^2 - 2\left(-\frac{5}{2}\right) = \frac{9}{4} + 5 = \frac{29}{4}$$

$$\alpha^2\beta^2 = (\alpha\beta)^2 = \left(-\frac{5}{2}\right)^2 = \frac{25}{4}$$

∴ the new equation is

$$x^2 - \frac{29}{4}x + \frac{25}{4} = 0$$

or

$$4x^2 - 29x + 25 = 0$$

(iv) If $2x^2 + 4x - 3 = 0$ has roots α and β, find, without solving the equation, the values of $\alpha^2 + \beta^2$, $\alpha^3 + \beta^3$, and $(\alpha - \beta)^2$.

$$\alpha + \beta = -\frac{4}{2} = -2; \quad \alpha\beta = -\frac{3}{2}$$

$$\alpha^2 + \beta^2 = (\alpha + \beta)^2 - 2\alpha\beta = (-2)^2 - 2\left(-\frac{3}{2}\right) = 4 + 3 = \underline{7}$$

$$\alpha^3 + \beta^3 = (\alpha + \beta)(\alpha^2 + \beta^2 - \alpha\beta) = (-2)\left[7 - \left(-\frac{3}{2}\right)\right]$$
$$= -\underline{17}$$

$$(\alpha - \beta)^2 = \alpha^2 + \beta^2 - 2\alpha\beta = 7 - 2(-\tfrac{3}{2}) = \underline{10}$$

2.2 Quadratic Functions

Consider the function $f(x) = ax^2 + bx + c$. The sign of the function may vary as x varies.

(i) Let the roots of $ax^2 + bx + c = 0$ be *real* and *different* (α and β), i.e. $\Delta > 0$,

$f(x) = a(x - \alpha)(x - \beta)$. Let α be less than β, i.e. $\alpha < \beta$.

When $x < \alpha$, $(x - \alpha)$ and $(x - \beta)$ are both *negative*

$\therefore \qquad\qquad f(x)$ has the same sign as a.

When $\alpha < x < \beta$, $(x - \alpha)$ and $(x - \beta)$ have *opposite signs*

$\therefore \qquad\qquad f(x)$ has the opposite sign to a.

When $x > \beta$, $(x - \alpha)$ and $(x - \beta)$ are both *positive*

$\therefore \qquad\qquad f(x)$ has the same sign as a.

(ii) Let the roots of $ax^2 + bx + c = 0$ be *equal* (α, α), i.e. $\Delta = 0$,
$$f(x) = a(x - \alpha)(x - \alpha) = a(x - \alpha)^2$$

$(x - \alpha)^2$ is positive for real values of x

$\therefore \qquad\qquad f(x)$ has the same sign as a for all real x.

(iii) Let the roots of $ax^2 + bx + c = 0$ be *complex*, i.e $\Delta < 0$,

$$f(x) = ax^2 + bx + c = a\left(x^2 + \frac{b}{a}x + \frac{c}{a}\right)$$

$$= a\left[x^2 + \frac{b}{a}x + \left(\frac{b}{2a}\right)^2 + \frac{c}{a} - \left(\frac{b}{2a}\right)^2\right]$$

$$= a\left[\left(x + \frac{b}{2a}\right)^2 + \frac{4ac - b^2}{4a^2}\right]$$

but $b^2 - 4ac$ is *negative*, $\therefore$ $4ac - b^2$ is *positive* and $\left(x + \frac{b}{2a}\right)^2$ is always *positive* for real x,

$\therefore$ $\qquad$ $f(x)$ has the same sign as a for all real x.

These results may all be included in the following summary:

$ax^2 + bx + c$ *has always the same sign as a for all values of x, except when the roots of* $ax^2 + bx + c = 0$ *are real and different and x lies between them.*

2.2.1 Graphical illustrations

The results obtained above may be illustrated as in the sketch (Figure 2.1).

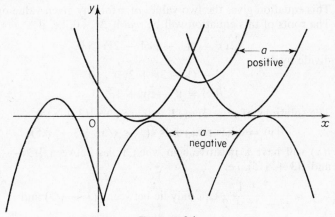

FIGURE 2.1

Examples:

(i) Find the range of values of c in order that $2x^2 + cx + 3$ shall always be positive for real x.

$$f(x) = 2x^2 + cx + 3 = 2\left(x^2 + \frac{c}{2}x + \frac{3}{2}\right)$$

$$= 2\left[x^2 + \frac{c}{2}x + \left(\frac{c}{4}\right)^2 + \frac{3}{2} - \left(\frac{c}{4}\right)^2\right]$$

$$= 2\left[\left(x + \frac{c}{4}\right)^2 + \frac{24 - c^2}{16}\right]$$

Now $\left(x + \frac{c}{4}\right)^2$ is always *positive* for real x except when $x = -\frac{c}{4}$,

$\therefore$ $f(x)$ will always be *positive* if $24 - c^2$ is *positive*, i.e. $c^2 < 24$,

$\therefore$ c must lie between $-\sqrt{24}$ and $\sqrt{24}$, or c lies between $-2\sqrt{6}$ and $2\sqrt{6}$.

(ii) Find the range of values of $\dfrac{x^2 + 1}{x^2 - 2x + 2}$ for real values of x.

Let $\dfrac{x^2 + 1}{x^2 - 2x + 2} = y$; $x^2 + 1 = yx^2 - 2yx + 2y$

$$x^2(1 - y) + 2yx + 1 - 2y = 0$$

This equation gives the two values of x for any given value of y. The roots of this equation will be *real* if $\Delta \geqslant 0$, i.e. if

$$(2y)^2 - 4(1 - y)(1 - 2y) \geqslant 0,$$

divide by 4,

$$y^2 - (1 - 3y + 2y^2) \geqslant 0$$

or $f(y) = y^2 - 3y + 1 \leqslant 0$

The solutions of $y^2 - 3y + 1 = 0$ are $y = \frac{1}{2}(3 \pm \sqrt{5})$,

$\therefore$ $f(y) = 0$ has real roots $\frac{1}{2}(3 - \sqrt{5})$, $\frac{1}{2}(3 + \sqrt{5})$

$f(y)$ will have a negative sign when y lies between $\frac{1}{2}(3 - \sqrt{5})$ and $\frac{1}{2}(3 + \sqrt{5})$, i.e.

$$y = \frac{x^2 + 1}{x^2 - 2x + 2} \text{ can only lie between } \tfrac{1}{2}(3 - \sqrt{5}) \text{ and}$$

$$\tfrac{1}{2}(3 + \sqrt{5}) \text{ for real } x.$$

2.3 Miscellaneous Equations

Many equations may, by using a suitable substitution, be reduced to quadratic equations and hence solved. Simultaneous equations may give rise to quadratic equations after eliminations. No additional theory is required for the solution of such equations. Two illustrations are given below.

Examples:

(i) Solve the equation $2^{2x} = 2^{x+3} - 13$.

Now $2^{2x} = (2^x)^2$; $2^{x+3} = 2^x \times 2^3 = 8 \times 2^x$.
Let $u = 2^x$, then the equation reduces to
$u^2 = 8u - 13$; $u^2 - 8u + 13 = 0$

$$u = \frac{8 \pm \sqrt{(64 - 52)}}{2} = \frac{8 \pm \sqrt{12}}{2} = 4 \pm \sqrt{3}$$
$$= 4 \pm 1 \cdot 732 = \underline{5 \cdot 732 \text{ or } 2 \cdot 268}$$

$\therefore \quad 2^x = 5 \cdot 732$; $x \log 2 = \log 5 \cdot 732$

$$x = \frac{0 \cdot 7584}{0 \cdot 3010} = \underline{2 \cdot 52}$$

or $\quad 2^x = 2 \cdot 268$; $x \log 2 = \log 2 \cdot 268$

$$x = \frac{0 \cdot 3556}{0 \cdot 3010} = \underline{1 \cdot 18(1)}; \quad \underline{x = 2 \cdot 52 \text{ or } 1 \cdot 18}$$

(ii) Solve the equations

$$2x^2 + 3xy - 4y^2 = 5 \tag{1}$$

$$x^2 + 4xy + y^2 = 6 \tag{2}$$

Multiply (1) by 6: $\quad 12x^2 + 18xy - 24y^2 = 30 \tag{3}$

Multiply (2) by 5: $\quad 5x^2 + 20xy + 5y^2 = 30 \tag{4}$

Subtract (4) from (3): $\quad 7x^2 - 2xy - 29y^2 = 0$

This equation may be treated as a quadratic to find x:

$$x = \frac{2y \pm \sqrt{[(2y)^2 + 4(7)(29)y^2]}}{14} = \frac{y}{7}(1 \pm \sqrt{204})$$

$$= \frac{y}{7}(1 \pm 14 \cdot 28) = \underline{2 \cdot 183y \text{ or } -1 \cdot 897y}$$

Substitute $x = 2 \cdot 183y$ in (2), then

$$y^2[(2 \cdot 183)^2 + 4(2 \cdot 183) + 1] = 6$$
$$y^2[4 \cdot 765 + 8 \cdot 732 + 1] = 6$$
$$14 \cdot 497y^2 = 6$$

$$\therefore \quad y^2 = 6/14 \cdot 497 = 0 \cdot 4138; \quad y = \pm\sqrt{(0 \cdot 4138)} = \underline{\pm 0 \cdot 6433}$$

Substitute $x = -1 \cdot 897y$ in (2), then

$$y^2[(1 \cdot 897)^2 - 4(1 \cdot 897) + 1] = 6$$
$$y^2[3 \cdot 600 - 7 \cdot 588 + 1] = 6$$
$$-2 \cdot 988y^2 = 6$$

$$\therefore \qquad y^2 = -6/2 \cdot 988 = -2 \cdot 008; \quad y = \pm j1 \cdot 417$$

The corresponding values of x will be:

$$x = +2 \cdot 183(\pm 0 \cdot 6433) = \pm 1 \cdot 405$$
$$\text{or} \qquad x = -1 \cdot 897(\pm j1 \cdot 417) = \mp j2 \cdot 688$$

There are four pairs of solutions (2 real pairs, 2 imaginary pairs) as follows:

$$x: \; +1 \cdot 405, \quad -1 \cdot 405, \quad 2 \cdot 688j, \; -2 \cdot 688j$$
$$y: \quad 0 \cdot 6433, \; -0 \cdot 6433, \; -1 \cdot 417j, \; +1 \cdot 417j$$

WORKED EXAMPLES

(a) The equation $Lm^2 + Rm + \dfrac{1}{C} = 0$ occurs in electrical theory. Write down the conditions that the values of m will be (i) real and unequal, (ii) equal, (iii) complex.

The discriminant of the quadratic is $R^2 - 4L\left(\dfrac{1}{C}\right) = R^2 - \dfrac{4L}{C}$.

$\therefore$ Roots will be real, unequal if $R^2 - \dfrac{4L}{C} > 0$; or $R^2 > \dfrac{4L}{C}$

Roots will be real, equal if $R^2 - \dfrac{4L}{C} = 0$; or $R^2 = \dfrac{4L}{C}$

Roots will be complex if $R^2 - \dfrac{4L}{C} < 0$; or $R^2 < \dfrac{4L}{C}$

(b) The roots of the equation $ax^2 + bx + c = 0$ are α and β. Form the equation whose roots are α^2 and β^2.

$\alpha + \beta = -b/a; \quad \alpha\beta = c/a$

$\alpha^2 + \beta^2 = (\alpha + \beta)^2 - 2\alpha\beta = (-b/a)^2 - 2(c/a) = (b^2 - 2ac)/a^2$

$\alpha^2\beta^2 = (\alpha\beta)^2 = (c/a)^2 = c^2/a^2$

The new equation may be written as

$$(x - \alpha^2)(x - \beta^2) = 0, \quad \text{i.e. } x^2 - (\alpha^2 + \beta^2)x + \alpha^2\beta^2 = 0$$

Substituting for $\alpha^2 + \beta^2$ and $\alpha^2\beta^2$, the equation becomes

$$x^2 - (b^2 - 2ac)x/a^2 + c^2/a^2 = 0$$

or
$$a^2x^2 - (b^2 - 2ac)x + c^2 = 0$$

(c) Show that the function $\dfrac{x^2 + 2x + 2}{x - 1}$ cannot assume values between $4 - 2\sqrt{5}$ and $4 + 2\sqrt{5}$ for real values of x.

Let $\dfrac{x^2 + 2x + 2}{x - 1} = K; \quad x^2 + 2x + 2 = Kx - K,$

$\therefore \qquad\qquad x^2 + x(2 - K) + 2 + K = 0$

is the equation to find x.

For real values of x, $\Delta \geqslant 0$, i.e. $(2 - K)^2 - 4(2 + K) \geqslant 0$,

$$4 - 4K + K^2 - 8 - 4K \geqslant 0, \quad \text{i.e. } K^2 - 8K - 4 \geqslant 0$$

$$f(K) = K^2 - 8K - 4; \quad K^2 - 8K - 4 = 0$$

when

$$K = [8 \pm \sqrt{(64 + 16)}]/2 = (8 \pm \sqrt{80})/2 = \underline{4 \pm 2\sqrt{5}}$$

$f(K)$ will only be positive when K lies outside this range, i.e. K cannot lie between $4 - 2\sqrt{5}$ and $4 + 2\sqrt{5}$ for real x.

(d) Find the co-ordinates of the points of intersection of the line $y = 2x + 1$ with the circle $x^2 + y^2 - 6x - 8y - 11 = 0$. The line meets the circle when

$$x^2 + (2x + 1)^2 - 6x - 8(2x + 1) - 11 = 0$$
$$x^2 + 4x^2 + 4x + 1 - 6x - 16x - 8 - 11 = 0$$
$$5x^2 - 18x - 18 = 0$$

$$x = \frac{18 \pm \sqrt{[18^2 + 4(5)(18)]}}{10} = \frac{18 \pm \sqrt{(324 + 360)}}{10}$$

$$= \frac{18 \pm \sqrt{684}}{10} = \frac{18 \pm 26 \cdot 15}{10} = 4 \cdot 415 \text{ or } -0 \cdot 815$$

$$y = 2x + 1$$

$$\therefore \qquad y = 9 \cdot 830 \text{ or } -0 \cdot 630.$$

The points of intersection of the line and circle are

$$(4 \cdot 415, 9 \cdot 830); \quad (-0 \cdot 815, -0 \cdot 630)$$

(e) Solve the equation $\frac{1}{2}(e^x + e^{-x}) = 1 \cdot 5$.

$$e^x + e^{-x} = 3; \quad e^x + 1/e^x = 3$$

multiply by e^x

$$e^{2x} + 1 = 3e^x; \quad e^{2x} - 3e^x + 1 = 0$$

Let $u = e^x$, then

$$u^2 - 3u + 1 = 0$$

$$u = [3 \pm \sqrt{(9 - 4)}]/2 = (3 \pm \sqrt{5})/2 = (3 \pm 2 \cdot 236)/2$$
$$= 5 \cdot 236/2 \text{ or } 0 \cdot 764/2 = 2 \cdot 618 \text{ or } 0 \cdot 382$$

$$\therefore \qquad u = e^x = 2 \cdot 618; \quad x = \log_e 2 \cdot 618 = \underline{\quad 0 \cdot 9624}$$

or $\qquad u = e^x = 0 \cdot 382; \quad x = \log_e 0 \cdot 382 = \underline{-0 \cdot 9624}$

(f) Using the method of completion of the square find:

(i) the minimum value of $3x^2 - 5x + 4$ and the value of x at which it occurs,

(ii) the maximum value of $17 + 3x - 4x^2$ and the value of x at which it occurs.

(i) $f(x) = 3x^2 - 5x + 4 = 3\left(x^2 - \frac{5}{3}x + \frac{4}{3}\right)$

$$= 3\left[\left(x - \frac{5}{6}\right)^2 + \frac{4}{3} - \left(\frac{5}{6}\right)^2\right] = 3\left[\left(x - \frac{5}{6}\right)^2 + \frac{23}{36}\right]$$

Now $(x - 5/6)^2$ is positive and always increases as x increases away from 5/6, but is least when $x = 5/6$ (zero).

$\therefore$ *Minimum* value of $f(x) = 3 \times 23/36 = \underline{23/12}$ when $\underline{x = 5/6}$.

(ii) $f(x) = 17 + 3x - 4x^2 = 4\left[\dfrac{17}{4} + \dfrac{3}{4}x - x^2\right]$

$\qquad = 4\left[\dfrac{17}{4} - \left(\dfrac{3}{8} - x\right)^2 + \left(\dfrac{3}{8}\right)^2\right] = 4\left[\dfrac{17}{4} + \dfrac{9}{64} - \left(\dfrac{3}{8} - x\right)^2\right]$

$\qquad = 4\left[\dfrac{272}{64} + \dfrac{9}{64} - \left(\dfrac{3}{8} - x\right)^2\right] = 4\left[\dfrac{281}{64} - \left(\dfrac{3}{8} - x\right)^2\right]$

Now the least value of $(\tfrac{3}{8} - x)^2$ is zero,

$\therefore$ $f(x)$ is a maximum when $\tfrac{3}{8} - x = 0$, i.e. $x = \tfrac{3}{8}$

$\therefore$ *Maximum* value of $f(x)$ is $4 \times \dfrac{281}{64} = \underline{17\tfrac{9}{16}}$ when $\underline{x = \tfrac{3}{8}}$.

Examples 2

1. Express the roots of the following equations in the form $a + bj$:
   (i) $x^2 + x + 1 = 0$, (ii) $4(z^2 + 1) = z(2 - z)$.
2. Solve the equations:
   (i) $3^{x^2-3} = 7^x$ (ii) $5^{2x} = 7(5^x) - 12$.
3. (i) Solve the equation $\dfrac{1}{x^4} - \dfrac{5}{x^2} + 4 = 0$.

   (ii) Solve the equation $2\sqrt{x} - \sqrt{(3x - 8)} = 2$.
4. (i) Solve the simultaneous equations:
$$x - 2y = 3, \qquad x^2 + y^2 = 26.$$
   (ii) Solve the equation $\tfrac{1}{2}(e^x - e^{-x}) = 1\cdot 7$.
5. Determine the coordinates of the points of intersection of the circle $x^2 + y^2 + 2x + 3y - 5 = 0$ and the line $y = 3x - 1$.
6. Find the coordinates of the points of intersection of the curves
$$x^2 - xy + y^2 = 3 \text{ and } 2x^2 - xy - y^2 = 5.$$
7. The roots of the equation $5x^2 + 3x - 1 = 0$ are α and β. Without solving the equation show that the roots are real and calculate the values of $\alpha^2 + \beta^2$, $(\alpha - \beta)^2$, $(\alpha^3 + \beta^3)$.
8. Form the quadratic equation whose roots are the squares of the roots of the equation $2x^2 - 3x - 5 = 0$.
9. Find the range of values of K for the function $3x^2 + Kx + 4$ to be positive for all real values of x.

10. Find the range of values of the function

$$\frac{3x + 5}{1 + 2x - 3x^2} \text{ for real values of } x.$$

11. By using the method of completion of the square, show that the function $4x^2 - 4x + 7$ has a minimum value of 6 when $x = \frac{1}{2}$.

12. By using the method of completion of the square, show that the function $13 + 12x - 9x^2$ has a maximum value of 17 when $x = \frac{2}{3}$.

13. The bending moment M in a beam of length $2l$ and weight per unit length w which is simply supported at both ends and carries a point load of W at $3l/2$ from one end is given by $M = \left(\dfrac{W}{4} + wl\right) x - \dfrac{wx^2}{2}$ at a distance x from one end. If W = weight of the beam show that the maximum bending moment is $9wl^2/8$ when $x = 3l/2$.

14. A template is in the form of a rectangle of height x surmounted by a semi-circle of radius r. If the perimeter of the template is 100 inches, find an expression for the area of the template in terms of r and π. Taking $\pi = 22/7$, show that the area is $25(28r - r^2)/7$, and hence, by completing the square show that the maximum area is 700 in^2 and occurs when $r = x = 14$ in.

15. The expression $T = 2\pi \sqrt{\left(\dfrac{K^2 + x^2}{xg}\right)}$ sec gives the period of vibration of a compound pendulum. By considering the function $(K^2 + x^2)/x$ and writing it in the form of a complete square plus a constant, show that the period is least when $x = K$. What is the least period?

3

Permutations and Combinations— Binomial Expansions

3.1 Permutations

Consider the three letters A, B, C. These may be arranged in the following ways: ABC, ACB, BAC, BCA, CAB, CBA, i.e. six ways. These six ways are referred to as the different *permutations* of the three letters A, B, C. In such a permutation the *order* of arrangement of the letters is important. The *number* of permutations could have been found as follows:

Consider three *locations* (positions), ① ② ③. To fill the first position with one of the letters there are three choices. Once a letter is placed in location ①, two letters remain, and to fill the second location ② with a different letter there are two choices. Having placed two different letters in locations ① and ②, only one choice is left to fill location ③. Hence the total number of ways of arranging the three (different) letters, taking all three is $3 \times 2 \times 1 = 6$, or there are six *permutations*.

The product of the first three natural numbers is abbreviated as follows: $1 \times 2 \times 3 = 3!$ (*factorial 3*). Similarly $1 \times 2 \times 3 \times 4 \times 5 = 5!$ and generally $1 \times 2 \times 3 \times 4 \times \ldots \times n = n!$ (*factorial n*).

3.1.1 Permutations of n unlike objects

(i) Using all n objects.

The total number of permutations will, using the method of 3.1, be $n \times (n - 1) \times (n - 2) \times \ldots \times 3 \times 2 \times 1 = n!$

(ii) Using r objects at a time.

The number of *locations* to be filled is r. The number of different ways of filling these r locations will be $n(n-1)(n-2)\ldots(n-r+1)$ (r factors). This is abbreviated by using the following notation: Number of permutations of n unlike things taking r at a time $= {}^nP_r \text{ (or } {}_nP_r) = n(n-1)\ldots(n-r+1)$

$$= \frac{n(n-1)(n-2)\ldots(n-r+1)(n-r)\ldots3\times2\times1}{(n-r)(n-r-1)\ldots3\times2\times1} = \frac{n!}{(n-r)!}$$

3.1.2 Permutations of objects when some are alike

When some of the objects are *alike* the number of *different permutations* is reduced.

Consider the permutations of $AABBBC$ using all six letters. Let N be the number of *different* six letter permutations.

Consider $AABBBC$:

The A's may be permuted in $2\times1 = 2!$ ways without changing the permutation and the B's may be permuted $3\times2\times1 = 3!$ ways, without changing the permutation, $\therefore$ each of the N different arrangements may be permuted $2! \times 3!$ times without changing the permutation, $\therefore N \times 3! \times 2! = $ Total number of six permutations $= 6!$

$$N = \frac{6!}{2!3!} = 60$$

Similarly if n unlike objects contain s alike of one kind, t alike of another, etc., then the number of different permutations will be given by $\frac{n!}{s!t!\ldots}$.

3.2 Combinations (or Selections)

Consider the *group* of letters (ABC). This is a *selection* of letters or *combination* or *group*. This combination (or group) can be permuted in $3!$ ways. Hence, when selections are made from a number of objects only the grouping is important—not the arrangement or order.

Example: How many combinations (or selections) can be made from the three letters ABC taking two at a time?

There are only three distinct groups, i.e. (AB), (BC), (AC). Hence the number of selections is three. Each of these combinations or selections can be permuted in $2 \times 1 = 2!$ ways.

$\therefore$ Number of permutations taking two at a time

$$= 3 \times 2! = 3 \times 2 \times 1 = 3!, \text{ Number of selections} = 3.$$

(*Note:* The number of permutations = Number of combinations × Number of permutations of one group.)

3.2.1 Combinations (selections)—general formula

Consider a group of n distinct objects. Let a selection (group) of r be chosen from the n objects. This selection may be permuted in $r!$ ways. Let the number of distinct combinations (selections or groups) be denoted by $^nC_r \left[\text{or } _nC_r \text{ or } \left(\dfrac{n}{r} \right) \right]$, then

$$^nC_r \times r! = \text{Total number of permutations of } n \text{ unlike objects}$$
$$r \text{ at a time}$$
$$= {^nP_r}$$

$$\therefore \quad ^nC_r = {^nP_r} \div r! = \frac{n!}{(n-r)!r!}$$

$$= \frac{n(n-1)(n-2) \ldots (n-r+1)}{1 \times 2 \times 3 \times \ldots \times r}$$

Examples:

(i) Find the number of different arrangements of the letters of the word PERMUTE taking all the letters.

Number of items = 7 (2 alike, E)

Total number of arrangements = 7!

But in each distinct arrangement the two E's can be permuted in 2! ways *without changing the permutation.*

$\therefore$ Number of distinct arrangements = $7! \div 2!$

$$= \frac{7 \times 6 \times 5 \times 4 \times 3 \times 2 \times 1}{2 \times 1} = \underline{2,520}$$

2

(ii) How many samples of three letters may be selected from the set *ABCDEFG*?

The number of samples (groups) of size 3

$$= {}^nC_r = {}^7C_3 = \frac{7!}{4!3!} = \frac{7 \times 6 \times 5 \times 4 \times 3 \times 2 \times 1}{4 \times 3 \times 2 \times 1 \times 3 \times 2 \times 1}$$
$$= 7 \times 5 = 35$$

(iii) A bag contains 5 balls of equal size, 2 white, 3 red. Estimate the chance of drawing, at random, (1) 2 red balls at one drawing, (2) 1 red and 1 white at one drawing.

The chance may be taken as the ratio:

$$\frac{\text{Number of pairs of the right kind}}{\text{Total number of pairs possible}}$$

(1) Number of pairs of reds $= {}^3C_2 = \dfrac{3 \times 2}{1 \times 2} = 3$

Total number of pairs in $5 = {}^5C_2 = \dfrac{5 \times 4}{1 \times 2} = 10$

∴ an estimate of the chance of drawing two reds $= 3/10$.

(2) Number of pairs, 1 white, 1 red $= 2 \times 3 = 6$

∴ an estimate of the chance of drawing 1 red and 1 white at one selection $= 6/10 = 3/5$.

3.3 Binomial Expansions

These have already been dealt with in Volume I, but the ${}^nC_r$ notation was not used.

Consider the expansion of $(a + x)^n$, where n is a positive integer. Clearly the first term of the expansion will be a^n, and the last term x^n, if the expansion is carried out in ascending powers of x.

$$(a + x)^n = (a + x)(a + x) \ldots n \text{ times}$$

The coefficient of x will be $a^{n-1} \times {}^nC_1$, the coefficient of x^2 will be $a^{n-2} \times {}^nC_2$, and so on. The coefficient of x^r will be $a^{n-r} \times {}^nC_r$, i.e.

$$(a + x)^n = a^n + {}^nC_1 a^{n-1}x + {}^nC_2 a^{n-2}x^2 + \ldots + {}^nC_r a^{n-r}x^r + \ldots + x^n$$

and there are $(n + 1)$ terms.

This may be written as follows

$$(a + x)^n = a^n + na^{n-1}x + \frac{n(n-1)}{1 \times 2} a^{n-2}x^2 + \ldots$$

$$+ \frac{n(n-1)(n-2)\ldots 3 \times 2 \times 1}{1 \times 2 \times 3 \times \ldots \times r} a^{n-r}x^r + \ldots + x^n$$

If $a = 1$ then

$$(1 + x)^n = 1 + nx + \frac{n(n-1)}{1 \times 2} x^2 + \frac{n(n-1)(n-2)}{1 \times 2 \times 3} x^3 + \ldots + x^n$$

When n is not a positive integer, the expansion may be written down in a similar way, but the expansion will be *infinite*, i.e. any number of terms may be written down, i.e.

$$(1 + x)^n = 1 + nx + \frac{n(n-1)}{1 \times 2} x^2 + \frac{n(n-1)(n-2)}{1 \times 2 \times 3} x^3$$

$$+ \ldots \text{ to infinity}$$

The expansion only *converges* (i.e. is *valid*) when x is numerically less than 1, i.e. $|x| < 1$, or $-1 < x < 1$. This means that it can only be used for numerical evaluation when n is *fractional* or *negative* if x lies in this range.

Examples:

(i) $(2x + y)^5 = (2x)^5 + {}^5C_1(2x)^4y + {}^5C_2(2x)^3y^2 + {}^5C_3(2x)^2y^3$
$$+ {}^5C_4(2x)y^4 + {}^5C_5y^5$$

   (*Note:* ${}^5C_1 = {}^5C_4$; ${}^5C_2 = {}^5C_3$, and generally ${}^nC_r = {}^nC_{n-r}$.)

$$\therefore (2x + y)^5 = 2^5x^5 + 5(2^4)x^4y + \frac{5 \times 4}{1 \times 2}(2^3)x^3y^2$$

$$+ \frac{5 \times 4 \times 3}{1 \times 2 \times 3}(2^2)x^2y^3 + \frac{5 \times 4 \times 3 \times 2}{1 \times 2 \times 3 \times 4}(2)xy^4$$

$$+ \frac{5 \times 4 \times 3 \times 2 \times 1}{1 \times 2 \times 3 \times 4 \times 5}y^5$$

$$= 32x^5 + 80x^4y + 80x^3y^2 + 40x^2y^3 + 10xy^4 + y^5$$

(ii) $(m - 3n)^4$
$$= m^4 - {}^4C_1m^3(3n) + {}^4C_2m^2(3n)^2 - {}^4C_3m(3n)^3 + {}^4C_4(3n)^4$$

(*Note:* Alternation in signs.)

$$\therefore \quad (m - 3n)^4$$

$$= m^4 - 4 \times 3m^3n + \frac{4 \times 3}{1 \times 2} \times 9m^2n^2$$

$$- \frac{4 \times 3 \times 2}{1 \times 2 \times 3} \times 27mn^3 + 81n^4$$

$$= m^4 - 12m^3n + 54m^2n^2 - 108mn^3 + 81n^4$$

(iii) $\dfrac{1}{(1 + 2x)^3}$

$$= (1 + 2x)^{-3}$$

$$= 1 + (-3)(2x) + \frac{(-3)(-4)}{1 \times 2}(2x)^2$$

$$+ \frac{(-3)(-4)(-5)}{1 \times 2 \times 3}(2x)^3 + \ldots$$

$$= 1 - 6x + 24x^2 - 80x^3 + \ldots \text{ to infinity}$$

The expansion will only be *valid* for $|2x| < 1$ or $|x| < \frac{1}{2}$, i.e.

$$-\tfrac{1}{2} < x < \tfrac{1}{2}$$

(iv) $\sqrt{(0\cdot98)} = (1 - 0\cdot02)^{\frac{1}{2}}$

$$= 1 + \tfrac{1}{2}(-0\cdot02) + \frac{\frac{1}{2}(-\frac{1}{2})}{1 \times 2}(-0\cdot02)^2$$

$$+ \frac{\frac{1}{2}(-\frac{1}{2})(-\frac{3}{2})}{1 \times 2 \times 3}(-0\cdot02)^3 + \ldots$$

$$= 1 - 0\cdot01 - \tfrac{1}{8}(0\cdot0004) - \tfrac{1}{16}(0\cdot000008) + \ldots$$

$$= 1 - 0\cdot01 - 0\cdot00005 - 0\cdot0000005 + \ldots$$

$$= 0\cdot9899495 = 0\cdot9899 \text{ to 4 s.f.}$$

3.3.1 Successive terms of a binomial expansion

(Recurrence or recursion formula.)

The term of $(a + x)^n$ containing x^r is $^nC_r a^{n-r} x^r$, the term of $(a + x)^n$

containing x^{r+1} is $^nC_{r+1}a^{n-r-1}x^{r+1}$. If these terms be called T_r and T_{r+1}, then

$$\frac{T_{r+1}}{T_r} = \frac{n!}{(n-r-1)!(r+1)!} \times \frac{(n-r)!r!}{n!} \times \frac{a^{n-r-1}x^{r+1}}{a^{n-r}x^r}$$

$$= \frac{n-r}{r+1} \times \frac{x}{a}$$

This formula is useful when evaluating binomial *probabilities* to be dealt with in a later chapter. A formula of this type is referred to as a *recurrence* or *recursion* formula as it can be used in a similar form with r increased in steps of 1.

Examples:

(i) Evaluate the first 3 terms of $(0{\cdot}7 + 0{\cdot}3)^{10}$.

The first term $T_0 = (0{\cdot}7)^{10} = \underline{0{\cdot}02825}$ (using 4 figure tables).

Using the recurrence formula

$$T_{r+1} = \frac{n-r}{r+1}\frac{x}{a}T_r = \frac{10-r}{r+1} \times \frac{3}{7}T_r \quad (n=10, x=0{\cdot}3, a=0{\cdot}7)$$

Put $r=0$; $T_1 = \frac{10}{1} \times \frac{3}{7} \times T_0 = 10 \times \frac{3}{7} \times 0{\cdot}02825 = \underline{0{\cdot}12102}$

Put $r=1$; $T_2 = \frac{10-1}{1+1} \times \frac{3}{7} \times T_1 = \frac{9}{2} \times \frac{3}{7} \times 0{\cdot}12102$

$$= \underline{0{\cdot}23328}$$

The first three terms are: $\underline{0{\cdot}02825, 0{\cdot}12102, 0{\cdot}23328.}$

(ii) In a binomial 'distribution', the chances of obtaining $0, 1, 2, 3, \ldots$ successes in n trials are given by the successive terms of the expansion of $(q+p)^n$, where $q = 1 - p$. Find the chances of $0, 1, 2, 3$ successes in 6 trials when $p = 0{\cdot}1$ $(q = 0{\cdot}9)$. The chance of 0 successes $= T_0 = (0{\cdot}9)^6 = \underline{0{\cdot}5315}$ (4 s.f.). Using the recurrence formula of 3.3.1

$$T_{r+1} = \frac{n-r}{r+1}\frac{p}{q}T_r = \frac{6-r}{r+1} \times \frac{0{\cdot}1}{0{\cdot}9}T_r = \frac{6-r}{r+1} \times \frac{1}{9}T_r$$

The chance of 1 success

$$= T_1 = \frac{6-0}{1+0} \times \frac{1}{9} \times T_0 = \frac{2}{3} \times 0 \cdot 5315 = \underline{0 \cdot 3544}$$

The chance of 2 successes

$$= T_2 = \frac{6-1}{1+1} \times \frac{1}{9} \times T_1 = \frac{5}{18} \times 0 \cdot 3544 = \underline{0 \cdot 0984}$$

The chance of 3 successes

$$= T_3 = \frac{6-2}{2+1} \times \frac{1}{9} \times T_2 = \frac{4}{27} \times 0 \cdot 0984 = \underline{0 \cdot 0146}$$

(*Note:* The value of the chance falls very sharply.)

WORKED EXAMPLES

(*a*) How many permutations of the letters of the word IMPACT can be made using 3 letters at a time?

Number of permutations $= {}^6P_3 = 6 \times 5 \times 4 = \underline{120 \text{ (all unlike)}}$

(*b*) Find the number of different groups of 3 and 4 letters which can be selected from the letters of the word CONQUEST.

Number of 3 letter groups $= {}^8C_3$ (since all letters are different)

$$= \frac{8 \times 7 \times 6}{1 \times 2 \times 3} = \underline{56}$$

Number of 4 letter groups $= {}^8C_4 = \frac{8 \times 7 \times 6 \times 5}{1 \times 2 \times 3 \times 4} = \underline{70}$

(*c*) In how many ways can a working party of 4 men be chosen from 10 men?

Number of ways = Number of groups of 4 selected from 10

$$= {}^{10}C_4 = \frac{10 \times 9 \times 8 \times 7}{1 \times 2 \times 3 \times 4} = \underline{210}$$

(*d*) In a batch of 10 articles there are known to be 4 faulty. If a random sample of 3 is made from the 10, estimate the chances (i) that all 3 will be faulty, (ii) that only 2 will be faulty.

(i) Number of possible samples of 3 from 10

$$= {}^{10}C_3 = \frac{10 \times 9 \times 8}{1 \times 2 \times 3} = \underline{120}$$

Number of possible samples of 3 faulty from the 4 faulty

$$= {}^4C_3 = \frac{4 \times 3 \times 2}{1 \times 2 \times 3} = \underline{4}$$

∴ an estimate of the chance that all three will be faulty

= Ratio of possible samples (all faults) to the total number of samples of 3 which can be taken

$= 4/120 = \underline{1/30}$

(ii) Number of samples of 3 containing 2 faulty, 1 not

= (No. of ways of selecting 2 from 4) × 6, since with each pair of faulty articles, 1 not faulty can be chosen in 6 ways

∴ Number of samples of 3 containing 2 faulty, 1 not

$$= {}^4C_2 \times 6 = \frac{4 \times 3}{1 \times 2} \times 6 = \underline{36}$$

∴ an estimate of chance = $36/120 = \underline{3/10}$

(e) Expand $(1 - 3x)^{\frac{3}{2}}$ as far as the term in x^3. For what range of values of x is the expansion valid?

$$(1 - 3x)^{\frac{3}{2}} = 1 + \frac{3}{2}(-3x) + \frac{\frac{3}{2} \times \frac{1}{2}}{1 \times 2}(-3x)^2$$

$$+ \frac{\frac{3}{2} \times \frac{1}{2} \times (-\frac{1}{2})}{1 \times 2 \times 3}(-3x)^3 + \dots$$

$$= 1 - \frac{9}{2}x + \frac{27}{8}x^2 + \frac{27}{16}x^3 + \dots$$

and is *valid* only for $|3x| < 1$; $\underline{|x| < \frac{1}{3} \text{ or } -\frac{1}{3} < x < \frac{1}{3}}$

(f) Calculate the first three terms of the expansion of $(0.95 + 0.05)^8$.

First term $T_0 = (0.95)^8 = \underline{0.6631}$ (using 4 figure tables)

$$T_{r+1} = \frac{n-r}{r+1} \times \frac{x}{a} T_r = \frac{n-r}{r+1} \times \frac{0.05}{0.95} T_r = \frac{n-r}{r+1} \times \frac{1}{19} T_r$$

$$= \frac{8-r}{r+1} \times \frac{1}{19} T_r \quad (n = 8, x = 0.05, a = 0.95)$$

Second term T_1 $(r = 0) = \dfrac{8 - 0}{0 + 1} \times \dfrac{1}{19} T_0 = \dfrac{8}{19} \times 0\cdot6631 = \underline{0\cdot2794}$

Third term T_2 $(r = 1) = \dfrac{8 - 1}{1 + 1} \times \dfrac{1}{19} T_1 = \dfrac{7}{38} \times 0\cdot2794 = \underline{0\cdot0514}$,

or to 3 significant figures the first three terms are

$$\underline{0\cdot663,\ 0\cdot279,\ 0\cdot051}$$

Examples 3

1. Evaluate (i) $^7P_3$, $^9P_4$, $^6P_5$, (ii) $^5C_3$, $^7C_5$, $^9C_7$.
2. By evaluating ^nC_r in each case show that:

   (i) $^7C_3 + ^7C_2 = ^8C_3$, (ii) $^8C_4 + ^8C_3 = ^9C_4$,
   (iii) $^5C_2 + ^5C_1 = ^6C_2$, (iv) $^nC_r + ^nC_{r-1} = ^{n+1}C_r$.

3. Write down the expansion for $(1 + x)^8$ using the general coefficients ^nC_r. By using $x = +1$, then $x = -1$, show that:

   (i) $^8C_1 + ^8C_2 + ^8C_3 + \ldots + ^8C_8 = 255$,
   (ii) $^8C_1 + ^8C_3 + ^8C_5 + ^8C_7 = ^8C_2 + ^8C_4 + ^8C_6 + ^8C_8 + 1$.

4. Calculate the number of permutations of the letters of the word DYNAMIC taking:

   (i) 2 letters only, (ii) 3 letters only, (iii) all 7 letters.

5. Calculate the number of distinct permutations of 8 coloured discs, using all 8, if 1 is black, 1 white, 2 red, 4 yellow.

6. How many distinct 3 digit numbers can be formed from the digits 1, 2, 5, 7, 8, 9? How many of these begin with 1 and how many will contain the digits 1 and 2.

7. A bag contains 10 balls of equal size, 6 white, 4 red. Estimate the chance of drawing, at random, a sample of 3 balls which will include:

   (i) 3 whites, (ii) 3 reds, (iii) 2 whites, 1 red, (iv) 1 white, 2 reds.

8. For a certain job there are 5 men and 4 boys available. In how many ways can a working party of 5 be made up consisting of:

   (i) 3 men and 2 boys, (ii) 2 men and 3 boys.

9. From a normal pack of 52 playing cards, 4 cards are drawn at random. What are the chances that (i) all four cards are black? (ii) all four cards are aces? (The four cards should be regarded as a random sample in each case.)

10. Write down the expansions of.

    (i) $(2a + 3b)^5$, (ii) $(x - 2y)^6$.

11. Write down the following functions as far as the term in x^5 in each case. State the range of values of x for which each expansion is valid.

    (i) $(1 + x)^{3/2}$, (ii) $\sqrt{(1 - 3x)}$, (iii) $1/(1 - x)^3$.

12. Using a binomial expansion evaluate $\sqrt[3]{0\cdot94}$ to 4 decimal places.
13. By writing $1 + x + x^2$ in the form $[1 + x(1 + x)]$ find the expansion of $(1 + x + x^2)^4$ in ascending powers of x.

14. By using the recursion (or recurrence) formula $T_{r+1} = \dfrac{n - r}{r + 1} \times \dfrac{x}{a} \, T_r$, connecting successive terms of $(a + x)^n$ in ascending powers of x, evaluate exactly all the terms of the expansion of $(\frac{1}{2} + \frac{1}{2})^6$. Verify that the sum of all the terms is unity.

15. Using four figure tables, evaluate the first term of the expansion of $(0\cdot75 + 0\cdot25)^{10}$. Using the recursion formula deduce the next three terms to four significant figures and find the sum of the first four terms.

4

Binomial Expansions—Other Expansions—Convergence of Infinite Series

4.1 Binomial Expansions (Any Index)

The expansion of a binomial to any power (positive on negative integral, or fractional) has been dealt with in Chapter 3. In order to write down any binomial expansion it is sufficient to apply the expansions,

(i) $(1 + x)^n = 1 + nx + \dfrac{n(n - 1)}{1 \times 2} x^2$
$$+ \frac{n(n - 1)(n - 2)}{1 \times 2 \times 3} x^3 + \ldots \text{etc.}$$

When n is a *positive integer* there will be $(n + 1)$ terms, but when n is *negative* or *fractional* the resulting expansion will be an *infinite series*, i.e. any required number of terms may be written down. The expansion is only valid (i.e. can only be used for numerical evaluation) when $|x| < 1$ or $-1 < x < 1$.

(ii) $(a + x)^n = \left[a \left(1 + \dfrac{x}{a} \right) \right]^n$

$= a^n \left[1 + n \dfrac{x}{a} + \dfrac{n(n - 1)}{1 \times 2} \left(\dfrac{x}{a} \right)^2 + \dfrac{n(n - 1)(n - 2)}{1 \times 2 \times 3} \left(\dfrac{x}{a} \right)^3 + \ldots \right]$

As before, if n is a positive integer there will be $(n + 1)$ terms, but when n is negative or fractional the resulting expansion will be an infinite series, only valid when $|x/a| < 1$ or $|x| < |a|$.

Examples:

(i) Expand $(1 + 2x)^{\frac{1}{2}}$ as far as the term in x^5 and hence evaluate $\sqrt{1 \cdot 2}$ to 3 decimal places.

$$(1 + 2x)^{\frac{1}{2}} = 1 + \tfrac{1}{2}(2x) + \frac{\frac{1}{2}(-\frac{1}{2})}{1 \times 2}(2x)^2 + \frac{\frac{1}{2}(-\frac{1}{2})(-\frac{3}{2})}{1 \times 2 \times 3}(2x)^3$$

$$+ \frac{\frac{1}{2}(-\frac{1}{2})(-\frac{3}{2})(-\frac{5}{2})}{1 \times 2 \times 3 \times 4}(2x)^4$$

$$+ \frac{\frac{1}{2}(-\frac{1}{2})(-\frac{3}{2})(-\frac{5}{2})(-\frac{7}{2})}{1 \times 2 \times 3 \times 4 \times 5}(2x)^5 + \dots$$

$$= 1 + x - \tfrac{1}{2}x^2 + \tfrac{1}{2}x^3 - \tfrac{5}{8}x^4 + \tfrac{7}{8}x^5 + \dots$$

Let $x = 0 \cdot 1$, then $1 + 2x = 1 \cdot 2$,

$$\therefore \sqrt{1 \cdot 2} = (1 + 0 \cdot 2)^{\frac{1}{2}} = 1 + 0 \cdot 1 - \tfrac{1}{2}(0 \cdot 1)^2 + \tfrac{1}{2}(0 \cdot 1)^3$$
$$- \tfrac{5}{8}(0 \cdot 1)^4 + \tfrac{7}{8}(0 \cdot 1)^5$$

$$= 1 + 0 \cdot 1 - 0 \cdot 005 + 0 \cdot 0005 - \tfrac{5}{8} \times 0 \cdot 0001$$
$$+ \tfrac{7}{8} \times 0 \cdot 00001 + \dots$$

$$\simeq 1 \cdot 1005 - 0 \cdot 005 - 0 \cdot 0000625 + 0 \cdot 00000875$$
$$\simeq 1 \cdot 10050875 - 0 \cdot 00506250$$
$$\simeq 1 \cdot 09544625 \simeq \underline{1 \cdot 095 \text{ (3 decimal places)}}$$

(ii) Expand $1/(2 - x)^3$ as far as the term in x^4.

$$1/(2 - x)^3 = 1/[2^3(1 - x/2)]^3 = \tfrac{1}{8}(1 - x/2)^{-3}$$

$$= \frac{1}{8}\left[1 + (-3)\left(-\frac{x}{2}\right) + \frac{(-3)(-4)}{1 \times 2}\left(-\frac{x}{2}\right)^2\right.$$

$$+ \frac{(-3)(-4)(-5)}{1 \times 2 \times 3}\left(-\frac{x}{2}\right)^3$$

$$\left. + \frac{(-3)(-4)(-5)(-6)}{1 \times 2 \times 3 \times 4}\left(-\frac{x}{2}\right)^4\right]$$

$$= \frac{1}{8}\left[1 + \frac{3}{2}x + \frac{3}{2}x^2 + \frac{5}{4}x^3 + \frac{15}{16}x^4 + \dots\right]$$

$$= \frac{1}{8} + \frac{3}{16}x + \frac{3}{16}x^2 + \frac{5}{32}x^3 + \frac{15}{128}x^4 + \dots$$

4.2 Partial Fractions—Application to Expansions

The methods for reducing an algebraic fraction to partial fractions were dealt with adequately in Volume I. The following are summaries of the types of fractions which should be allowed:

(i) $\dfrac{ax + b}{(x + 2)(x + 3)} = \dfrac{A}{x + 2} + \dfrac{B}{x + 3}$

(ii) $\dfrac{ax^2 + bx + c}{(x + 1)^2(x + 2)} = \dfrac{A}{x + 1} + \dfrac{B}{(x + 1)^2} + \dfrac{C}{x + 2}$

(iii) $\dfrac{ax^2 + bx + c}{(x + 2)(2x^2 + x + 1)} = \dfrac{A}{x + 2} + \dfrac{Bx + C}{2x^2 + x + 1}$

(i) Deals with proper fractions with simple linear factors in the denominator.

(ii) Deals with cases when a repeated linear factor is present in the denominator.

(iii) Deals with cases when a quadratic factor is present in the denominator (the quadratic factor cannot be factorized).

If an *improper fraction* is encountered, it should be divided out until a polynomial plus a *proper* fraction is obtained.

Example:

Expand the function $\dfrac{3 + 2x}{(x - 1)(1 + 2x)}$ in ascending powers of x as far as the term in x^5.

Let $\dfrac{3 + 2x}{(x - 1)(1 + 2x)} = \dfrac{A}{x - 1} + \dfrac{B}{1 + 2x}$

Then
$$3 + 2x = A(1 + 2x) + B(x - 1)$$

let $x = 1$; $5 = A(3)$, $\therefore A = 5/3$

let $x = -\tfrac{1}{2}$; $2 = B(-3/2)$, $\therefore B = -4/3$

$\therefore \dfrac{3 + 2x}{(x - 1)(1 + 2x)} = \dfrac{5}{3(x - 1)} - \dfrac{4}{3(1 + 2x)}$

$$= \dfrac{-5}{3(1 - x)} - \dfrac{4}{3(1 + 2x)}$$

$$= -\tfrac{1}{3}[5(1-x)^{-1} + 4(1+2x)^{-1}]$$
$$= -\tfrac{1}{3}[5(1 + x + x^2 + x^3 + x^4 + x^5 + \ldots)$$
$$+ 4(1 - 2x + 4x^2 - 8x^3 + 16x^4 - 32x^5 + \ldots)]$$

$$= -\tfrac{1}{3}[9 - 3x + 21x^2 - 27x^3 + 69x^4 - 123x^5 + \ldots]$$
$$= -3 + x - 7x^2 + 9x^3 - 23x^4 + 41x^5 + \ldots$$

The expansion of $(1-x)^{-1}$ is only *valid* for $|x| < 1$, the expansion of $(1+2x)^{-1}$ is only *valid* for $|x| < \tfrac{1}{2}$. Hence the full expansion is only *valid* for $|x| < \tfrac{1}{2}$.

4.3 Exponential, Logarithmic, Cosine, and Sine Series

There are a number of expansions (other than binomial) which are used for the numerical evaluation of functions. Four such expansions are shown in 4.3.1, 4.3.2, 4.3.3, and 4.3.4.

4.3.1 Exponential series

This is an infinite power series for e^x. It can be derived by considering the binomial expansion of $\left(1 + \dfrac{x}{n}\right)^n$ as n tends to infinity. The series will be derived as an example of an application of the binomial expansion.

$$\left(1 + \frac{x}{n}\right)^n = 1 + n\left(\frac{x}{n}\right) + \frac{n(n-1)}{1 \times 2}\left(\frac{x}{n}\right)^2$$
$$+ \frac{n(n-1)(n-2)}{1 \times 2 \times 3}\left(\frac{x}{n}\right)^3 + \ldots$$

Consider the general term

$$u_r = \frac{n(n-1)\ldots(n-r+1)}{r!}\left(\frac{x}{n}\right)^r$$

u_r may be written as

$$u_r = \frac{\dfrac{n}{n}\left(1 - \dfrac{1}{n}\right)\left(1 - \dfrac{2}{n}\right)\ldots\left(1 - \dfrac{r-1}{n}\right)x^r}{r!}$$

If r is fixed, then, as n tends to infinity, i.e. as $n \to \infty$, $\dfrac{r-1}{n}$ approaches zero and hence $1 - \dfrac{r-1}{n}$ tends to 1.

$\therefore$ u_r tends to $\dfrac{1 \times 1 \times 1 \times 1 \ldots (r \text{ times})}{r!} x^r = \dfrac{x^r}{r!}$

$\therefore$ as n tends to infinity the expansion of $\left(1 + \dfrac{x}{n}\right)^n$ tends to

$$1 + \frac{x}{1!} + \frac{x^2}{2!} + \frac{x^3}{3!} + \ldots \text{ to infinity}$$

i.e. $e^x = \lim\limits_{n \to \infty} \left(1 + \dfrac{x}{n}\right)^n = 1 + \dfrac{x}{1!} + \dfrac{x^2}{2!} + \dfrac{x^3}{3!} + \ldots$ *ad inf.*

Similarly

$$e^{-x} = \lim\limits_{n \to \infty} \left(1 - \frac{x}{n}\right)^n = 1 - \frac{x}{1!} + \frac{x^2}{2!} - \frac{x^3}{3!} + \ldots \text{ } ad\text{ } inf.$$

These two series are of great practical importance. In particular $e^1 = e$ was derived in Chapter 1. The series are *valid* for all real values of x.

4.3.2 Logarithmic series

This may be derived in various ways but will only be stated at this stage:

$\log_e (1 + x) = x - \frac{1}{2}x^2 + \frac{1}{3}x^3 - \frac{1}{4}x^4 + \frac{1}{5}x^5 - \frac{1}{6}x^6 + \ldots$
and so on to infinity

$\log_e (1 - x) = -x - \frac{1}{2}x^2 - \frac{1}{3}x^3 - \frac{1}{4}x^4 - \frac{1}{5}x^5 - \frac{1}{6}x^6 + \ldots$
and so on to infinity

These series are only valid for $|x| < 1$, although the first series could be used when $x = 1$.

4.3.3 Cosine series

This series may be used to evaluate the cosine of an angle x (in radians).

$$\cos x = 1 - \frac{x^2}{2!} + \frac{x^4}{4!} - \frac{x^6}{6!} + \frac{x^8}{8!} \ldots \text{to infinity}$$

This series is only valid for $|x| < 1$.

4.3.4 Sine series

This series may be used to evaluate the sine of an angle x (in radians).

$$\sin x = x - \frac{x^3}{3!} + \frac{x^5}{5!} - \frac{x^7}{7!} + \frac{x^9}{9!} + \ldots \text{to infinity}$$

This series is only valid for $|x| < 1$.

(*Note:* When x is *very small*:

$$\cos x \simeq 1 - \frac{x^2}{2!}; \ \sin x \simeq x - \frac{x^3}{3!}.$$

These series will be used later.)

4.4 Convergence of Infinite Series

Reference has been made to the *validity* of expansions and ranges of values of x for which series are *valid* (i.e. ranges of values of x for which the series can be used for numerical evaluation). These ranges arise from the *conditions of convergence* of infinite series of terms.

In general a series may be said to be convergent if:

(i) The rth term tends to zero as r tends to infinity.
(ii) The remainder after r terms is less than an arbitrarily small numerical value.

Most theorems on the *convergence* of series are based on the *convergence* of the *Geometric Progression* (or *series*)

$$S = a + ar + ar^2 + ar^3 + \ldots \text{ad inf.}$$

which in Volume I was shown to possess a sum to infinity of $a/(1 - r)$ when r is numerically less than 1, i.e. the *geometric series converges* to a finite sum when $|r| < 1$.

4.4.1 Absolute convergence

A series of terms $S = u_1 + u_2 + u_3 + \ldots ad\ inf. = \sum_1^\infty u_r$ ($\Sigma =$ Sum of terms like—) is said to *converge absolutely*, if, when all the numerical values of the terms are taken, the series is *convergent*. If a series is *absolutely convergent* it will also be *convergent* in its original form.

4.4.2 Conditional convergence

If a series of terms *converges*, but does *not converge absolutely* it is said to be *conditionally convergent*.

A series which is not convergent is said to be *divergent*. In such cases the sum tends to infinity as the number of terms increases.

4.4.3 Tests for convergence

There are a number of *tests* which can be applied to decide whether a series is *Absolutely convergent*. The proofs of these tests are outside the scope of this volume. Two tests, which can be applied in many cases are as follows:

(i) *Comparison test for absolute convergence*

If $\sum_{r=0}^\infty u_r$ is a series of terms *known* to be absolutely convergent, and if $\sum_{r=0}^\infty v_r$ is another series of terms, such that for all values of r greater than a finite number $|v_r| \leqslant K|u_r|$, where K is a constant, then $\sum v_r$ is also absolutely convergent [see Worked Example (a)].

(ii) *Ratio test for absolute convergence*

If an infinite series of positive terms is denoted by

$$S = u_0 + u_1 + u_2 + u_3 + \ldots + u_r + u_{r+1} + \ldots ad\ inf.$$

then $\sum\limits_{0}^{\infty} u_r$ *converges*, if, after a finite number of terms the ratio of any term to the previous is always less than 1, i.e. if $\dfrac{u_{r+1}}{u_r} < 1$ for all $r >$ a finite number r_0.

Limit form of ratio test (series of positive terms)

$$\sum_{r=0}^{\infty} u_r \text{ is convergent if } \lim_{r \to \infty} \frac{u_{r+1}}{u_r} < 1.$$

Forms of test for any series

$$\sum_{0}^{\infty} u_r \text{ is absolutely convergent if}$$

(1) $\left| \dfrac{u_{r+1}}{u_r} \right| < 1$ for all $r >$ a finite value r_0

or (2) $\lim\limits_{r=\infty} \left| \dfrac{u_{r+1}}{u_r} \right| < 1$, where $|x|$ means the numerical value of x.
[See Worked Examples (b), (c), and (d).]

WORKED EXAMPLES

(a) The series $\sum\limits_{0}^{\infty} x^r$ is known to be absolutely convergent for $|x| < 1$. Show that the series $\sum\limits_{0}^{\infty} \dfrac{x^r}{r}$ is also absolutely convergent for the same range of values of x.

Let $u_r = x^r, v_r = \dfrac{x^r}{r}$

$$\text{when } r \geqslant 1, |v_r| \leqslant \frac{1}{r} |u_r|$$

Now $\dfrac{1}{1} > \dfrac{1}{2} > \dfrac{1}{3} > \dfrac{1}{4} \ldots$

$\therefore \qquad\qquad |v_r| \leqslant |u_r|$ for all $r > 1$

$\therefore$ by the comparison test $\sum v_r$ must be absolutely convergent for the same range of values of x, i.e. $|x| < 1$.

(b) Show that the series $\sum\limits_{0}^{\infty} x^r$ is absolutely convergent for $|x| < 1$.

Let $u_r = x^r$, then $u_{r+1} = x^{r+1}$

$$\left| \frac{u_{r+1}}{u_r} \right| = \left| \frac{x^{r+1}}{x^r} \right| = |x|$$

By the ratio test if this is less than 1 for all r then the series is absolutely convergent, i.e. A.C. if $|x| < 1$.

(c) Apply the ratio test to show that the series for e^x is absolutely convergent for all real values of x.

$$e^x = 1 + \frac{x}{1!} + \frac{x^2}{2!} + \frac{x^3}{3!} + \ldots = \sum\limits_{0}^{\infty} \frac{x^r}{r!}$$

$$u_r = \frac{x^r}{r!}, \quad u_{r+1} = \frac{x^{r+1}}{(r+1)!}, \quad \therefore \frac{u_{r+1}}{u_r} = \frac{x}{r+1}$$

$$\therefore \qquad \left| \frac{u_{r+1}}{u_r} \right| = \left| \frac{x}{r+1} \right| = \frac{|x|}{r+1}$$

For a fixed (finite) value of x $\lim\limits_{r \to \infty} \dfrac{|x|}{r+1} = 0 < 1$,

$\therefore$ by the limit form of the ratio test the series is absolutely convergent for any finite value of x, i.e. it is absolutely convergent for *all real values of x*.

(d) Apply the ratio test to find the range of values of x for which the binomial expansion $(1 + x)^n$ is absolutely convergent (n negative or a fraction).

$$(1 + x)^n = 1 + nx + \frac{n(n-1)}{1 \times 2} x^2 + \ldots \textit{ ad infinitum}$$

Let

$$u_r = \frac{n(n-1) \ldots (n-r+1)}{r!} x^r;$$

$$u_{r+1} = \frac{n(n-1) \ldots (n-r)}{(r+1)!} x^{r+1};$$

then

$$\frac{u_{r+1}}{u_r} = \frac{n-r}{r+1} x = \frac{\dfrac{n}{r} - 1}{1 + \dfrac{1}{r}} \cdot x$$

$$\therefore \qquad \left|\frac{u_{r+1}}{u_r}\right| = \left|\frac{n/r - 1}{1 + 1/r}\right| |x| \text{ taking numerical values.}$$

For a fixed value of n, as $r \to \infty$, $\dfrac{n}{r} \to 0$, $\dfrac{1}{r} \to 0$

$$\therefore \lim_{r \to \infty} \left|\frac{u_{r+1}}{u_r}\right| = \left|\frac{-1}{1}\right| |x| = |x|$$

$\therefore$ By the ratio test the series converges absolutely if $|x| < 1$.

(e) Obtain the expansion of $(1 + x + x^2)e^x$ as far as the term in x^4.

$$(1 + x + x^2)\, e^x$$
$$= (1 + x + x^2)\left(1 + \frac{x}{1!} + \frac{x^2}{2!} + \frac{x^3}{3!} + \frac{x^4}{4!} + \ldots\right)$$

(There is no need to expand e^x beyond the x^4 term.)

Multiply out, then

$$(1 + x + x^2)e^x$$
$$= 1 + \left(x + \frac{x}{1!}\right) + \left(x^2 + x \times \frac{x}{1!} + \frac{x^2}{2!}\right)$$
$$+ \left(\frac{x^3}{3!} + \frac{x^2}{2!} \times x + \frac{x}{1!} \times x^2\right) + \left(\frac{x^4}{4!} + \frac{x^3}{3!} \times x + \frac{x^2}{2!} \times x^2\right)$$
$$= 1 + 2x + x^2(1 + 1 + \tfrac{1}{2}) + x^3(\tfrac{1}{6} + \tfrac{1}{2} + 1) + x^4(\tfrac{1}{24} + \tfrac{1}{6} + \tfrac{1}{2})$$
$$= \underline{1 + 2x + \tfrac{5}{2}x^2 + \tfrac{5}{3}x^3 + \tfrac{17}{24}x^4 + \ldots}$$

(f) Express $\frac{1}{2}(e^x + e^{-x})$ as a power series in x.

$$e^x = 1 + \frac{x}{1!} + \frac{x^2}{2!} + \frac{x^3}{3!} + \frac{x^4}{4!} + \frac{x^5}{5!} + \ldots \, ad \, \infty$$

$$e^{-x} = 1 - \frac{x}{1!} + \frac{x^2}{2!} - \frac{x^3}{3!} + \frac{x^4}{4!} - \frac{x^5}{5!} + \ldots \, ad \, \infty$$

$$\therefore \; e^x + e^{-x} = 2\left[1 + \frac{x^2}{2!} + \frac{x^4}{4!} + \frac{x^6}{6!} + \frac{x^8}{8!} + \ldots \, ad \, \infty\right]$$

$$\therefore \qquad \tfrac{1}{2}(e^x + e^{-x}) = 1 + \frac{x^2}{2!} + \frac{x^4}{4!} + \frac{x^6}{6!} + \frac{x^8}{8!} + \ldots \, ad \, \infty$$

(g) Obtain the series for $\log_e\left(\dfrac{1+x}{1-x}\right)$ for $|x| < 1$.

Using $x = 0\cdot2$ evaluate $\log_e 1\cdot5$ approximately.

$$\log_e\left(\frac{1+x}{1-x}\right) = \log_e(1+x) - \log_e(1-x)$$

$$= (x - \tfrac{1}{2}x^2 + \tfrac{1}{3}x^3 - \tfrac{1}{4}x^4 \ldots)$$
$$- (-x - \tfrac{1}{2}x^2 - \tfrac{1}{3}x^3 - \tfrac{1}{4}x^4 \ldots)$$

$$= 2(x + \tfrac{1}{3}x^3 + \tfrac{1}{5}x^5 + \tfrac{1}{7}x^7 + \ldots)$$

When $x = 0\cdot2$, $\log_e[(1+x)/(1-x)] = \log_e 1\cdot2/0\cdot8 = \log_e 1\cdot5$,

$$\therefore \log_e 1\cdot5 = 2[(0\cdot2) + \tfrac{1}{3}(0\cdot2)^3 + \tfrac{1}{5}(0\cdot2)^5 + \tfrac{1}{7}(0\cdot2)^7 + \ldots]$$
$$= 2[0\cdot2 + \tfrac{1}{3} \times 0\cdot008 + \tfrac{1}{5} \times 0\cdot00032$$
$$+ \tfrac{1}{7} \times 0\cdot0000128 \ldots]$$
$$= 2[0\cdot2 + 0\cdot0026667 + 0\cdot0000640 + 0\cdot0000018 \ldots]$$
$$= 0\cdot4054650 = \underline{0\cdot4055 \text{ to 4 decimal places}}$$

Examples 4

1. Write down the rth and $(r + 1)$th terms of the expansion of $(1 + x)^n$ for any value of n in terms of n, r, and x. Hence, using the ratio test, show that the expansion is absolutely convergent (or valid) for $-1 < x < 1$.

2. Write down the rth term and the $(r + 1)$th terms of the logarithmic series for $\log_e(1 + x)$. Hence, by using the limit form of the ratio test, show that the series is absolutely convergent for $|x| < 1$.

3. The series $\sum\limits_{1}^{\infty}\dfrac{1}{n^2}$, where n has integral (whole number) values 1, 2, 3, 4, $\ldots$ is known to converge. Apply the comparison test to show that $\sum\limits_{1}^{\infty}\dfrac{1}{(n + s)^2}$ converges for $s > 0$, and that the series $\sum\limits_{1}^{\infty}\dfrac{x^n}{n^2}$ converges absolutely for $|x| < 1$.

4. Write down the expansion of $(4 - x)^{3/2}$ as far as the term containing x^3 For what range of values of x is the series absolutely convergent.

5. Express the function $\dfrac{2x + 3}{(1 + x)(1 - 3x)}$ in partial fractions. Hence expand the function as far as the term in x^4. For what range of values of x is the infinite expansion valid?

6. Express $\dfrac{4x - 5}{(1 + x)^2(1 - 2x)}$ in partial fractions and hence find the expansion of the function as far as the term in x^4. For what range of values of x is the expansion valid?

7. Write down the expansion of e^{-x} as far as the term in x^4. Hence obtain the expansion of $(1 + 2x)e^{-x}$ as far as the term in x^4. What would be the coefficient of x^r in the expansion?

8. Express $\frac{1}{2}(e^x - e^{-x})$ as a power series as far as the term in x^7. What would be the coefficient of x^{2r+1} when r is a positive integer?

9. If $|x| < 1$, obtain the expansion of $e^x/(1 + x)$ as far as the term in x^3.

10. Using the series for e^x, evaluate $e^{0\cdot 1}$ and $e^{-0\cdot 2}$ correct to 4 decimal places.

11. Factorize the quadratic function $1 - 3x + 2x^2$ and hence write down the expansion of $\log_e (1 - 3x + 2x^2)$ in ascending powers of x as far as the term in x^4. What would be the coefficient of x^n? State the range of values of x for which the expansion is valid.

12. Write down the expansion of $\log_e \left(\dfrac{1 + x}{1 - 2x} \right)$ as far as the term in x^5. Find the value of x to satisfy $\dfrac{1 + x}{1 - 2x} = 1\cdot 375$ and use this value of x in the expansion obtained to evaluate $\log_e 1\cdot 375$ to 4 decimal places.

13. Write down the series for $\cos x$ and e^x as far as the term in x^4. Hence derive the expansion of $e^x \cos x$ as far as the term in x^4.

14. By reducing $\sqrt[3]{(7\cdot 84)}$ to the form $a(1 - x)^{1/3}$, evaluate the cube root to 3 decimal places using a binomial expansion.

15. Using a suitable test, discuss the absolute convergence of the following series:

(i) $\displaystyle\sum_{r=1}^{\infty} \frac{1}{2^r}$, (ii) $\displaystyle\sum_{r=0}^{\infty} \frac{x^{2r}}{(2r)!}$,

(iii) $\displaystyle\sum_{1}^{\infty} \left(\frac{r + 1}{r^3} \right) x^r$, (iv) $\displaystyle\sum_{1}^{\infty} \frac{2^r x^r}{r}$.

5

Location of Roots of Equations—
Approximations—Iteration

5.1 Location of Real Roots of an Equation

By considering the graph of a function $f(x)$, i.e. the graph $y = f(x)$ (see Figure 5.1), it is clear that if the function is continuous and $f(a)$ and $f(b)$ have *opposite signs* then the graph crosses the x-axis at *least once* between $x = a$ and $x = b$ and hence the equation $f(x) = 0$ will have at least one real root between $x = a$ and $x = b$. From the figure it is seen that there might be an odd number of roots between a and b.

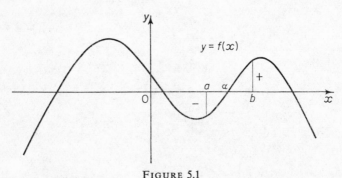

FIGURE 5.1

If there is a single root between a and b, let it be α then α satisfies the equation $f(\alpha) = 0$.

44

5.1.1 Rollé's theorem

Consider the sketch (Figure 5.2) which shows how a polynomial function $f(x)$ varies as x varies.

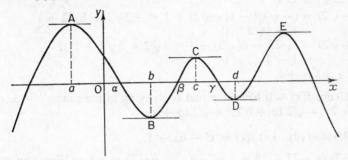

FIGURE 5.2

The real roots of the equation $f(x) = 0$ are given by the values of x where the graph crosses the axis of x. At certain values of x given by $dy/dx = 0$ (or $f'(x) = 0$) the function has turning values (e.g. maximum at A, minimum at B). If these two successive turning values have *opposite signs*, then clearly the graph crosses the x axis between A and B at a point where $x = \alpha$ and this value of x is a real root of $f(x) = 0$. In addition there is a minimum at D and clearly between A and D the graph crosses the x-axis at 3 points where $x = \alpha, \beta,$ and γ.

From these results, the following theorem (*Rollé's theorem*) follows:

Theorem; If α and β are *consecutive* real roots of the equation $f(x) = 0$ then the equation $f'(x) = 0$ has an odd number of real roots between α and β.

(*Notes:* (i) The roots of $f'(x) = 0$ separate the roots of $f(x) = 0$. (ii) Between two consecutive real roots of $f'(x) = 0$ there is at most one real root of $f(x) = 0$. (iii) The results of 5.1 and 5.1.1 may be used to locate real roots of both *polynomial* and *transcendental* equations.)

Example:

Show that the equation $x^3 - 6x + 1 = 0$ has three real roots and find ranges in which the roots lie.

Method (i). Let $f(x) = x^3 - 6x + 1$, then $\dfrac{dy}{dx} = f'(x) = 3x^2 - 6$.

$f'(x) = 0$ when $3x^2 - 6 = 0$ or $\underline{x = \pm\sqrt{2}}$

$f(-\infty) = -\infty$

$f(-\sqrt{2}) = (-\sqrt{2})^3 - 6(-\sqrt{2}) + 1 = -2\sqrt{2} + 6\sqrt{2} + 1$
$\qquad = 1 + 4\sqrt{2} > 0$

$f(+\sqrt{2}) = (\sqrt{2})^3 - 6(\sqrt{2}) + 1 = 2\sqrt{2} - 6\sqrt{2} + 1$
$\qquad = 1 - 4\sqrt{2} < 0$

$f(+\infty) = +\infty$

Hence $f(x) = 0$ has three real roots lying in the ranges: $-\infty$ to $-\sqrt{2}$; $-\sqrt{2}$ to $+\sqrt{2}$; $+\sqrt{2}$ to $+\infty$.

Method (ii). Let $f(x) = x^3 - 6x + 1$.

$f(-3) = -27 + 18 + 1 = -8 \qquad f(-2) = -8 + 12 + 1 = +5$
$f(0) = +1 \qquad\qquad\qquad f(1) = 1 - 6 + 1 = -4$
$f(2) = 8 - 12 + 1 = -3 \qquad f(3) = 27 - 18 + 1 = +10$

Hence $f(x)$ *changes sign* between $x = -3$ and -2, between $x = 0$ and $+1$, between $x = 2$ and 3, so it follows that there are three real roots of $f(x) = 0$ occurring in the ranges shown.

5.2 Repeated Approximation to Roots

By applying method (ii) of the example above, the range in which a root lies can be narrowed even further.

Consider the equation $f(x) = x^3 - 6x + 1 = 0$. As shown in the example of 5.1.1, $f(2) = -3$, $f(3) = +10$, i.e. a real root of $f(x) = 0$ lies between $x = 2$ and $x = 3$.

$f(2\cdot5) = (2\cdot5)^3 - 6(2\cdot5) + 1 = 15\cdot625 - 15 + 1 = +1\cdot625$

$\therefore f(2)$ and $f(2\cdot5)$ have opposite signs and hence the root lies between 2 and 2·5.

$f(2\cdot3) = (2\cdot3)^3 - 6(2\cdot3) + 1 = 12\cdot167 - 13\cdot8 + 1 = -0\cdot633$

$\therefore f(2\cdot3)$ and $f(2\cdot5)$ have opposite signs and hence the root lies between 2·3 and 2·5.

By continuing this process, smaller and smaller ranges can be found in which the root lies.

However, once an approximate value has been found there are *more efficient* methods of approximating closer to the roots. Two of these methods will be discussed in 5.3.1 and 5.4. These are:

(i) Approximation to roots of a polynomial equation using binomial approximation.

(ii) Newton's method of approximation (applicable to polynomial and transcendental equations).

5.3 Binomial Approximations

From previous work

$$(1 + x)^n = 1 + nx + \frac{n(n - 1)}{2!} x^2 + \frac{n(n - 1)(n - 2)}{3!} x^3 + \ldots \, ad \, \infty$$

for any value of n provided $-1 < x < 1$.

When x is *very small* only a few terms are required to evaluate $(1 + x)^n$ to a near enough accuracy. Depending on the number of terms used, so the approximation is of different *orders*, e.g.

$$(1 + x)^n \simeq 1 + nx \text{ (1st order approximation)}$$

$$(1 + x)^n \simeq 1 + nx + \frac{n(n - 1)}{2!} x^2 \text{ (2nd order approximation)}$$

and so on.

Examples:

(i) Using a first order approximation evaluate $(0{\cdot}9)^3\sqrt{(4{\cdot}3)}$.

$$(0{\cdot}9)^3\sqrt{(4{\cdot}3)} = (1 - 0{\cdot}1)^3\sqrt{(4 + 0{\cdot}3)} = (1 - 0{\cdot}1)^3 2\sqrt{(1 + 0{\cdot}075)}$$
$$= 2(1 - 0{\cdot}1)^3(1 + 0{\cdot}075)^{\frac{1}{2}}$$
$$\simeq 2(1 - 3 \times 0{\cdot}1 + \ldots)(1 + \tfrac{1}{2} \times 0{\cdot}075)$$
$$\simeq 2(1 - 3 \times 0{\cdot}1 + \tfrac{1}{2} \times 0{\cdot}075)$$
$$\simeq 2(1 - 0{\cdot}3 + 0{\cdot}0375)$$
$$\simeq 2 \times 0{\cdot}7375 \simeq \underline{1{\cdot}475}$$

(ii) What is the first order approximation for small values of x, y, and z for $(1 + x)^m(1 + y)^n(1 + z)^p$.

$$(1 + x)^m(1 + y)^n(1 + z)^p \simeq (1 + mx)(1 + ny)(1 + pz)$$
$$\simeq \underline{1 + mx + ny + pz}$$

neglecting powers above the first.

(iii) Evaluate $\dfrac{\sqrt{(8 \cdot 82)} \times (1 \cdot 02)^4}{(1 \cdot 95)^3} = E$ using first order binomial approximation.

$$E = \frac{(9 - 0 \cdot 18)^{\frac{1}{2}}(1 + 0 \cdot 02)^4}{(2 - 0 \cdot 05)^3} = \frac{\sqrt{9}(1 - 0 \cdot 02)^{\frac{1}{2}}(1 + 0 \cdot 02)^4}{2^3(1 - 0 \cdot 025)^3}$$

$$= \tfrac{3}{8}(1 - 0 \cdot 02)^{\frac{1}{2}}(1 + 0 \cdot 02)^4(1 - 0 \cdot 025)^{-3}$$

$$\simeq \tfrac{3}{8}(1 - \tfrac{1}{2} \times 0 \cdot 02 + 4 \times 0 \cdot 02 + 3 \times 0 \cdot 025)$$

$$\simeq \tfrac{3}{8}(1 - 0 \cdot 01 + 0 \cdot 08 + 0 \cdot 075) \simeq \tfrac{3}{8} \times 1 \cdot 145$$

$$\simeq 3 \cdot 435/8 \simeq 0 \cdot 429375 \text{ or } \underline{0 \cdot 429} \text{ to 3 decimal places}$$

5.3.1 Approximation to the root of a polynomial equation

Consider again the equation $f(x) = x^3 - 6x + 1 = 0$. It has already been shown that the equation has a root between 2 and 2·5. Let the root be $2 + h_1$, then the value of the function when $2 + h_1$ is substituted will be zero, i.e.

$$(2 + h_1)^3 - 6(2 + h_1) + 1 = 0$$

$\therefore \qquad 8 + 12h_1 + 6h_1{}^2 + h_1{}^3 - 12 - 6h_1 + 1 = 0$

Using the first order approximation, i.e. neglecting powers of h_1 above the first:

$$8 + 12h_1 - 12 - 6h_1 + 1 = 0$$

$$6h_1 = 3, \quad h_1 = 0 \cdot 5$$

$\therefore \qquad \underline{x = 2 \cdot 5 \text{ is a first approximation to the root}}$

This process is now repeated, i.e. let $x = 2 \cdot 5 + h_2$. Then

$$(2 \cdot 5 + h_2)^3 - 6(2 \cdot 5 + h_2) + 1 = 0$$

Neglecting powers of h_2 above the first,

$$[(2 \cdot 5)^3 + 3(2 \cdot 5)^2 h_2 + \ldots] - 15 - 6h_2 + 1 = 0$$

$$15 \cdot 625 + 18 \cdot 75h_2 - 15 - 6h_2 + 1 = 0$$

$$12 \cdot 75h_2 = -1 \cdot 625$$

$\therefore \qquad\qquad\qquad h_2 = -1 \cdot 625/12 \cdot 75$

$$= -0 \cdot 1275$$

i.e. a closer value of the root is $2 \cdot 5 - 0 \cdot 1275 = \underline{2 \cdot 3725}$, or $\underline{2 \cdot 37}$

to 2 decimal places. This is a *second approximation*. This process could be repeated further, thus obtaining *more significant figures* for the root.

5.4 Newton's Method of Approximation to a Root

Consider the graph of the function $y = f(x)$ and let it cross the

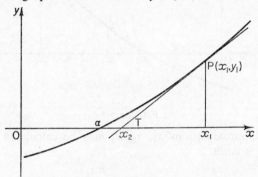

FIGURE 5.3

x-axis at $x = \alpha$ (Figure 5.3), then α is a real root of $f(x) = 0$. Let $x = x_1$ be an *estimate* of the root. Let the tangent to the graph at the point P $[x_1, f(x_1)]$ meet the x-axis at the point T. If the value of x at the point T is x_2, then x_2 is a nearer value to α than x_1, i.e. x_2 is a closer approximation to the root.

The slope of the tangent at P $= \dfrac{\mathrm{d}y}{\mathrm{d}x} = f'(x_1)$,

$\therefore$ the equation of the tangent PT is $y - y_1 = m(x - x_1)$,

where $y_1 = f(x_1)$, m (slope) $= f'(x_1)$,

$\therefore$ the equation of the tangent is $y - f(x_1) = f'(x_1)[x - x_1]$.

This meets the x-axis where $y = 0$, i.e.

$$-f(x_1) = f'(x_1)[x - x_1]; \quad x - x_1 = \frac{-f(x_1)}{f'(x_1)}$$

$$x = x_1 - \frac{f(x_1)}{f'(x_1)}; \quad \text{i.e. } x_2 = x_1 - \frac{f(x_1)}{f'(x_1)}.$$

This is the *Newton (or Newton–Raphson) approximation formula*.

5.4.1 Iteration

Consider Figure 5.4. By repeating the process described in 5.4,

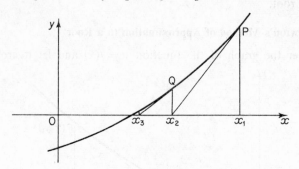

FIGURE 5.4

successive approximations x_1, x_2, x_3 may be calculated using the *recurrence formula*

$$x_{n+1} = x_n - \frac{f(x_n)}{f'(x_n)}$$

Each approximation is derived from the previous one by the same formula. This method of approximation to the root of an equation is one of a number which are *iterative*, i.e. successive approximations are obtained by repeating a standard process.

Newton's method may be used with any type of equation in a single variable, i.e. *polynomial or transcendental equations*.

Example:

Show that the equation $2x^3 - 4x - 1 = 0$ has three real roots. Find ranges in which the roots lie and using Newton's approximation, find the largest root to two decimal places.

Let $f(x) = 2x^3 - 4x - 1$; $f'(x) = 6x^2 - 4 = 0$, when $x = \pm\sqrt{\tfrac{2}{3}}$.

$\therefore f'(x) = 0$ has real roots $x = \pm\sqrt{0.6667} = \pm 0.8165$

$$f(-\infty) = -\infty \qquad f(-0.8165) = +1.177$$
$$f(0.8165) = -3.177 \qquad f(+\infty) = +\infty$$

hence $f(x) = 0$ has real roots in the ranges $-\infty$ to -0.8165, -0.8165 to 0.8165, 0.8165 to $+\infty$.

Now by consideration of Figure 5.5 it is clear that the largest root α is greater than 0·8165.

$$f(1) = 2(1)^3 - 4(1) - 1 = -3; \quad f(2) = 2(2)^3 - 4(2) - 1 = +7$$

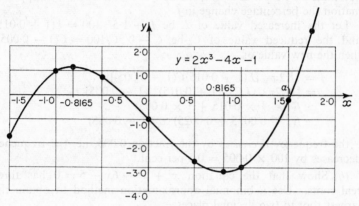

FIGURE 5.5

Since these values are of opposite signs, the required root lies between 1 and 2.

$$f(1·5) = 2(1·5)^3 - 4(1·5) - 1 = 6·75 - 6 - 1 = -0·25$$

Since $f(1·5)$ and $f(2)$ are of opposite signs, α lies between 1·5 and 2. Now

$$\frac{dy}{dx} = f'(x) = 6x^2 - 4; \quad f'(1·5) = 9·5$$

If $x_1 = 1·5$ is an approximate value of the root, then by Newton's approximation $x_2 = x_1 - \dfrac{f(x_1)}{f'(x_1)}$ is a closer value.

$$x_2 = 1·5 - f(1·5)/f'(1·5) = 1·5 + 0·25/9·5 = 1·5 + 0·026 = \underline{1·526}$$

A further approximation is

$$x_3 = x_2 - f(x_2)/f'(x_2) = 1·526 - f(1·526)/f'(1·526)$$
$$= 1·526 - 0·008/9·98 = 1·526 - 0·0008 = \underline{1·5252}$$

or to two decimal places, root α = $\underline{1·53}$.

WORKED EXAMPLES

(a) The frequency f of vibration of an electrical oscillatory circuit is given by $f = 1/[2\pi\sqrt{(LC)}]$. If L increases by 1·5 per cent and C decreases by 0·5 per cent, find, using a first order binomial approximation the percentage change in f.

Let the increased value of L be $L + 1·5L/100 = L(1 + 0·015)$ and the reduced value of C be $C - 0·5C/100 = C(1 - 0·005)$, then the new value of

$$f = 1/\{2\pi\sqrt{[L(1 + 0·015)C(1 - 0·005)]}\}$$
$$= [1/\{2\pi\sqrt{(LC)}\}](1 + 0·015)^{-\frac{1}{2}}(1 - 0·005)^{-\frac{1}{2}}$$
$$\simeq f(1 - \tfrac{1}{2} \times 0·015 + \tfrac{1}{2} \times 0·005)$$
$$\simeq f(1 - 0·0075 + 0·0025) \simeq f(1 - 0·005)$$

$\therefore$ the frequency *decreases* by the fraction 0·005, i.e. the frequency decreases by $100 \times 0·005 = 0·5$ per cent.

(b) Show that the equation $x^3 + x^2 - 6x - 5 = 0$ has three real roots. Use a binomial approximation method to obtain the largest root to two decimal places.

Let $f(x) = x^3 + x^2 - 6x - 5$

$$f(-\infty) = -\infty, f(-1) = +1, f(0) = -5, f(2) = -5, f(3) = +13$$

Hence the equation $f(x) = 0$ has real roots in the ranges $-\infty$ to -1, -1 to 0, 2 to 3.

Let the largest root be $2 + h$, then

$$(2 + h)^3 + (2 + h)^2 - 6(2 + h) - 5 = 0$$
$$(8 + 12h + \ldots) + (4 + 4h) + \ldots) - 12 - 6h - 5 = 0$$

[neglecting h^2, h^3]

$$10h = 5, \quad h = 0·5$$

$\therefore$ a closer value of the root is $2·5$

Let the root be $2·5 + h_1$

$$(2·5 + h_1)^3 + (2·5 + h_1)^2 - 6(2·5 + h_1) - 5 = 0$$

$$(15·625 + 18·75h_1 + \ldots) + (6·25 + 5h_1 + \ldots) - 15 - 6h_1 - 5 = 0$$

$$17·75h_1 + 1·875 = 0, \quad h_1 = -1·875/17·75 = -0·1056$$

$\therefore$ a closer value of the root is $2·5 - 0·1056 = 2·3944$

Let the root be $2 \cdot 3944 + h_2$

$$(2 \cdot 3944 + h_2)^3 + (2 \cdot 3944 + h_2)^2 - 6(2 \cdot 3944 + h_2) - 5 = 0$$

$$13 \cdot 73 + 17 \cdot 19 h_2 + \ldots + 5 \cdot 73 + 4 \cdot 7888 h_2 - 14 \cdot 3664 - 6 h_2 - 5 = 0$$

$$15 \cdot 98 h_2 + 0 \cdot 0936 = 0, \quad h_2 = -0 \cdot 00586 \simeq -0 \cdot 0059$$

$\therefore$ a closer value of the root is $2 \cdot 3944 - 0 \cdot 0059 = \underline{2 \cdot 3885}$, i.e. root is 2·39 to two decimal places.

(c) Show that the equation $\mathrm{e}^{-x} = x$ has a real root between $x = 0 \cdot 5$ and $0 \cdot 7$. Use the Newton approximation to evaluate the root to three decimal places.

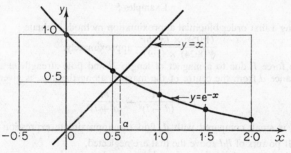

FIGURE 5.6

From Fig. 5.6, the root of $\mathrm{e}^{-x} = x$ is given by the value of x at the intersection of the graphs $y = \mathrm{e}^{-x}$ and $y = x$. Clearly the equation will only have one real root. Let $f(x) = \mathrm{e}^{-x} - x$,

$f(0 \cdot 5) = \mathrm{e}^{-0 \cdot 5} - 0 \cdot 5 = 0 \cdot 6065 - 0 \cdot 5 = +0 \cdot 1065$, i.e. positive,
$f(0 \cdot 7) = \mathrm{e}^{-0 \cdot 7} - 0 \cdot 7 = 0 \cdot 4966 - 0 \cdot 7 = -0 \cdot 2034$, i.e. negative.

Since $f(0 \cdot 5)$ and $f(0 \cdot 7)$ have opposite signs, the equation $f(x) = 0$ has a real root between $x = 0 \cdot 5$ and $0 \cdot 7$.

Using Newton's approximation $x_2 = x_1 - f(x_1)/f'(x)$. Let the first approximation to the root be $x_1 = 0 \cdot 5$

$$f'(x) = -\mathrm{e}^{-x} - 1 \quad \left(\textit{Note:} \ \frac{\mathrm{d}}{\mathrm{d}x}(\mathrm{e}^{Kx}) = K\mathrm{e}^{Kx} \right)$$

$\therefore \quad f'(0 \cdot 5) = -\mathrm{e}^{-0 \cdot 5} - 1 = -0 \cdot 6065 - 1 = -1 \cdot 6065$
$ f(0 \cdot 5) = \mathrm{e}^{-0 \cdot 5} - 0 \cdot 5 = 0 \cdot 1065$

$\therefore \quad x_2 = 0 \cdot 5 - 0 \cdot 1065/(-1 \cdot 6065) = 0 \cdot 5 + 0 \cdot 0663 = \underline{0 \cdot 5663}$

Let the second approximation be $x_2 = 0.5663$

$$f'(0.5663) = -e^{-0.5663} - 1 = -0.5676 - 1 = -1.5676$$
$$f(0.5663) = e^{-0.5663} - 0.5663 = 0.5676 - 0.5663 = 0.0013$$

$$\therefore \quad x_3 = x_2 - f(x_2)/f'(x_2) = 0.5663 - 0.0013/(-1.5676)$$
$$= 0.5663 + 0.00083 = 0.56713$$

Hence, to three decimal places the root is 0·567.

(*Note:* $f(0.567) = e^{-0.567} - 0.567 = 0.5672 - 0.5670$
$$= 0.0002, \text{ i.e. almost zero.})$$

Examples 5

1. Using a first order binomial approximation method, evaluate
$$\frac{(1.05)^3 \times (3.96)^2}{\sqrt[3]{(8.24)} \times (1.97)^4} \text{ approximately.}$$

2. The force F due to a magnet of length $2l$ and pole strength m at a point distance d from the centre of the magnet, along the axis, is given by

$$F = \frac{m}{(d-l)^2} - \frac{m}{(d+l)^2}.$$

If l is small compared with d, find the approximate expression for F:

(i) If powers of l/d above the first are neglected,
(ii) If powers of l/d above the third are neglected.

3. The frequency f of torsional oscillations of a wire is given by the formula $f = Kl/(t^2 r^4)$, where l, t, and r are observed quantities and K is a known constant.

Find, using a first order binomial approximation, the percentage error in the calculation of f if errors are made of 0·3 per cent in l (too high), 1·5 per cent in t (too low) and 0·8 per cent in r (too low).

4. In calculating the period of vibration T seconds of a simple pendulum, the formula $T = 2\pi\sqrt{(l/g)}$ is used, where g is the acceleration due to gravity in ft/s². If l is increased by 1·5 per cent and g decreases by 0·8 per cent calculate the percentage change in T, stating whether the change is an increase or decrease.

5. The frequency f, in cycles per second, of oscillation of an electrical circuit is given by the formula $f = 1/[2\pi\sqrt{(LC)}]$, where L is the inductance in henries and C the capacitance in farads. If L is 2 per cent too large and C is 0·75 per cent too small, estimate the percentage error in the calculation of f.

6. Use the binomial expansion to expand $\sqrt[3]{(8 + x)}$ in a series of ascending powers of x as far as the term in x^3. Hence find the value of $\sqrt[3]{(8.4)}$ correct to three decimal places.

7. Given $V = 2d^2\sqrt{P}$, use the binomial approximation to find the percentage change in V when d increases by 3 per cent and P decreases by 2·5 per cent.

8. The velocity c of sound in a gas is given by $c = \sqrt{\left(\dfrac{p\gamma}{d}\right)}$ where γ = ratio of specific heats, p = pressure. d = density. If both p and d rise by 2 per cent and γ falls by 1 per cent, find the percentage rise or fall in c using a first order approximation.

9. Neglecting powers of x above the second, determine x from the equation $\left(1 + \dfrac{x}{3}\right)^4 + \left(1 - \dfrac{x}{3}\right)^4 = 2{\cdot}35$.

10. Show that the equation $2x^3 + x^2 - 8x - 10 = 0$ has a real root between $x = 2$ and $x = 3$. Use a binomial approximation process to determine the root to two decimal places.

11. Using Rollé's theorem, show that the equation $2x^3 - 5x + 2 = 0$ has three real roots. Using variation in the sign of the function, find ranges of values of x in which the roots lie and estimate the middle root correct to two decimal places.

12. Show that the equation $x^3 - 8x + 5 = 0$ has three real roots in the range -4 to $+4$. Determine the largest root correct to three significant figures using the Newton method of approximation.

13. Using Rollé's theorem, show that the equation $x^3 = 9x - 4$ has three real roots. Use an approximation method (binomial or Newton) to estimate the middle root to two decimal places.

14. By considering the sign of the function $e^x - 3x$, show that the equation $e^x - 3x = 0$ has a root between $x = 0{\cdot}6$ and $0{\cdot}8$. Use the Newton approximation method to obtain the root correct to two decimal places. $\left(\text{Note: } \dfrac{d}{dx}(e^x) = e^x.\right)$

15. By considering the sign of a suitable function, show that the equation $4\log_e x = x$ has a root between $x = 1$ and $1{\cdot}5$. Use a suitable approximate method to show that the root is about $1{\cdot}424$. $\left(\text{Note: } \dfrac{d}{dx}(\log_e x) = \dfrac{1}{x}.\right)$

3

6

Graphical Solution of Equations—Determination of Laws—Curve Sketching

6.1 Graphical Solution of Equations

This topic was dealt with fully in Volume I (Chapter 16). A summary of the methods will be given here together with an example.

(i) Let it be required to obtain solutions of an equation which may be represented as $f(x) = 0$, where $f(x)$ is a given function of x. The values of $y = f(x)$ are tabulated for values of x in a given range. This range may be determined by applying the principles laid down in Chapter 5, i.e. using Rollé's theorem or change of sign of the function. Having plotted the graph $y = f(x)$, the solutions of $f(x) = 0$ are the values of x at the points where the graph crosses the x-axis. In order to estimate the intersection values (or *roots*) more closely it may be necessary to insert additional values in the function table.

(ii) *Two graph method of solution*

In a number of cases, if an approximate idea only of the root is required, the process of graphical solution may be speeded up by writing the equation in the form $f_1(x) = f_2(x)$. It may be found that the two graphs $y = f_1(x)$ and $y = f_2(x)$ can be plotted rapidly, then the required solutions of $f(x) = 0$ will be the values of x at *the points of intersection* of the two graphs.

(iii) *Enlargements*

Having estimated that a root of $f(x) = 0$ is approximately α, by enlarging a portion of the single graph over a *small range* of x values, including α, it will be found that a much closer value of the root may be obtained rapidly.

(iv) *Simultaneous equations*

If it be required to solve simultaneously two equations of the forms $f_1(x, y) = 0$, $f_2(x, y) = 0$ it may be done by plotting graphs of the two functions and reading off the corresponding (x, y) values (or co-ordinates) of the points of intersection of the graphs.

Example:

Determine a range of values of x in which the three real roots of the equation $x^3 - 4x - 2 = 0$ lie. Hence estimate graphically the values of the roots using (1) a single graph, (2) two graphs. Use an enlargement to determine a closer value for the middle root.

Let $f(x) = x^3 - 4x - 2$

$$f(-3) = -27 + 12 - 2 = -17 \qquad f(-2) = -8 + 8 - 2 = -2$$
$$f(-1) = -1 + 4 - 2 = +1 \qquad f(0) = -2$$
$$f(1) = 1 - 4 - 2 = -5 \qquad f(2) = 8 - 8 - 2 = -2$$
$$f(3) = 27 - 12 - 2 = 13$$

Hence $f(x)$ changes sign between $(-2, -1)$, $(-1, 0)$, $(2, 3)$, therefore there are three real roots of $f(x) = 0$ lying in these ranges. Clearly it will be sufficient to plot the graph (or graphs) for values of x between -2 and $+3$.

(1) $y = f(x) = x^3 - 4x - 2$

Table of values:

x	-2	-1	0	1	2	3	$-1\frac{1}{2}$	$-\frac{1}{2}$	$2\frac{1}{2}$
$f(x)$	-2	1	-2	-5	-2	13	$\frac{5}{8}$	$-\frac{1}{8}$	$3\frac{5}{8}$

(*Note:* The insertion of intermediate values as it is already known in which ranges the roots lie.)

The function is plotted as shown in Figure 6.1 and the solutions of the equation $x^3 - 4x - 2 = 0$ read off.

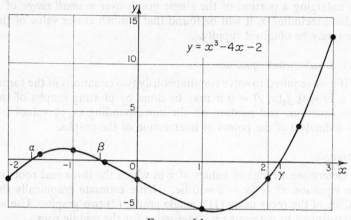

FIGURE 6.1

From the graph the solutions are approximately $-1\cdot66$, $-0\cdot55$, $2\cdot20$.

(2) Write the equation as $x^3 = 4x + 2$. Plot the graphs of $y = x^3$ and $y = 4x + 2$. These can be fairly rapidly plotted as shown in Figure 6.2.

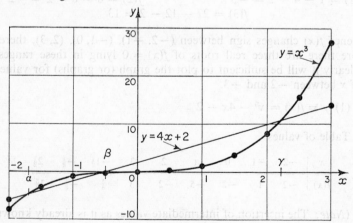

FIGURE 6.2

From the graphs the solutions are approximately $-1·7$, $-0·6$, $2·2$. It will be noted that, using this method, whilst a rapid approximate solution is effected, the values cannot be read off as closely as when method (1) is used.

Enlargement of $y = x^3 - 4x - 2$, near $x = -0·6$

x	$-0·7$	$-0·6$	$-0·5$
y	$0·457$	$0·184$	$-0·125$

(see Figure 6.3)

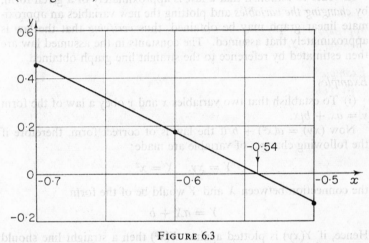

FIGURE 6.3

For the enlargement three points were used, but it can be seen that the portion of the graph is approximately straight, hence if the range chosen is *very small*, two points would be sufficient. From the enlargement the root is about $-0·54$.

6.2 Graphical Determination of Laws—Linear Form

This topic was also dealt with carefully in Volume I (Chapters 17 and 18). A summary of the techniques required is given here together with one example.

(i) Linear law

If values of x are plotted horizontally and the graph of y against x is approximately linear (due allowance being made for variation), then the law connecting x and y is of the form $y = ax + b$, where $a = $ slope of the graph, $b = $ intercept. The values of a and b may be read off from the graph, or alternatively *two points on the graph* are taken, the values of x and y read off and substituted in $y = ax + b$ and the two resulting simultaneous equations solved for a and b.

(ii) Reduction to linear law

In order to establish that a law is approximately of a given form, by *changing the variables* and plotting the new variables an approximate linear graph may be obtained, thus *verifying* that the law is approximately that assumed. The constants in the assumed law are then estimated by reference to the straight line graph obtained.

Examples:

(i) To establish that two variables x and y obey a law of the form $y = ax + b/x$.

Now $(xy) = a(x^2) + b$ if the law is of correct form, therefore if the following changes of variable are made:

$$Y = xy, \quad X = x^2$$

the connection between X and Y would be of the form

$$Y = aX + b$$

Hence, if $Y(xy)$ is plotted against $X(x^2)$ then a straight line should result and from it likely values for a and b may be estimated.

(ii) To establish that two variables T and θ approximately obey a law of the form $T = a\theta^n$.

If this law is approximately correct, then

$$\log_{10} T = \log_{10} a + n \log_{10} \theta$$

Let $\log_{10} T = Y$ and $\log_{10} \theta = X$, then

$$Y = \log_{10} a + nX$$

Hence, if Y ($\log_{10} T$) be plotted against X ($\log_{10} \theta$) then a straight line will result and from it estimates may be made of a and n.

(The reader should refer to Chapters 17 and 18, Volume I, at this stage for further examples.)

Example:

The torque T (lbf in.) and angle of twist (θ degrees) obtained experimentally in a torsion test are given as:

T (lbf in.)	850	900	950	1000	1050	1100	1150	1200
θ (degrees)	11·6	14·0	16·6	19·8	22·9	26·5	30·6	35·5

Show that the law connecting T and θ is approximately of the form $T = a\theta^n$ and estimate likely values for a and n.

If the law is as stated, then $\log_{10} T = \log_{10} a + n \log_{10} \theta$,

$\log_{10} T$	2·9294	2·9542	2·9777	3·0000	3·0212	3·0414	3·0607	3·0792
$\log_{10} \theta$	1·0645	1·1461	1·2201	1·2967	1·3598	1·4232	1·4857	1·5502

The plot of these values is shown in Figure 6.4.

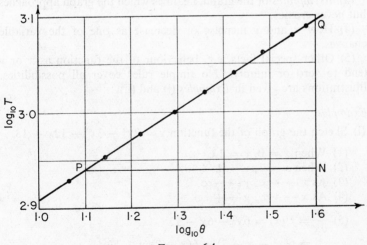

FIGURE 6.4

Since the graph is approximately linear, the assumed law is of correct form and

$$n = \text{Slope of graph} = \frac{QN}{PN} = \frac{3\cdot0950 - 2\cdot9400}{0\cdot5} = 0\cdot31$$

at the point P,

$$\log_{10} T = 2\cdot9400, \quad \log_{10} \theta = 1\cdot1$$

$\therefore$ substituting in the equation $\log T = \log a + n \log \theta$,

$$2\cdot9400 = \log_{10} a + 0\cdot31 \times 1\cdot1$$

$$\log_{10} a = 2\cdot9400 - 0\cdot3410 = 2\cdot5990, \quad a = \text{antilog } 2\cdot5990 = \underline{397\cdot2}$$

$\therefore$ The law is approximately $\underline{T = 397\theta^{0\cdot31}}$.

6.3 Principles of Curve Sketching

In order to *sketch* the graph of a function $y = f(x)$, certain special points may be quickly obtained in order to *sketch* the graph.

Examples of these points are as follows:

(1) Values of y when $x = 0$, and of x when $y = 0$.
(2) Turning values of the function.
(3) *Asymptotes* of the graph, i.e. lines which the graph approaches but never crosses.
(4) How x and y increase or decrease as one of the variable changes.
(5) Other special facts, e.g. behaviour of the function as x or y tend to zero or infinity. No simple rules cover all possibilities. Illustrations are given in *Examples* (i) and (ii).

Examples:

(i) Sketch the graph of the function $y = 2x^3 - 3x^2 - 12x + 13$.

(1) When $x = 0$, $y = 13$.
(2) When $x = 1$, $y = 0$.
(3) As $x \to -\infty$, $y \to -\infty$.
(4) As $x \to +\infty$, $y \to +\infty$.
(5) $\dfrac{dy}{dx} = f'(x) = 6x^2 - 6x - 12$

$\qquad = 6(x^2 - x - 2) = 6(x - 2)(x + 1)$.

$$\therefore \ \frac{dy}{dx} = 0 \text{ when } x = -1 \text{ or } 2$$

i.e. the function has turning values when $\underline{x = -1 \text{ or } 2.}$

When $x = -1$, $y = -2 - 3 + 12 + 13 = \underline{20}$

When $x = 2$, $y = 16 - 12 - 24 + 13 = \underline{-7}$

A *sketch* of the graph of the function is shown in Figure 6.5

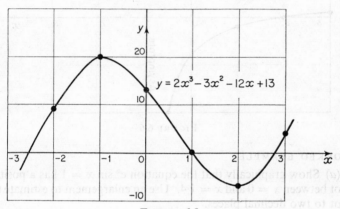

$$y = 2x^3 - 3x^2 - 12x + 13$$

FIGURE 6.5

(ii) Sketch the curve $y = \dfrac{x+2}{x-1} = 1 + \dfrac{3}{x-1}$.

(1) When $x = -2$, $y = 0$.

(2) When $x = 0$, $y = -2$.

(3) As $x \to 1$, $x - 1 \to 0$, $\therefore \dfrac{x+2}{x-1} \to +\infty$,

$\therefore x = 1$ is an *asymptote*.

(4) When x is just >1, y will be positive (large).

(5) When x is just <1, y will be negative (large).

(6) As x increases >1, $3/(x-1) \to 0$, $\therefore y \to 1$.

(7) As $x \to -\infty$, $3/(x-1) \to 0$, $\therefore y \to 1$.

Both $x = 1$ and $y = 1$ are *asymptotes* of the graph.

For sketch see Figure 6.6.

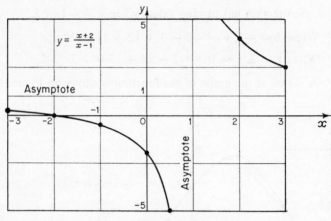

FIGURE 6.6

WORKED EXAMPLES

(a) Show graphically that the equation $e^x \sin x = 1$ has a positive root between $x = 0$ and $x = 1\cdot4$. Use an enlargement to estimate the root to two decimal places.

Write the equation as $\sin x = e^{-x}$ (x in radians). The required values are tabulated as follows in order to plot the graphs $y = \sin x$ and $y = e^{-x}$.

x (rad)	0	0·2	0·4	0·6	0·8	1·0	1·2	1·4
x (degrees)	0	11°28′	22°55′	34°23′	45°50′	57°18′	68°45′	80°13′
$\sin x$	0	0·1988	0·3894	0·5647	0·7173	0·8415	0·9320	0·9854
e^{-x}	1	0·8187	0·6703	0·5488	0·4493	0·3679	0·3012	0·2466

The graphs are plotted as in Figure 6.7.

From the graphs the root of $e^x \sin x = 1$ is seen to be approximately 0·58.

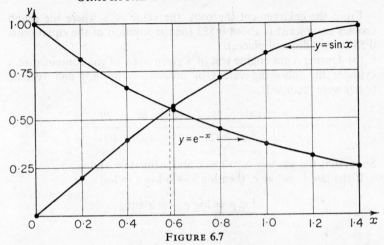

FIGURE 6.7

Enlargement: an enlargement of the graph of $y = e^x \sin x - 1$
near to $x = 0.58$ is plotted using the following values (Figure 6.8):

x	0·5	0·6	0·7
e^x	1·6487	1·8221	2·0138
$\sin x$	0·4794	0·5647	0·6442
$e^x \sin x - 1$	−0·2095	0·0290	0·2970

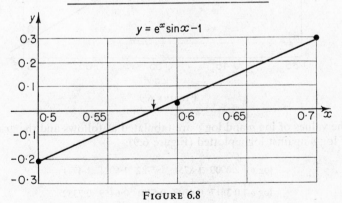

FIGURE 6.8

From the enlargement the root (the value of x where the graph crosses the x-axis) is about 0·583 (or the solution of the equation is 0·58 to two decimal places).

(b) During a gas engine test of a given mass of gas contained in a cylinder the following values of pressure (p lbf/in.^{2}) and volume (v ft^3) were recorded:

p	100	75	60	40	30
v	2·300	2·824	3·312	4·425	5·435

Show that the gas law is $pv^n = c$ and estimate n and c.

If the law is $pv^n = c$, then $\log p + n \log v = \log c$ or

$$\log p = \log c - n \log v$$

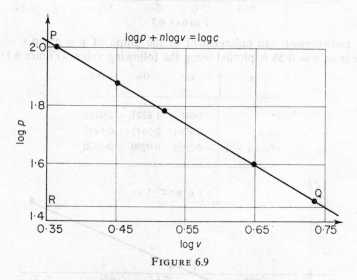

FIGURE 6.9

The values of $\log p$ and $\log v$ are tabulated as follows and the graph of $\log p$ against $\log v$ plotted (Figure 6.9).

$\log p$	2·0000	1·8751	1·7782	1·6021	1·4771
$\log v$	0·3617	0·4508	0·5201	0·6459	0·7352

Since the graph of $\log p$ against $\log v$ is linear, the assumed law is approximately of correct form. From the graph the slope is negative

$$= -\frac{PR}{RQ} = -\frac{2\cdot02 - 1\cdot45}{0\cdot4} = -\frac{0\cdot57}{0\cdot4} = -1\cdot425 = -n,$$

$$\therefore n = 1\cdot425$$

Substituting the values at P,

$$\log c = 2\cdot02 + 1\cdot425 \times 0\cdot35 = 2\cdot02 + 0\cdot4987 = 2\cdot5187$$
$$c = \text{antilog } 2\cdot5187 = 330\cdot1$$

Hence the law is approximately $pv^{1\cdot43} = 330$.

(c) Sketch the graph of the function given by the equation $\dfrac{x^2}{9} + \dfrac{y^2}{4} = 1$.

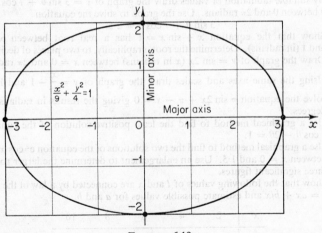

FIGURE 6.10

(1) When $y = 0$, $x^2/9 = 1$, $x = \pm3$.
(2) When $x = 0$, $y^2/4 = 1$, $y = \pm2$.
(3) $y^2/4 = 1 - x^2/9 = \frac{1}{9}(9 - x^2)$.

$$y^2 = \tfrac{4}{9}(9 - x^2), \quad y = \pm\tfrac{2}{3}\sqrt{(9 - x^2)}$$

When x lies between -3 and $+3$, y has equal and opposite values, being greatest when $x = 0$, i.e. $y = \pm\frac{2}{3} \times 3 = \pm 2$. When $x = \pm 3$, $y = 0$ as above. In between these values, for any value of x, y has equal and opposite values. The curve is an *oval* as shown in Figure 6.10.

The curve is an *ellipse* with *major axis* of length 6 units and *minor axis* of length 4 units.

Examples 6

1. Using variation in the sign of a function for real values of x, show that the equation $x^3 - 8x = 5$ has three real roots between $x = -4$ and $x = 4$. By plotting a suitable graph, solve the equation and use an enlargement to determine the largest root correct to two decimal places.

2. Show that the equation $2x^3 - x^2 = 7x - 6$ has three real roots between $x = -3$ and $+3$. By plotting a suitable graph, find the three roots of the equation.

3. Draw the graph of $y = \tan x$ for values of x from 0 to $\pi/3$ radians. By drawing another suitable graph to intersect the first, find a solution of the equation $2 \tan x = 3 - 2x$ to the nearest minute.

4. By suitable tabulation of values draw the graph of $y = 3 \sin \theta + 7 \cos \theta$ for θ between 0 and 2π radians. Use the graph to solve the equation
$$3 \sin \theta + 7 \cos \theta = 2 \cdot 5.$$

5. Show that the equation $x + \sin x = 1$ has a real root between $x = 0$ and 1 (in radians). Determine the root, graphically, to two places of decimals.

6. Draw the graph of $y = \sin 2x$ (x in radians) between $x = 0$ and 2π radians.

Using the same axes and scales draw the graph of $y = \dfrac{x}{\pi} - 1$ and hence solve the equation $\pi \sin 2x + \pi - x = 0$ giving the results in radians and degrees.

7. Use a graphical method to find the least positive solution of the equation $3 \cos \theta - 2\theta = 1$.

8. Use a graphical method to find the two solutions of the equation $e^x \cos x = 1$ between $x = 0$ and $1 \cdot 5$. Use an enlargement to determine the larger root to three significant figures.

9. Show that the following values of I and x are connected by a law of the form $I = ax + b/x$ and estimate possible values for a and b.

x	2	4	6	8	10
I	55·0	72·5	98·3	126·2	155·0

10. By plotting a suitable graph, show that the following values of x and y satisfy a law of the form $y = ax^n$ and estimate possible values for a and n.

x	1·380	1·905	3·020	4·571	8·318
y	7·079	12·59	28·18	56·23	158·3

11. The luminosity I of a lamp for varying voltage v is as follows:

v	60	80	100	120	140	160
I	10	31·6	88	184	322	580

Show that the law connecting I and v is of the form $I = Av^m$ and estimate values for A and m.

12. The head of water h in feet, and the velocity of water flow v in ft/s are noted as follows:

h (feet)	5	7	10	12	15
v (ft/s)	3·43	4·0	4·8	5·2	5·8

Show that the law connecting v and h is of the form $v = Kh^n$ and find K and n.

13. A law of the form $y = ae^{bx}$ is presumed to connect x and y as tabulated. Show that this is so and find likely values for a and b.

x	1·0	1·5	2·0	2·5	3·0	3·5	4·0	4·5
y	13·28	15·04	17·53	19·80	23·11	26·00	30·50	34·40

14. The tensions T and T_0, in a rope wrapped round a drum when the angle of lap is θ radians, are known to be connected by a law of the form $T = T_0 e^{\mu\theta}$ where μ is the coefficient of friction. Show that the following values of T and θ verify this law and estimate T_0 and μ.

T	19·00	30·84	50·59	105·5
θ	3·14	7·91	13·70	21·99

15. Sketch the graphs represented by the following functional relationships:

(i) $xy = 25$, (ii) $y^2 = 8x$, (iii) $x^2 = 16y$, (iv) $x^2/16 + y^2/9 = 1$, (v) $x^2/16 - y^2/9 = 1$, (vi) $(x - 2)(y - 3) = 1$, (vii) $y = 2x^3 + 3x^2 - 36x + 17$, (viii) $y = (x - 3)/(x + 2)$.

7

Co-ordinate systems—Co-ordinate Geometry—Parameters

Co-ordinate System

In order to trace the paths which are followed by moving points, some form of *co-ordinate system* is required. A co-ordinate system

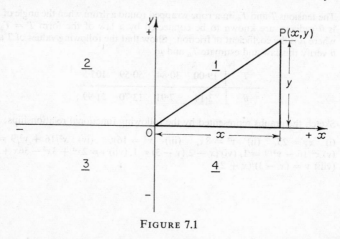

FIGURE 7.1

allows the position of any point to be specified or located. In two dimensional work the following *systems* are most frequently used: (i) rectangular cartesian, (ii) polar, (iii) parametric.

7.1 Rectangular Co-ordinates

This system has been used for all graphical work done so far. Two *axes of reference* are taken, at right angles to each other, their point of intersection being the *origin of co-ordinates*. The magnitudes of the co-ordinates are measured from the origin. Using this system, if an equation of some form connects the two co-ordinates, then this equation represents the *locus* of a moving point.

Example (see Figure 7.1):

The position of any point P is specified by two numbers (x, y). The numbers may be positive or negative, the combination of signs will specify the *quadrant* in which a point will lie, e.g.

the point $(3, 4)$ lies in quadrant 1,

the point $(-3, 4)$ lies in quadrant 2,

the point $(-3, -4)$ lies in quadrant 3,

and the point $(3, -4)$ lies in quadrant 4.

7.1.1 The equation of a locus

A locus is represented by an *algebraic* equation connecting the two co-ordinates x and y. For example, the equation $y = mx + c$, where m and c are constants *represents* the equation of a *straight line*. This locus has been well illustrated and $m = gradient$ (or *slope*) of the line with reference to the axis of x and $c = intercept$ of the line on the axis of y.

Circle: Consider the distance OP of the point P (see Figure 7.1) from the origin.

$$OP^2 = ON^2 + NP^2 = x^2 + y^2$$

Let $OP = r$, then

$$x^2 + y^2 = r^2$$

If the point P moves on a circle whose centre is O, with radius r, then OP is constant. Hence, for all points on the circle the two co-ordinates always obey the equation $x^2 + y^2 = r^2$ (constant).

This is the *equation of a circle*, radius r, centre O.

7.1.2 Distance between two points

If the co-ordinates of two points P and Q are denoted by (x_1, y_1) and (x_2, y_2) the distance between P and Q is readily found by using the theorem of *Pythagoras* (see Figure 7.2).

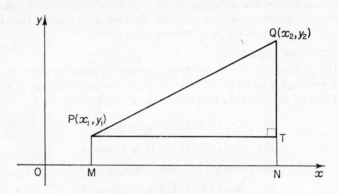

FIGURE 7.2

$$PQ^2 = PT^2 + TQ^2$$
$$= MN^2 + TQ^2$$
$$= (NO - MO)^2 + (QN - TN)^2$$
$$= (NO - MO)^2 + (QN - PM)^2$$

But $OM = x_1$, $ON = x_2$, $QN = y_2$, $PM = y_1$.

Hence $PQ^2 = (x_2 - x_1)^2 + (y_2 - y_1)^2$
or $\underline{PQ = \sqrt{[(x_2 - x_1)^2 + (y_2 - y_1)^2]}}$

This is an algebraic expression and may be used whatever the signs of the co-ordinates.

Examples:

(i) Find the distance between the points P(7, 10) and Q(2, 3).

$$PQ^2 = (x_2 - x_1)^2 + (y_2 - y_1)^2 = (7 - 2)^2 + (10 - 3)^2$$
$$= 25 + 49 = 74$$
$$\therefore \ PQ = \sqrt{74} = \underline{8 \cdot 602 \text{ units}}$$

(ii) Find the distance between the points P(5, 8) and Q(−3, −5).

$$PQ^2 = (x_2 - x_1)^2 + (y_2 - y_1)^2$$
$$= [5 - (-3)]^2 + [8 - (-5)]^2$$
$$= (5 + 3)^2 + (8 + 5)^2 = 8^2 + 13^2$$
$$= 64 + 169 = 233$$
$$\therefore \quad PQ = \sqrt{233} = \underline{15 \cdot 26 \text{ units}}$$

7.2 Polar Co-ordinates

The position of a point is specified by its distance r from a fixed point (the POLE) and the angle θ between OP and an *initial line*, i.e. P is specified as (r, θ) (see Figure 7.3).

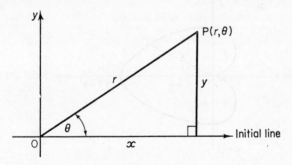

FIGURE 7.3

If the x-axis of a rectangular cartesian system be used as the *initial line*, then the polar co-ordinate system and rectangular system can be connected as follows:

$$r^2 = x^2 + y^2, \quad \tan \theta = y/x,$$
also
$$x = r \cos \theta, \quad y = r \sin \theta.$$

7.2.1 The polar equation of a locus

A locus may be represented by an equation connecting r and θ. The method of sketching a locus given by a *polar equation* is illustrated as follows.

Example:

Sketch the locus represented by the polar equation $r = a(1 + \cos \theta)$, where a is a constant.

A table of values may be drawn up to *plot* the locus in detail. The curve is *sketched* as follows:

When $\theta = 0$, $r = a(1 + 1) = 2a$. As θ increases from 0 to $\pi/2$, $\cos \theta$ decreases from 1 to 0 and r decreases gradually from $2a$ to a. For $\pi/2 < \theta < \pi$, $\cos \theta$ decreases from 0 to -1 and hence r decreases from a to $a(1 - 1) = 0$. It can be shown that, for $\pi < \theta < 2\pi$ similar conditions apply and a curve symmetrical about the initial line is obtained (see Figure 7.4).

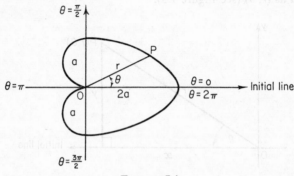

FIGURE 7.4

The curve is referred to as a *cardioid*.

7.3 Conversion Between Cartesian and Polar Equations

By using the relations in 7.2 it is possible to convert between Cartesian and polar equations. This will be illustrated with the following two examples.

Examples:

(i) Find the polar equation of the curve $y^2 = 4ax$. (This is the equation of a *parabola*—see Figure 7.5.)

$$x = r \cos \theta, \quad y = r \sin \theta$$

Hence the polar equation of the curve is given by

$$(r \sin \theta)^2 = 4a(r \cos \theta)$$
$$r^2 \sin^2 \theta = 4ar \cos \theta$$

or,

$$r = 4a \cos \theta / \sin^2 \theta = 4a \cot \theta \operatorname{cosec} \theta$$

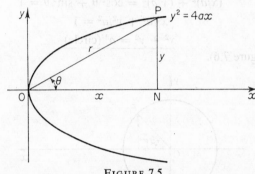

FIGURE 7.5

(ii) Find the Cartesian equation of the curve whose polar equation is $r^2(b^2 \cos^2 \theta + a^2 \sin^2 \theta) = a^2b^2$.

This may be written as $b^2(r^2 \cos^2 \theta) + a^2(r^2 \sin^2 \theta) = a^2b^2$ but $x = r \cos \theta$, $y = r \sin \theta$,

$$\therefore \qquad b^2x^2 + a^2y^2 = a^2b^2$$

or

$$\frac{x^2}{a^2} + \frac{y^2}{b^2} = 1$$

is the Cartesian equation of the curve. (The curve is an ellipse.)

7.4 Parametric Co-ordinates

Many curves may be represented by a third system of co-ordinates, i.e. *parametric co-ordinates*. A parametric co-ordinate is such that both x and y may be expressed as functions of the parameter. The curve is then traced out by the *parameter* varying *continuously*. The parameter involved may be *time*, an *angle*, or some other variable like the length of *arc* along a curve.

Examples:

(i) If the parametric equations of a curve are $x = a \cos \theta$, $y = a \sin \theta$, $0 \leqslant \theta \leqslant 2\pi$, what is the Cartesian equation of the curve and what is the locus?

The *parameter* is eliminated as follows:

$$x/a = \cos \theta, \quad y/a = \sin \theta$$

$\therefore \qquad\qquad (x/a)^2 + (y/a)^2 = \cos^2 \theta + \sin^2 \theta = 1$

or $\qquad\qquad\qquad x^2/a^2 + y^2/a^2 = 1$

$\therefore \qquad\qquad\qquad \underline{x^2 + y^2 = a^2} \text{ (circle)}$

(see Figure 7.6).

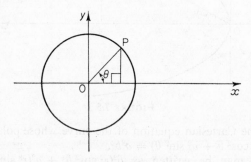

FIGURE 7.6

(*Note:* The parameter in this case is the angle θ shown and as θ increases from 0 to 2π the full circle is swept out.)

(ii) Eliminate the parameter t between the parametric equations $x = at^2$, $y = 2at$. What locus is represented by these equations?

$$t = y/2a, \quad \therefore \quad t^2 = (y/2a)^2 = x/a$$

$\therefore \qquad\qquad y^2/(4a^2) = x/a \quad \text{or} \quad \underline{y^2 = 4ax}$

i.e. the locus is a *parabola*.

7.5 Standard Cartesian and Parametric Equations

A number of standard loci, with appropriate Cartesian and parametric equations are stated here for reference. These equations refer to the straight line, circle, parabola, ellipse, hyperbola, and rectangular hyperbola.

(i) *Straight line.* The equation may be stated in various ways, although all forms are equivalent to $y = mx + c$ or $x = m_1y + c_1$. The general equation may be written as $ax + by + c = 0$. Generally parametric forms are not used but such equations could be obtained as follows (see Figure 7.7):

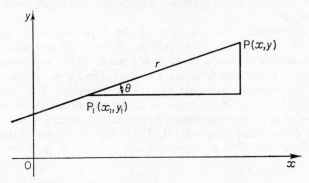

FIGURE 7.7

Let P_1 be a fixed point (x_1, y_1) on the line and $P(x, y)$ a variable point. Let $PP_1 = r$, θ is a constant for the line

$$x - x_1 = r \cos \theta, \quad x = x_1 + r \cos \theta \qquad (1)$$
$$y - y_1 = r \sin \theta, \quad y = y_1 + r \sin \theta \qquad (2)$$

$\therefore$ r may be treated as a parameter and the equations (1) and (2) may be treated as parametric equations of the line $(-\infty < r < \infty)$.

(ii) *Circle.* The simple equation of a circle of radius r, centre at the origin has already been derived as $x^2 + y^2 = r^2$ and the parametric equations of the curve are $x = r \cos \theta, y = r \sin \theta$ $(0 \leqslant \theta \leqslant 2\pi)$.

The next illustrations are of *conic sections*, which are the shapes of curves obtained by taking plane sections of a cone at various angles to the axis of the cone.

(iii) *Parabola.* (*a*) With its *axis* along Ox, *vertex* at the origin, the equation may be written as $y^2 = 4ax$, with parameteric equations $x = at^2, y = 2at, -\infty < t < \infty$.

(b) With its *axis* along Oy, *vertex* at the origin, the equation may be written as $x^2 = 4ay$, with parametric equations $x = 2at$, $y = at^2$, $-\infty < t < \infty$.

(c) With its *axis* parallel to the y-axis the general equation may be written as $y = ax^2 + bx + c$.

(iv) *Ellipse.* The equation of an ellipse with its *axes* (major and minor) along the axes Ox and Oy respectively may be written as $x^2/a^2 + y^2/b^2 = 1$ with parametric equations $x = a\cos\theta$, $y = b\sin\theta$, $0 \leqslant \theta \leqslant 2\pi$.

(v) *Hyperbola.* The equation of the hyperbola with its *axis* along the axis Ox may be written as $x^2/a^2 - y^2/b^2 = 1$ with parametric equations $x = a\sec\theta$, $y = b\tan\theta$. The hyperbola has *asymptotes* whose equations are $y = +bx/a$ and $y = -bx/a$.

(vi) *Rectangular hyperbola* (1). This hyperbola has the two *asymptotes* at *right angles* and $a = b$. Its equation may be written as $x^2 - y^2 = a^2$ with parametric equations $x = a\sec\theta$, $y = a\tan\theta$.

(vii) *Rectangular hyperbola* (2). Using the asymptotes as the axes of Ox and Oy. This equation reduces to $xy = c^2$ (c a constant)

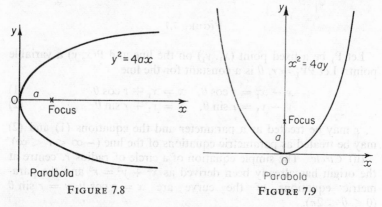

FIGURE 7.8   FIGURE 7.9

with parametric equations $x = ct$, $y = c/t$.
Figures 7.8 to 7.12 show sketches of some of these loci.

WORKED EXAMPLES

(a) Find: (i) the distance between the two points $(-2, -4)$ and $(3, 6)$. (ii) The equation of the line joining these two points.

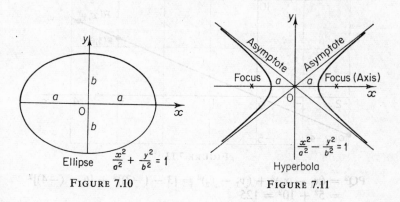

Ellipse $\dfrac{x^2}{a^2} + \dfrac{y^2}{b^2} = 1$

FIGURE 7.10

Hyperbola

$\dfrac{x^2}{a^2} - \dfrac{y^2}{b^2} = 1$

FIGURE 7.11

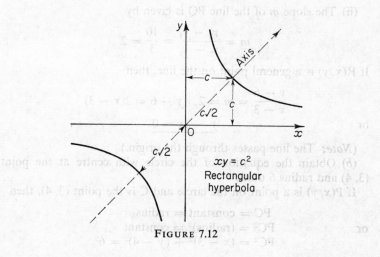

$xy = c^2$

Rectangular hyperbola

FIGURE 7.12

(i) Let the points be P and Q (Figure 7.13).

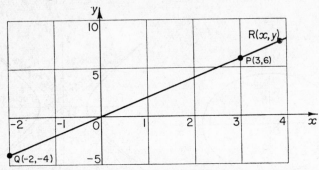

FIGURE 7.13

$$PQ^2 = (x_1 - x_2)^2 + (y_1 - y_2)^2 = [3 - (-2)]^2 + [6 - (-4)]^2$$
$$= 5^2 + 10^2 = 125$$

$\therefore \qquad PQ = \sqrt{125} = 5\sqrt{5} = 5 \times 2\cdot236 = \underline{11\cdot18 \text{ units}}$

(ii) The slope m of the line PQ is given by

$$m = \frac{y_2 - y_1}{x_2 - x_1} = \frac{10}{5} = \underline{2}$$

If R(x, y) is a general point on the line, then

$$\frac{y - 6}{x - 3} = m = 2, \quad y - 6 = 2(x - 3)$$

or $\qquad \underline{y - 2x = 0}$

(*Note:* The line passes through the origin.)

(*b*) Obtain the equation of the circle with centre at the point (3, 4) and radius 6 units.

If P(x, y) is a point on the circle and C is the point (3, 4), then

PC = constant = radius

or $\qquad$ PC^2 = (radius)^2 = constant

$\qquad$ PC$^2 = (x - 3)^2 + (y - 4)^2 = 6^2$

$\therefore \qquad x^2 - 6x + 9 + y^2 - 8y + 16 = 36$

or $\qquad \underline{x^2 + y^2 - 6x - 8y - 11 = 0}$

is the equation of the circle.

(c) The equation of an ellipse is $x^2/4 + y^2/9 = 1$. Find the co-ordinates of the points where the ellipse cuts the two co-ordinate axes and convert the equation to its polar form.

The curve meets the x-axis where $y = 0$, i.e.

$$x^2/4 = 1, \quad x^2 = 4; \quad \underline{x = \pm 2}$$

or the curve meets the x-axis at $(-2, 0)$, $(2, 0)$.

The curve meets the y-axis where $x = 0$, i.e.

$$y^2/9 = 1, \quad y^2 = 9; \quad \underline{y = \pm 3}$$

or the curve meets the y-axis at $(0, 3)$, $(0, -3)$.

The polar equation is derived by putting $x = r \cos \theta$, $y = r \sin \theta$, i.e.

$$(r \cos \theta)^2/4 + (r \sin \theta)^2/9 = 1$$

or $$\underline{r^2(9 \cos^2 \theta + 4 \sin^2 \theta) = 36}$$

(d) Find the condition that the line $y = mx + c$ shall be a tangent to the ellipse $x^2 + 2y^2 = 2$.

The line meets the curve where:

$$x^2 + 2(mx + c)^2 = 2; \quad x^2 + 2(m^2x^2 + 2cmx + c^2) = 2$$

i.e. $$x^2(1 + 2m^2) + 4cmx + 2(c^2 - 1) = 0$$

Now the line will be a *tangent* if it meets the curve in *two coincident points*. This will be so if the roots of the quadratic equation for x are *equal*, i.e. if

$$(4cm)^2 - 4(1 + 2m^2) \times 2(c^2 - 1) = 0$$
$$8[2c^2m^2 - (c^2 - 1)(1 + 2m^2)] = 0$$
$$2c^2m^2 - c^2 - 2c^2m^2 + 1 + 2m^2 = 0$$

or $$\underline{c^2 = 1 + 2m^2}$$

is the condition that the line should be a tangent.

(e) The parametric equations of the rectangular hyperbola $xy = c^2$ are $x = ct$, $y = c/t$. Find the equation of the chord joining the two points whose parameters are t_1 and t_2. Deduce the equation of the tangent to the curve at the point t (see Figure 7.14).

The slope of the chord PQ

$$= \frac{y_1 - y_2}{x_1 - x_2} = \frac{c/t_1 - c/t_2}{ct_1 - ct_2} = \frac{c(t_2 - t_1)}{ct_1 t_2(t_1 - t_2)} = -\frac{1}{t_1 t_2}$$

The equation of the chord may be written as

$$y - y_1 = m(x - x_1); \quad y - c/t_1 = (-1/t_1 t_2)(x - ct_1)$$

$$t_1 t_2 y - ct_2 = -x + ct_1$$

or $\underline{t_1 t_2 y + x = c(t_1 + t_2)}$

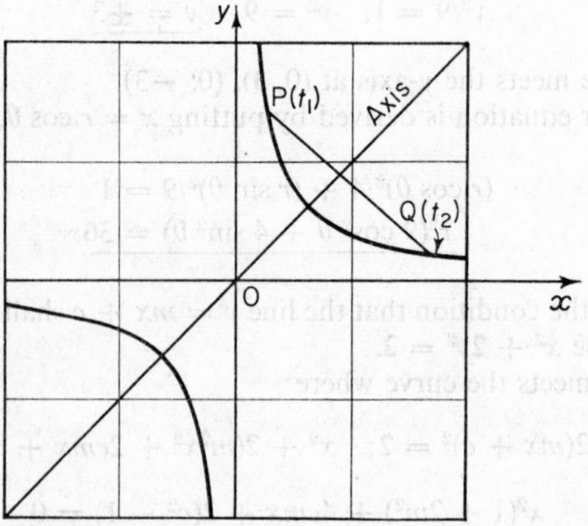

FIGURE 7.14

For the tangent at t, make $t_1 = t_2 = t$, then the equation of the tangent at t is given by

$$(t \times t)y + x = c(t + t) \quad or \quad \underline{x + t^2 y = 2ct}$$

(f) Sketch the curve whose polar equation is $4/r = 1 + \frac{1}{4} \cos \theta$. The equation may be written as

$$\frac{16}{r} = 4 + \cos \theta \quad or \quad r = \frac{16}{4 + \cos\theta}$$

When $\theta = 0$, $\cos \theta = 1$,

$$r = 16/5$$

as θ increases from 0 to $\pi/2$, $\cos \theta$ decreases from 1 to 0,

∴ r increases from 16/5 to 16/4, i.e. 16/5 to 4

When $\theta = \pi$, $\cos \theta = -1$, $r = 16/3$,

∴ r increases from 4 to 16/3

The curve will be symmetrical about the initial line.

The curve is an ellipse whose *focus* is at O (Figure 7.15).

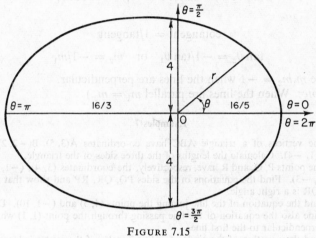

FIGURE 7.15

(g) Prove that two lines are perpendicular when $m_1 m_2 = -1$, where m_1 and m_2 are the gradients of the lines.

Consider the sketch in Figure 7.16.

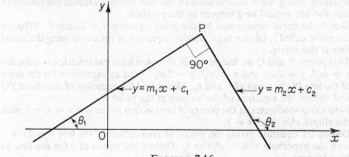

FIGURE 7.16

Let the two lines meet at point P. If the lines make angles θ_1 and θ_2 with Ox, then, with equal scales, $m_1 = \tan\theta_1$, $m_2 = \tan\theta_2$.

If the lines are *perpendicular*, then

$$\theta_2 = 90° + \theta_1$$

$$\therefore \quad \tan\theta_2 = \tan(90° + \theta_1) = -\tan(180° - 90° - \theta_1)$$

$$\therefore \quad \tan\theta_2 = -\tan(90° - \theta_1) = -\cot\theta_1$$

but

$$\text{cotangent} = 1/\text{tangent}$$

$$\therefore \quad \tan\theta_2 = -1/\tan\theta_1 \quad \text{or} \quad m_2 = -1/m_1$$

Hence $m_1 m_2 = -1$ when the lines are perpendicular.

(*Note:* When the lines are parallel $m_1 = m_2$.)

Examples 7

1. The vertices of a triangle ABC have co-ordinates A(3, 5), B(−2, 2) and C(1, −4). Calculate the lengths of the three sides of the triangle.
2. The points P, Q, and R have, respectively, the coordinates (3, 4), (−1, +1), (2, −3). Find the equations of the sides PQ, QR, RP and show that angle PQR is a right angle.
3. Find the equation of the line joining the points (4, 5) and (−1, 10). Determine also the equation of the line passing through the point (1, 1) which is perpendicular to the first line.
4. Find the equation of the circle of radius 7 inches if the centre is located at the point whose co-ordinates are (3, 4) inches.
5. Find the co-ordinates of the points of intersection of the circle $x^2 + y^2 = 25$ and the line $x - y + 3 = 0$.
6. The line $y = mx + c$ meets the parabola $y^2 = 4ax$. Find the quadratic equation giving the x co-ordinates of the two points. Deduce the condition that the line should be a tangent to the parabola.
7. Sketch the curve represented by the polar equation $r = 2a\cos\theta$. What is the curve called? Obtain the Cartesian equation of the curve using the initial line as the x-axis.
8. Two points P and Q on the parabola $y^2 = 4ax$ have parametric co-ordinates $x = at_1^2$, $y = 2at_1$, and $x = at_2^2$, $y = 2at_2$. Find an expression for the slope of the line PQ in terms of t_1 and t_2. Hence find the equation of the chord PQ and deduce the equation of the tangent at the point t.
9. Find the co-ordinates of the points of intersection of the line $y = x - 1$ with the ellipse $x^2/6 + y^2/4 = 1$.
10. Obtain the equation giving the points of intersection of the line $y = 2x + c$ with the hyperbola $x^2/6 - y^2/4 = 1$. Deduce the values of c for the line to be a tangent.

11. By drawing up a suitable table of values plot the graph of $r = a\cos 2\theta$. (*Note*: r is positive only.)

12. The normal to a curve is the line through the point of contact of the tangent, at right angles to the tangent.
    (i) If the equation of the tangent to the curve $y^2 = 4ax$ at the point $(at^2, 2at)$ is $y = x/t + at$ find the equation of the normal.
    (ii) If the equation of the tangent to $xy = c^2$ at the point $(ct, c/t)$ is $x + t^2y = 2ct$ find the equation of the normal.

13. Find the co-ordinates of the point of intersection of the rectangular hyperbola $xy = 2$ and the parabola $y^2 = 4x$.

14. The points A and B lie on the x-axis. A is the point $(4, 0)$ and B the point $(-6, 0)$. If P is any point (x, y) find the locus of P if (i) PA = PB, (ii) PA = 2PB.

15. Find the equation of the circle which has the line AB as a diameter where A is the point $(5, 0)$ and B is the point $(0, 3)$. (*Note:* The angle in a semicircle is a right angle.)

8

Trigonometry Revision—Periodic Functions—Compound Angles

8.1 Trigonometry Revision

Trigonometry was dealt with in some detail in Volume I. Readers are recommended to read the relevant chapters again. However, for revision purposes, some of the more important results are reproduced here.

8.1.1 Angles of any magnitude

Given a suitable value of any trigonometric ratio (sine, cosine, tangent, cosecant, secant, cotangent), since these functions are *periodic*, i.e. repeat regularly, many angles may be written down.

Examples:

(i) If $\sin \theta = \sin \alpha$, where α is an angle between $0°$ and $360°$ (*or* 0 and 2π radians), then

$$\theta = n180° + (-1)^n\alpha \quad or \quad \theta = n\pi + (-1)^n\alpha \text{ (radians)}.$$

(ii) If $\cos \theta = \cos \alpha$, where $0 < \alpha < 180°$ (*or* $0 < \alpha < \pi$ radians),

$$\theta = n360° \pm \alpha \quad or \quad \theta = 2n\pi \pm \alpha \text{ (radians)}.$$

(iii) If $\tan \theta = \tan \alpha$, where $0 < \alpha < 180°$ (*or* $0 < \alpha < \pi$ radians),

$$\theta = n180° + \alpha \quad or \quad \theta = n\pi + \alpha \text{ (radians)},$$

where, in each case, n is a positive *integer*. The ASTC rule will be found useful when dealing with angles of any magnitude.

8.1.2 Identities

A number of standard identities were established in Volume I, some of which are as follows (θ is any angle):

$$\operatorname{cosec} \theta = 1/\sin \theta, \quad \sec \theta = 1/\cos \theta, \quad \cot \theta = 1/\tan \theta$$

$$\sin^2 \theta + \cos^2 \theta = 1, \quad 1 + \tan^2 \theta = \sec^2 \theta, \quad 1 + \cot^2 \theta = \operatorname{cosec}^2 \theta$$

$$\tan \theta = \sin \theta/\cos \theta, \quad \cot \theta = \cos \theta/\sin \theta$$

Many other identities may be established by making use of the rules of algebra and a few standard identities.

8.1.3 Rules for the solution of triangles

(To be used in Chapter 9.)
Basically, most trigonometric problems may be solved by breaking the problem down into right-angled triangles. However, as was shown in Volume I, the labour of numerical solution may be reduced by applying rules for the general solution of triangles.

Results for a triangle ABC, generally used when the triangle is not right angled (the angles are normally labelled A, B, and C and the opposite sides a, b, and c) are as follows:

Sine rule:

$$\frac{a}{\sin A} = \frac{b}{\sin B} = \frac{c}{\sin C} = 2R$$

(R is the radius of the circumcircle.)

Cosine rule:

$$a^2 = b^2 + c^2 - 2bc \cos A$$
and $$\cos A = (b^2 + c^2 - a^2)/(2bc)$$

It will be remembered that when an angle is *obtuse* (i.e. between 90° and 180°) then if the angle is A,

$$\sin A = \sin (180° - A); \quad \cos A = - \cos (180° - A)$$

8.1.4 Formulae for the area of a triangle ($\triangle$)

$$\triangle = \tfrac{1}{2}bc \sin A \ (= \tfrac{1}{2}ac \sin B = \tfrac{1}{2}ab \sin C)$$
$$\triangle = \sqrt{\{s(s-a)(s-b)(s-c)\}}$$

where s = semi-perimeter = $\tfrac{1}{2}(a + b + c)$.

8.2 Periodic Functions

A general periodic function of *period* T may be defined as $f(x) = f(x + T)$, i.e. the function repeats exactly at intervals of T (the period). The trigonometric functions are special cases of periodic functions.

Consider the function $y = a \sin (x + \alpha)$, where a and α are constants, then a is called the *amplitude* of the function and α the *phase angle*. Note that α may be *positive* or *negative*.

When α is *positive* the *phase* angle is a *lead* angle.

When α is *negative* the *phase* angle is a *lag* angle.

The *frequency* of a periodic function is the number of *cycles* which occur, per *second*. An additional term used with periodic functions is the '*angular velocity*'. This arises originally in connection with the

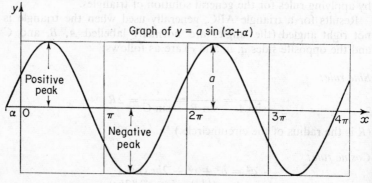

FIGURE 8.1

angular velocity in a circular motion. If ω is the *angular velocity* and f the frequency in cycles per second, then $\omega = 2\pi f$ and $f = \omega/2\pi$.

When the periodic function has a *maximum* (positive) and a

minimum (negative), the values of these maxima and minima are often referred to as the positive and negative *peak values*.

Figure 8.1 illustrates 2 *cycles* of a periodic function of period 2π, amplitude *a*, phase angle α.

Notice that the *phase* angle is a measure of the amount the function is *out of step* (or out of *phase*) with the function $a \sin x$. The *period* of this function is 2π.

8.2.1 General expression for a periodic function

If a periodic function has an *angular velocity* ω, period *T*, and frequency *f*, then

$$T = 2\pi/\omega, \quad f = 1/T, \quad \omega = 2\pi f$$

Using a *sine function*, then if *t* is the time in seconds, $x = \omega t$, hence a periodic function $f(x)$ of period *T* may be represented as:

$$f(x) = a \sin (x + \alpha) = a \sin (\omega t + \alpha) = a \sin (2\pi f t + \alpha)$$

If t_1 is the *time phase*, then $\alpha = \omega t_1 = 2\pi f t_1$,

$$\therefore \qquad f(x) = a \sin (\omega t + \omega t_1) = a \sin \omega(t + t_1)$$
$$= a \sin (2\pi f t + 2\pi f t_1) = \underline{a \sin 2\pi f(t + t_1)}$$

Examples:

(i) An alternating voltage has a peak value of 150 volts, and a frequency of 50 cycles per second. Express the voltage *V* as a time function if, at time $t = 0$ the voltage is 75 volts.

The voltage *V* may be represented as a sine function as follows:

$$V = 150 \sin (2\pi f t + \alpha)$$

where α is a phase angle.

$f = 50$ c/s,

$$\therefore \qquad V = 150 \sin (2\pi \times 50t + \alpha) = 150 \sin (100\pi t + \alpha)$$

at $t = 0$,

$$75 = 150 \sin \alpha; \quad \sin \alpha = 75/150 = 0 \cdot 5$$

$\therefore \alpha$ could be 30° or $\pi/6$ radians,

$$\therefore \qquad \underline{V = 150 \sin (100\pi t + \pi/6)}$$

(ii) The displacement s inches of a particle during a vibration, at time t seconds is given by $s = 25 \sin (50t + 0.03)$. State the amplitude, period, frequency, and time phase.

Amplitude = 25 inches; angular velocity $\omega = 50$ rad/sec.

Period $T = 2\pi/\omega = 2\pi/50 = \underline{\pi/25}$ second.

Frequency $f = \omega/2\pi = \underline{25/\pi}$ cycles per second.

Time phase = $0.03/\omega = 0.03/50 = \underline{0.0006}$ second.

8.3 Compound Angle Formulae

These formulae were derived in Volume I and are quoted here for reference:

(i) $\sin (A + B) = \sin A \cos B + \cos A \sin B$

(ii) $\cos (A + B) = \cos A \cos B - \sin A \sin B$

(iii) $\sin (A - B) = \sin A \cos B - \cos A \sin B$

(iv) $\cos (A - B) = \cos A \cos B + \sin A \sin B$

(v) $\tan (A + B) = \dfrac{\tan A + \tan B}{1 - \tan A \tan B}$

(vi) $\tan (A - B) = \dfrac{\tan A - \tan B}{1 + \tan A \tan B}$

It should be remembered that most of the formulae may be established using basic formulae.

Example:

Express $\tan (A + B)$ in terms of $\tan A$ and $\tan B$.

$$\tan (A + B) = \frac{\sin (A + B)}{\cos (A + B)} = \frac{\sin A \cos B + \cos A \sin B}{\cos A \cos B - \sin A \sin B}$$

Divide numerator and denominator by $\cos A \cos B$, then

$$\tan (A + B) = \frac{\sin A/\cos A + \sin B/\cos B}{1 - \sin A \sin B/(\cos A \cos B)} = \frac{\tan A + \tan B}{1 - \tan A \tan B}$$

8.3.1 Double angle formulae

$\sin 2\theta = 2 \sin \theta \cos \theta$

$\cos 2\theta = \cos^2 \theta - \sin^2 \theta = 2 \cos^2 \theta - 1 = 1 - 2 \sin^2 \theta$

$\tan 2\theta = 2 \tan \theta/(1 - \tan^2 \theta)$

These formulae may be readily established using the formulae in 8.3.

Example:

Express $\cos 2\theta$ in terms of $\cos \theta$.

$$\cos 2\theta = \cos (\theta + \theta) = \cos \theta \cos \theta - \sin \theta \sin \theta$$
$$= \cos^2 \theta - \sin^2 \theta$$

but $\cos^2 \theta + \sin^2 \theta = 1, \quad \therefore \; \sin^2 \theta = 1 - \cos^2 \theta$

Hence $\cos 2\theta = \cos^2 \theta - (1 - \cos^2 \theta) = \underline{2 \cos^2 \theta - 1}$

8.3.2 Sum and product formulae

(i) Consider

$$\sin (A + B) = \sin A \cos B + \cos A \sin B$$

and $$\sin (A - B) = \sin A \cos B - \cos A \sin B$$

By addition of these two *identities*

$$\sin (A + B) + \sin (A - B) = 2 \sin A \cos B$$

and by subtraction

$$\sin (A + B) - \sin (A - B) = 2 \cos A \sin B$$

If $P = A + B$, $Q = A - B$, then $A = \frac{1}{2}(P + Q)$, $B = \frac{1}{2}(P - Q)$,

then:
$$\sin P + \sin Q = 2 \sin \tfrac{1}{2}(P + Q) \cos \tfrac{1}{2}(P - Q)$$
$$\sin P - \sin Q = 2 \cos \tfrac{1}{2}(P + Q) \sin \tfrac{1}{2}(P - Q)$$

also:
$$\sin A \cos B = \tfrac{1}{2}[\sin (A + B) + \sin (A - B)]$$
$$\cos B \sin A = \tfrac{1}{2}[\sin (A + B) - \sin (A - B)]$$

These are the sum and product formulae using sines.

(ii) Similarly, by considering

$$\cos (A + B) = \cos A \cos B - \sin A \sin B$$
$$\cos (A - B) = \cos A \cos B + \sin A \sin B$$

the following formulae are derived:

$$\cos P + \cos Q = 2 \cos \tfrac{1}{2}(P + Q) \cos \tfrac{1}{2}(P - Q)$$
$$\cos P - \cos Q = 2 \sin \tfrac{1}{2}(P + Q) \sin \tfrac{1}{2}(Q - P)$$
$$\cos A \cos B = \tfrac{1}{2}[\cos (A + B) + \cos (A - B)]$$
$$\underline{\sin A \sin B = \tfrac{1}{2}[\cos (A - B) - \cos (A + B)]}$$

WORKED EXAMPLES

(a) Find all the angles θ between $0°$ and $360°$ which satisfy the equation $\cos 2\theta = 0.75$.

The smallest positive angle for 2θ is $41° 24'$,

$$\therefore \qquad 2\theta = n360° \pm 41°24' \quad (n = 0, 1, 2, 3, \ldots)$$
$$\theta = n180° \pm 20°42'$$
$$\underline{= 20°42', 159°18', 200°42', 339°18'}$$

(b) Using sums and products simplify
$(\cos 3\theta + \cos 5\theta)/(\sin 5\theta - \sin 3\theta)$.

$$\frac{\cos 3\theta + \cos 5\theta}{\sin 5\theta - \sin 3\theta} = \frac{2 \cos \tfrac{1}{2}(3\theta + 5\theta) \cos \tfrac{1}{2}(5\theta - 3\theta)}{2 \cos \tfrac{1}{2}(5\theta + 3\theta) \sin \tfrac{1}{2}(5\theta - 3\theta)}$$

$$= \frac{2 \cos 4\theta \cos \theta}{2 \cos 4\theta \sin \theta} = \frac{\cos \theta}{\sin \theta} = \underline{\cot \theta}$$

(c) Without using trigonometric tables evaluate $\cos 15°$.

Method (i)
$$\cos 15° = \cos (45° - 30°) = \cos 45° \cos 30° + \sin 45° \sin 30°$$

$$= \frac{1}{\sqrt 2} \frac{\sqrt 3}{2} + \frac{1}{\sqrt 2} \frac{1}{2} = \frac{\sqrt 3 + 1}{2\sqrt 2} = \frac{(\sqrt 3 + 1)\sqrt 2}{4}$$

$$= \frac{\sqrt 6 + \sqrt 2}{4} = \frac{2.449 + 1.4142}{4} = \underline{0.9658}$$

Method (ii)
$$\cos 2\theta = 2 \cos^2 \theta - 1, \text{ let } \theta = 15°$$
$$\cos 30° = 2 \cos^2 15° - 1$$
$$\cos^2 15° = \tfrac{1}{2}(1 + \cos 30°)$$
$$\cos^2 15° = \tfrac{1}{2}\left(1 + \frac{\sqrt 3}{2}\right) = \tfrac{1}{4}(2 + \sqrt 3) = \tfrac{1}{4}(3.7321) = 0.9330$$
$$\cos 15° = \sqrt{0.9330} = \underline{0.9659}$$

The small difference is explained by the fact that the square roots are given to the nearest 4th significant figure only.

(d) Two voltages E_1 and E_2 are given by $E_1 = 250 \sin (2\pi t + \alpha)$, $E_2 = 250 \sin (2\pi t - \alpha)$, find E when $E = E_1 + E_2$.
$$E = E_1 + E_2 = 250[\sin (2\pi t + \alpha) + \sin (2\pi t - \alpha)]$$
$$= 250 \times 2 \sin \tfrac{1}{2}[2\pi t + \alpha + 2\pi t - \alpha] \cos \tfrac{1}{2}[2\pi t + \alpha - (2\pi t - \alpha)]$$
$$= \underline{500 \sin 2\pi t \cos \alpha}$$

(e) If $\tan 2A = 1\cdot 8$, find the possible values for $\tan A$ without using trigonometric tables.

$$\tan 2A = 2 \tan A/(1 - \tan^2 A) = 1\cdot 8$$

$$1\cdot 8 - 1\cdot 8 \tan^2 A = 2 \tan A; \quad \tan^2 A + (2/1\cdot 8) \tan A - 1 = 0$$
$$\tan^2 A + 1\cdot 1111 \tan A - 1 = 0$$

$$\therefore \quad \tan A = [-1\cdot 1111 \pm \sqrt{(1\cdot 1111^2 + 4)}]/2$$
$$= [-1\cdot 1111 \pm \sqrt{5\cdot 235}]/2 = [-1\cdot 1111 \pm 2\cdot 288]/2$$
$$= -3\cdot 3991/2 \ or \ 1\cdot 1769/2 = -1\cdot 69955 \ or \ 0\cdot 58845$$
$$= \underline{-1\cdot 6996 \ or \ 0\cdot 5884 \text{ (to 4 decimal places)}}$$

(f) Express $5 \sin 2\theta + 8 \cos 2\theta$ in the form $R \sin (2\theta + \alpha)$.

Let
$$5 \sin 2\theta + 8 \cos 2\theta = R \sin (2\theta + \alpha)$$
$$= R[\sin 2\theta \cos \alpha + \cos 2\theta \sin \alpha]$$

Equating coefficients of $\cos 2\theta$ and $\sin 2\theta$

$$R \cos \alpha = 5, \quad R \sin \alpha = 8$$

$$R^2(\cos^2 \alpha + \sin^2 \alpha) = 5^2 + 8^2 = 89$$
$$R = \sqrt{89} = \underline{9\cdot 434}$$

$$\therefore$$

$$R \sin \alpha/R \cos \alpha = \tan \alpha = 8/5 = 1\cdot 6$$

$$\therefore \qquad \alpha = 58°$$

Hence
$$\underline{5 \sin 2\theta + 8 \cos 2\theta = 9\cdot 434 \sin (2\theta + 58°)}$$

(g) Given $\cos A = 3/5\,(0 < A < 90°)$, $\sin B = 5/13\,(90° < B < 180°)$ evaluate $\cos(A + B)$, $\sin(B - A)$.

$\cos(A + B) = \cos A \cos B - \sin A \sin B = (3/5)(-12/13) - (4/5)(5/13)$
$= -36/65 - 20/65 = \underline{-56/65}$

$\sin(B - A) = \sin B \cos A - \cos B \sin A = (5/13)(3/5) - (-12/13)(4/5)$
$= 15/65 + 48/65 = \underline{63/65}$

(h) Express $E = 50 \sin(20\pi t - \pi/3) + 100 \cos(20\pi t - \pi/4)$ in the form $R \sin(20\pi t + \alpha)$. What are the amplitude, period, and phase angle of the resultant wave form?

$50 \sin(20\pi t - \pi/3) + 100 \cos(20\pi t - \pi/4)$

$= 50[\sin 20\pi t \cos \pi/3 - \cos 20\pi t \sin \pi/3]$
$\qquad\qquad + 100[\cos 20\pi t \cos \pi/4 + \sin 20\pi t \sin \pi/4]$
$= 50[\sin 20\pi t(\tfrac{1}{2}) - \cos 20\pi t(\tfrac{1}{2}\sqrt{3})]$
$\qquad\qquad + 100[\cos 20\pi t(1/\sqrt{2}) + \sin 20\pi t(1/\sqrt{2})]$
$= \sin 20\pi t[25 + 50\sqrt{2}] + \cos 20\pi t[50\sqrt{2} - 25\sqrt{3}]$
$= 95{\cdot}7 \sin 20\pi t + 27{\cdot}4 \cos 20\pi t$
$= R \sin(20\pi t + \alpha) = R[\sin 20\pi t \cos \alpha + \cos 20\pi t \sin \alpha]$

$\therefore \qquad R \cos \alpha = 95{\cdot}7, \quad R \sin \alpha = 27{\cdot}4$

$R^2 = 95{\cdot}7^2 + 27{\cdot}4^2 = 9157 + 751 = 9908$
$R = \sqrt{9908} = \underline{99{\cdot}54}$
$\tan \alpha = 27{\cdot}4/95{\cdot}7 = 0{\cdot}2864$
$\alpha = 15° 29' = \underline{0{\cdot}279}$ (rad)

$\therefore \qquad \underline{E = 99{\cdot}54 \sin(20\pi t + 0{\cdot}279)}$

The amplitude of $E = 99{\cdot}54$, $\omega = 20\pi$
Period $= 2\pi/\omega = 2\pi/20\pi = 1/10 = 0{\cdot}1$ sec.
Phase angle $= \alpha = 0{\cdot}279$ rad (or $15° 59'$).

Examples 8

1. (i) Using the expansion for $\cos(A + B)$, show that $\cos 2\theta = 2\cos^2\theta - 1$ and hence find the value of $\cos 22\tfrac{1}{2}°$ without using trigonometric tables.
(ii) Given A is acute and $\sin A = 3/5$, evaluate $\sin 2A$, $\cos 2A$ and $\tan 2A$ without using trigonometric tables.

2. (i) Given $\tan A = 3/4$ (A acute), $\sin B = 5/13$ (B obtuse) calculate, without using trigonometric tables, the values of $\sin (A - B)$ and $\cos 2A - \cos 2B$.
   (ii) If $180° < A < 270°$ and $180° > B > 90°$, $\sin A = -12/13$, $\tan B = -4/3$ calculate $\sin (A - B)$, $\cos (A + B)$ and $\cos 2B$ without using trigonometric tables.

3. (i) Given $\cos \theta = 0{\cdot}66$ (θ acute), find the values of $\sin 2\theta$, $\cos 2\theta$ and $\tan 2\theta$ without evaluating θ.
   (ii) If $\tan(A + B) = 2{\cdot}1$ and $\tan A = 0{\cdot}56$ find $\tan B$.

4. Write down all the solutions between $0°$ and $360°$ of:
   (i) $\cos 2\theta = 0{\cdot}8$, (ii) $\sin 3x = 0{\cdot}52$, (iii) $\tan \frac{3}{2}\theta = 1{\cdot}8$,
   (iv) $\sin 4t = -0{\cdot}7$, (v) $\cos \frac{5}{2}x = 0{\cdot}58$.

5. (i) Express the following as products:
   (a) $\sin 50° + \sin 30°$, (b) $\cos 70° - \cos 20°$, (c) $\sin 110° - \sin 40°$,
   (d) $\sin 5x + \sin 3x$, (e) $\cos 5x - \cos x$, (f) $\sin 5\theta - \sin 3\theta$.

   (ii) Express the following as sums or differences:
   (a) $2 \sin 45° \cos 15°$, (b) $2 \cos 70° \cos 30°$, (c) $2 \sin 50° \sin 10°$,
   (d) $\cos 3x \cos 5x$, (e) $\sin 3t \sin t$, (f) $\sin 2\alpha \sin 3\alpha$.

6. Prove the following identities:
   (i) $\cot \theta + \tan \theta = 2 \operatorname{cosec} 2\theta$, (ii) $\operatorname{cosec} x - \cot x = \tan \frac{1}{2}x$.
   (iii) $\dfrac{\sin (2x + \alpha) + \sin(2x - \alpha)}{\cos (x + \alpha) + \cos (x - \alpha)} = 2 \sin x$,
   (iv) $\dfrac{\sin A + \sin B}{\sin A - \sin B} = \tan \frac{1}{2}(A + B) \cot \frac{1}{2}(A - B)$,
   (v) $\dfrac{\sin 3\theta + \sin \theta}{\cos 3\theta + \cos \theta} = \tan 2\theta$.

7. Prove the following identities:
   (i) $\sin 3\theta = 3 \sin \theta - 4 \sin^3 \theta$, (ii) $\cos 3\theta = 4 \cos^3 \theta - 3 \cos \theta$.

8. Using sums and differences expressed as products, simplify:
   (i) $(\cos 5\theta - \cos 7\theta)/(\sin 7\theta + \sin 5\theta)$,
   (ii) $(\cos 2\theta - \cos 5\theta)/(\sin 2\theta + \sin 5\theta)$.

9. Express $8 \cos \theta - 15 \sin \theta$ in the form $R \cos (\theta + \alpha)$ and hence find the maximum and minimum values of the function and the smallest positive values of θ at which these occur.

10. Express $6 \cos \omega t - 5 \sin \omega t$ in the form $R \cos (\omega t + \alpha)$ and hence find the maximum and minimum values of the function and the smallest values of t at which they occur (α in radians).

11. Two voltages $E_1 = 40 \sin 50\pi t$ and $E_2 = 30 \cos 50\pi t$ are combined. Express $E_1 + E_2$ and $E_1 - E_2$ in the compound angle forms $R_1 \sin (50\pi t + \alpha_1)$ and $R_2 \sin (50\pi t - \alpha_2)$. State the amplitude, period, and phase angle in each case.

12. Express $E = 100 \sin (40\pi t + \pi/6) + 100 \cos (40\pi t - \pi/3)$ in the form $R \sin (40\pi t + \alpha)$. What are the amplitude, period and phase angle of E?

13. The displacement s of a moving particle is given by the expression $s = 8 \sin (4\pi t + \pi/4)$. What are the amplitude, period, frequency, and phase angle of the vibration?

14. Prove that (i) $2 \tan A/(1 + \tan^2 A) = \sin 2A$,

(ii) $(1 - \tan^2 A)/(1 + \tan^2 A) = \cos 2A$.

15. Express $8 \cos (\theta + 30°) + 10 \sin (\theta + 20°)$ in the form $R \sin (\theta + \alpha)$.

9

Trigonometric Equations—Problems—Half-angle Formulae

9.1 Trigonometric Equations

The methods of solution of trigonometric equations require the use of trigonometric identities. Also, in order to write down all the solutions in a required range, the general solutions of equations of the types $\sin \theta = \sin \alpha$, $\cos \theta = \cos \alpha$, $\tan \theta = \tan \alpha$ are necessary. These general solutions (i.e. angles) were stated in 8.1.1.

In addition to analytic solutions, *graphical solutions* can also be used or approximation methods using Newton's approximation. Examples of analytic solutions are given in the following four examples.

Examples:

(i) Find all the solutions of the equation $3 \sec^2 \theta = 2 \tan \theta + 5$ which lie between $0°$ and $360°$.

Using the identity $\sec^2 \theta = 1 + \tan^2 \theta$ the equation reduces to:

$$3(1 + \tan^2 \theta) = 2 \tan \theta + 5$$

i.e. $$3 \tan^2 \theta - 2 \tan \theta - 2 = 0$$

This is a *quadratic equation* for $\tan \theta$ and is satisfied by

$$\tan \theta = \{2 \pm \sqrt{[(-2)^2 - 4(3)(-2)]}\}/6$$
$$= (2 \pm \sqrt{28})/6 = (2 \pm 2\sqrt{7})/6 = (1 \pm \sqrt{7})/3$$
$$= (1 \pm 2 \cdot 646)/3 = 3 \cdot 646/3 \text{ or } -1 \cdot 646/3$$
$$\tan \theta = 1 \cdot 2153 \text{ or } -0 \cdot 5487$$

$\therefore \quad \theta = \tan^{-1}(1 \cdot 2153) = n180° + 50°33' = \underline{50°33' \text{ or } 230°33'}$

or $\quad \theta = \tan^{-1}(-0 \cdot 5487) = n180° + (180° - 28°45')$
$= n180° + 151°15' = \underline{151°15' \text{ or } 331°15'}$

$\therefore$ the solutions between 0° and 360° are:

$$\theta = \underline{50°33', \ 151°15', \ 230°33', \ 331°15'}$$

(ii) Solve the equation $3 \cos 2\theta = 2 \sin \theta + 1$, giving possible solutions between 0° and 720°.

Using the identity $\cos 2\theta = 1 - 2 \sin^2 \theta$ the equation reduces to:

$$3(1 - 2 \sin^2 \theta) = 2 \sin \theta + 1$$
$$6 \sin^2 \theta + 2 \sin \theta - 2 = 0$$
or $\quad \underline{3 \sin^2 \theta + \sin \theta - 1 = 0}$

This is a *quadratic equation* for $\sin \theta$ and is satisfied by

$$\sin \theta = [-1 \pm \sqrt{(1 + 12)}]/6 = (-1 \pm \sqrt{13})/6$$
$$= (-1 \pm 3 \cdot 606)/6 = 2 \cdot 606/6 \text{ or } -4 \cdot 606/6$$
$$\underline{\sin \theta = 0 \cdot 4343 \text{ or } -0 \cdot 7677}$$

$\therefore \quad \theta = \sin^{-1}(0 \cdot 4343) = n180° + (-1)^n 25°44'$
$\quad \theta = 25°44', \ 154°16', \ 385°44', \ 514°16'$

or $\quad \theta = \sin^{-1}(-0 \cdot 7677) = n180° + (-1)^n(180° + 50°8')$
$\quad = 230°8', \ (-50°8'), \ 590°8', \ 309°52', \ 669°52'$

$\therefore$ the solutions between 0° and 720° are:

$$\theta = \underline{25°44', \ 154°16', \ 230°8', \ 309°52',}$$
$$\underline{385°44', \ 514°16', \ 590°8', \ 669°52'}$$

(iii) Find two solutions of the equation $5 \sin x + 9 \cos x = 7 \cdot 2$ between 0° and 360°.

The left-hand side of the equation is put in the form $R \sin (x + \alpha)$, i.e.

$5 \sin x + 9 \cos x = R \sin (x + \alpha) = R[\sin x \cos \alpha + \cos x \sin \alpha]$
$5 \sin x + 9 \cos x = (R \cos \alpha) \sin x + (R \sin \alpha) \cos x$
$\qquad R \cos \alpha = 5, \quad R \sin \alpha = 9$

Hence

$$R^2(\cos^2 \alpha + \sin^2 \alpha) = 5^2 + 9^2 = 25 + 81 = 106$$
$$R^2 = 106$$
$$R = \sqrt{106} = 10\cdot3$$

and

$$R \sin \alpha/(R \cos \alpha) = \sin \alpha/\cos \alpha = \tan \alpha = 9/5 = 1\cdot8$$
$$\alpha = \tan^{-1}(1\cdot8) = 60°57'$$

$\therefore$ the equation may be written as:

$$10\cdot3 \sin (x + 60°57') = 7\cdot2$$
$$\sin (x + 60°57') = 7\cdot2/10\cdot3 = 0\cdot6990$$

$\therefore \quad x + 60° 57' = \sin^{-1}(0\cdot6990) = n180° + (-1)^n 44°21'$
$$= 44°21', 135°39', 404°21'$$

or $x = 44°21' - 60°57', 135°39' - 60°57', 404°21' - 60°57'$

$$\underline{x = 74°42' \text{ or } 343°24'}$$

(iv) By considering radian measure, find the smallest positive value of t to satisfy the equation $5 \sin 100t - 12 \cos 100t = 7\cdot95$.

Express the left-hand side in the form $R \sin (100t - \alpha)$, where α is in radians,

$$R = \sqrt{(5^2 + 12^2)} = 13, \quad \tan \alpha = 12/5 = 2\cdot4$$
$$\alpha = \tan^{-1} 2\cdot4 = (67°23') = 1\cdot1761 \text{ radians}$$

$\therefore$ the equation reduces to:

$$13 \sin (100t - 1\cdot1761) = 7\cdot95$$
$$\sin (100t - 1\cdot1761) = 7\cdot95/13 = 0\cdot6115$$
$\therefore \qquad (100t - 1\cdot1761) = \sin^{-1}(0\cdot6115) = (37° 42')$
$$= 0\cdot6580 \text{ rad}$$
$\therefore \qquad\qquad 100t = 0\cdot6580 + 1\cdot1761 = 1\cdot8341$
$\therefore \qquad\qquad\qquad t = 0\cdot018341 = 0\cdot018 \text{ (2 sig. figs.)}$

9.2 Solution of Trigonometric Problems

The solution of most problems may be carried out by reducing the figure representing the problem to a set of right-angled triangles. However, in many cases this may lead to rather long and tedious solutions. In order to reduce the amount of calculation required, rules such as the *sine and cosine rules* are employed. The applications of these rules have been demonstrated in Volume I. Additional problems are illustrated in Examples (i) and (ii).

Examples:

(i) In the triangle ABC, AB = 7·2 in., BC = 8·5 in., AC = 13·1 in. Calculate the angles of the triangle, its area, and the radius of the circle to pass through A, B, and C (see Figure 9.1).

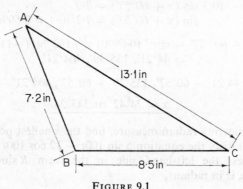

FIGURE 9.1

Using the cosine rule, angle B (the largest angle) is given by

$$\cos B = (a^2 + c^2 - b^2)/(2ac)$$
$$= (8\cdot5^2 + 7\cdot2^2 - 13\cdot1^2)/(2 \times 7\cdot2 \times 8\cdot5)$$
$$= (72\cdot25 + 51\cdot84 - 171\cdot5)/(17 \times 7\cdot2)$$
$$= -47\cdot41/122\cdot4$$
$$= -0\cdot3874$$
$$\therefore \quad B = 180° - 67°12' = \underline{112°48'}$$

For least error, the other angles should also be found using the cosine rule. However, provided the calculation is carried out carefully the *sine rule* will be sufficiently accurate.

Using the sine rule:

$$\frac{\sin A}{8\cdot5} = \frac{\sin C}{7\cdot2} = \frac{\sin 112°48'}{13\cdot1} = \frac{\sin 67°12'}{13\cdot1}$$

$$\therefore \qquad \sin A = (8\cdot5 \times \sin 67°12')/13\cdot1$$
$$\underline{A = 36° \ 45}$$

$$\sin C = (7\cdot2 \times \sin 67°12')/13\cdot1$$
$$\underline{C = 30°27'}$$

$A + B + C = 36°45' + 112°48' + 30°27' = 180°00'$ (check)
Area of triangle

$$= \tfrac{1}{2}ac \sin B = \tfrac{1}{2} \times 7\cdot2 \times 8\cdot5 \times \sin 112°48'$$
$$= 3\cdot6 \times 8\cdot5 \sin 67°12' = \underline{28\cdot21 \text{ in}^2}.$$

Radius of circumcircle,

$$R = a/(2 \sin A) = 8\cdot5/(2 \sin 36°45')$$
$$= 4\cdot25/\sin 36°45' = \underline{7\cdot104 \text{ in}}.$$

(ii) Two sides XY, XZ of a triangular plot of ground XYZ are respectively 125 yd and 180 yd. The angle $X = 71°34'$. Calculate the area of the plot and the length of fence required to surround the plot (see Figure 9.2).

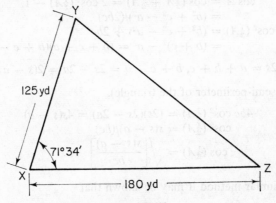

FIGURE 9.2

Area of plot $= \tfrac{1}{2}$XY $\times$ XZ $\sin 71°34'$
$$= \tfrac{1}{2} \times 125 \times 180 \times \sin 71°34'$$
$$= 90 \times 125 \sin 71°34' = \underline{10,680 \text{ yd}^2}$$

$$YZ^2 = XY^2 + XZ^2 - 2XY \times XZ \times \cos 71°34'$$
$$= 15,625 + 32,400 - 45,000 \times 0\cdot3162 = 48,025 - 14,229$$
$$= 33,796$$
$$YZ = \sqrt{33,796} = \underline{183\cdot8 \text{ yd}}$$

Perimeter $= 305 + \underline{183\cdot8 \text{ yd}} = 488\cdot8 \text{ yd}$

or length of fencing required $= \underline{489 \text{ yd}}$ (nearest yard)

9.3 Half-angle Formulae

These formulae are alternative formulae which can be used instead of the cosine rule for the solution of triangles. Only two of these are really necessary. These will be established and the others stated. The formulae are convenient for evaluations when logarithms are used, but the *error* involved in their use may be greater than when the cosine rule is used since an error in an angle is doubled. However, by using *interpolation* in the tables, errors may be reduced.

9.3.1 Formulae for cos $A/2$, sin $A/2$, tan $A/2$

$$\cos A = \cos (\tfrac{1}{2}A + \tfrac{1}{2}A) = 2 \cos^2 (\tfrac{1}{2}A) - 1$$
$$= (b^2 + c^2 - a^2)/(2bc)$$
$$\therefore \quad 4bc \cos^2 (\tfrac{1}{2}A) = (b^2 + c^2 - a^2) + 2bc$$
$$= (b + c)^2 - a^2 = (b + c + a)(b + c - a)$$

Let $\quad 2s = a + b + c, \; b + c - a = 2s - 2a = 2(s - a)$

(s is the semi-perimeter of the triangle),

$$\therefore \qquad\qquad 4bc \cos^2 (\tfrac{1}{2}A) = (2s)(2s - 2a) = 4s(s - a)$$
$$\therefore \qquad\qquad \cos^2 (\tfrac{1}{2}A) = s(s - a)/(bc)$$
$$\therefore \qquad\qquad \cos (\tfrac{1}{2}A) = \sqrt{\left[\frac{s(s - a)}{bc}\right]}$$

By a similar method it may be shown that

$$\sin (\tfrac{1}{2}A) = \sqrt{\left[\frac{(s - b)(s - c)}{bc}\right]}$$

hence

$$\tan (\tfrac{1}{2}A) = \sqrt{\left[\frac{(s - b)(s - c)}{s(s - a)}\right]}$$

One of these formulae may be used to calculate an angle when *three sides of the triangle are known.*

Example:

Calculate the angle B of the triangle ABC in Example (i) of 9.2.

$a = 8{\cdot}5$ in., $b = 13{\cdot}1$ in., $c = 7{\cdot}2$ in.

$s = (a + b + c)/2 = (8{\cdot}5 + 13{\cdot}1 + 7{\cdot}2)/2 = 28{\cdot}8/2 = 14{\cdot}4$

$s - b = 14{\cdot}4 - 13{\cdot}1 = 1{\cdot}3$

$$\cos\left(\tfrac{1}{2}B\right) = \sqrt{\left[\frac{s(s-b)}{ac}\right]} = \sqrt{\frac{14{\cdot}4 \times 1{\cdot}3}{8{\cdot}5 \times 7{\cdot}2}}$$

$\log\left(\cos\tfrac{1}{2}B\right) = \tfrac{1}{2}[\log 14{\cdot}4 + \log 1{\cdot}3 - \log 8{\cdot}5 - \log 7{\cdot}2]$

$\qquad\qquad = \tfrac{1}{2}[1{\cdot}1584 + 0{\cdot}1139 - 0{\cdot}9294 - 0{\cdot}8573]$

$\qquad\qquad = \tfrac{1}{2}[\bar{1}{\cdot}4856] = \bar{1}{\cdot}7428$

$\therefore \qquad\qquad \tfrac{1}{2}B = \text{antilog} \cos (\bar{1}{\cdot}7428) = 56°25'$

$\therefore \qquad\qquad B = 2 \times 56°25' = \underline{112°50'}$

(*Note:* This is $2'$ more than obtained in 9.2 (i).)

9.3.2 Formula for tan $\tfrac{1}{2}(B - C)$

From the sine rule

$$a/\sin A = b/\sin B = c/\sin C = 2R$$
$$a = 2R \sin A, \quad b = 2R \sin B, \quad c = 2R \sin C$$

consider

$$\frac{b - c}{b + c} = \frac{2R(\sin B - \sin C)}{2R(\sin B + \sin C)} = \frac{2 \cos \tfrac{1}{2}(B + C) \sin \tfrac{1}{2}(B - C)}{2 \sin \tfrac{1}{2}(B + C) \cos \tfrac{1}{2}(B - C)}$$
$$= \tan \tfrac{1}{2}(B - C)/\tan \tfrac{1}{2}(B + C)$$

But $A + B + C = 180°$, $B + C = 180° - A$,

$$\tfrac{1}{2}(B + C) = 90 - \tfrac{1}{2}A$$

$\therefore \qquad\qquad \tan \tfrac{1}{2}(B + C) = \cot \left(\tfrac{1}{2}A\right)$

Hence

$$\tan \tfrac{1}{2}(B - C) = \frac{b - c}{b + c} \cot \left(\tfrac{1}{2}A\right)$$

With similar formulae for tan $\tfrac{1}{2}(A - C)$, tan $\tfrac{1}{2}(A - B)$, etc.

One of these formulae may be used to solve a triangle when *two sides and the included angle are given*.

Example:

Use a suitable half-angle formula to find the angles B and C of a triangle ABC in which AB = 17·5 in., AC = 22·3 in., angle BAC = 73° 12′.

Since $b = $ AC $ = 22·3$, $c = $ AB $ = 17·5$, and $b > c$,

$$\therefore \qquad\qquad\qquad \text{angle } B > \text{angle } C$$

$$\tan \tfrac{1}{2}(B - C) = \frac{b - c}{b + c} \cot \tfrac{1}{2}A = \frac{22·3 - 17·5}{22·3 + 17·5} \cot 36°36'$$

$$= \frac{4·8}{39·8} \tan 53°24' \text{ (logarithms now used)}$$

$$\begin{aligned}
\log [\tan \tfrac{1}{2}(B - C)] &= \log 4·8 + \log (\tan 53°24') - \log 39·8 \\
&= 0·6812 + 0·1292 - 1·5999 = 0·8104 - 1·5999 \\
&= \bar{1}·2105
\end{aligned}$$

$$\therefore \qquad\qquad \tfrac{1}{2}(B - C) = 9°13\tfrac{1}{2}'$$
$$\underline{B - C = 18°27'}$$

But $A + B + C = 180°$, $B + C = 180° - 73°12' = \underline{106°48'}$.

Hence
$$2B = 125°15',$$
$$\therefore \qquad\qquad\qquad \underline{B = 62°37\tfrac{1}{2}'}$$
$$2C = 88°21'$$
$$\therefore \qquad\qquad\qquad \underline{C = 44°10\tfrac{1}{2}'}$$

(*Note:* Care must be taken to interpolate in the tables to estimate angles nearer than to the nearest minute.)

WORKED EXAMPLES

(*a*) Show that $\cot 2\theta = (1 - \tan^2 \theta)/(2 \tan \theta)$ and hence find all solutions of the equation $\cot 2\theta = 4 - \tan \theta$ for values of θ between 0° and 360°.

$$\sin 2\theta = 2 \sin \theta \cos \theta, \quad \cos 2\theta = \cos^2 \theta - \sin^2 \theta$$

$$\cot 2\theta = \frac{\cos 2\theta}{\sin 2\theta} = \frac{\cos^2 \theta - \sin^2 \theta}{2 \sin \theta \cos \theta}$$

$$= \frac{1 - \sin^2 \theta/\cos^2 \theta}{2 \sin \theta/\cos \theta} = \frac{1 - \tan^2 \theta}{2 \tan \theta}$$

Substituting for cot 2θ, the equation becomes

$$(1 - \tan^2 \theta)/(2 \tan \theta) = 4 - \tan \theta$$

$\therefore$ $1 - \tan^2 \theta = (4 - \tan \theta)\, 2 \tan \theta = 8 \tan \theta - 2 \tan^2 \theta$

$\therefore$ $\tan^2 \theta - 8 \tan \theta + 1 = 0$

$$\tan \theta = [8 \pm \sqrt{(64 - 4)}]/2 = (8 \pm \sqrt{60})/2$$
$$= (8 \pm 2\sqrt{15})/2 = 4 \pm \sqrt{15}$$
$$= 4 \pm 3{\cdot}873 = \underline{7{\cdot}873 \text{ or } +0{\cdot}127}$$

$\therefore$ $\theta = \tan^{-1}(7{\cdot}873) = n180° + 82°46' = \underline{82°46' \text{ or } 262°46'}$

or $\theta = \tan^{-1}(0{\cdot}1270) = n180° + 7°14' = \underline{7°14' \text{ or } 187°14'}$

        $\underline{\theta = 7°14',\ 82°\ 46',\ 187°\ 14',\ 262°46'}$

(b) Express $5 \cos 20t + 9 \sin 20t$ in the form $R \cos(20t - \alpha)$, where α is in radians. Hence find the least positive value of t to satisfy the equation $5 \cos 20t + 9 \sin 20t = 6{\cdot}35$.

$$5 \cos 20t + 9 \sin 20t = R \cos(20t - \alpha) = R(\cos 20t \cos \alpha + \sin 20t \sin \alpha)$$

$\therefore$ $R \cos \alpha = 5$, $R \sin \alpha = 9$; $R = \sqrt{(5^2 + 9^2)} = \sqrt{106} = \underline{10{\cdot}3}$

       $\tan \alpha = 9/5 = 1{\cdot}8$, $\alpha = (60°\ 57') = \underline{1{\cdot}0638}$ rad

$\therefore$ The equation becomes

   $10{\cdot}3 \cos(20t - 1{\cdot}0638) = 6{\cdot}35$

$\therefore$ $\cos(20t - 1{\cdot}0638) = 6{\cdot}35/10{\cdot}3 = 0{\cdot}6166$

       $20t - 1{\cdot}0638 = \cos^{-1}(0{\cdot}6166) = (51°56') = 0{\cdot}9064$ rad

           $20t = 0{\cdot}9064 + 1{\cdot}0638 = 1{\cdot}9702$

            $\underline{t = 0{\cdot}09851 = 0{\cdot}0985 \text{ (3 figures)}}$

(c) The distance between two points A and B is 175 yd. Observations are made from the points A and B to two points C and D on the same level as A and B and on the same side of the line AB. The following angles are recorded:

     angle BAC = 107°15',   angle BAD = 37°24',
     angle ABD = 112°17',   angle ABC = 44°11'.

Determine the distance CD and the area of ABDC (Figure 9.3).

In triangle ABC:

$$\text{angle ACB} = 180° - (107°15' + 44°11') = \underline{28°34'}$$

In triangle ABD:

$$\text{angle ADB} = 180° - (37°24' + 112°17') = \underline{30°19'}$$

$$\frac{BC}{\sin 107°15'} = \frac{175}{\sin 28°34'}$$

$$BC = \frac{175 \sin 107°15'}{\sin 28°34'} = \frac{175 \sin 72°45'}{\sin 28°34'}$$

$$\underline{BC = 349 \cdot 4 \text{ yd}}$$

$$\frac{BD}{\sin 37°24'} = \frac{175}{\sin 30°19'}$$

$$BD = \frac{175 \sin 37°24'}{\sin 30°19'} = \underline{210 \cdot 6 \text{ yd}}$$

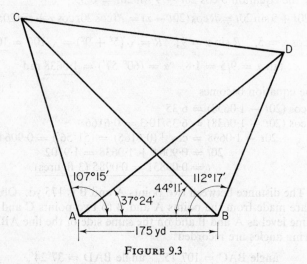

FIGURE 9.3

In triangle CBD,

$$\text{angle CBD} = 112°17' - 44°11' = 68°6'$$

$$CD^2 = BC^2 + BD^2 - 2BC \times BD \cos 68°6'$$
$$= (349\cdot4)^2 + (210\cdot6)^2 - 2 \times 349\cdot4 \times 210\cdot6 \cos 68°6'$$
$$= 122{,}100 + 44{,}350 - 54{,}880 = 166{,}450 - 54{,}880 = \underline{111{,}570}$$

$$\therefore \qquad CD = \sqrt{111{,}570} = \underline{334\cdot0 \text{ yd}}$$

Area of ABDC $= \frac{1}{2}AB \times BC \sin 44°11' + \frac{1}{2}BC \times BD \sin 68°6'$
$$= \frac{1}{2}BC[AB \sin 44°11' + BD \sin 68°6'] \text{ yd}^2$$
$$= 174\cdot7[175 \sin 44°11' + 210\cdot6 \sin 68°6'] \text{ yd}^2$$
$$= 174\cdot7[122\cdot0 + 195\cdot4] = 174\cdot7 \times 317\cdot4$$
$$= \underline{55{,}460 \text{ yd}^2}$$

(d) Use a half-angle formula to find the angle X of the triangle XYZ in which XY $= 11\cdot2$ cm, YZ $= 15\cdot8$ cm, XZ $= 8\cdot9$ cm.

$$s = \tfrac{1}{2}(x + y + z) = \tfrac{1}{2}(15\cdot8 + 8\cdot9 + 11\cdot2) = \tfrac{1}{2}(35\cdot9) = \underline{17\cdot95 \text{ cm}}$$

$$s - y = 17\cdot95 - 8\cdot9 = 9\cdot05, \quad s - z = 17\cdot95 - 11\cdot2 = 6\cdot75$$

$$\sin\tfrac{1}{2}X = \sqrt{\left[\frac{(s-y)(s-z)}{yz}\right]} = \sqrt{\left[\frac{9\cdot05 \times 6\cdot75}{8\cdot9 \times 11\cdot2}\right]}$$

$$\log(\sin\tfrac{1}{2}X) = \tfrac{1}{2}[\log 9\cdot05 + \log 6\cdot75 - \log 8\cdot9 - \log 11\cdot2]$$
$$= \tfrac{1}{2}[0\cdot9566 + 0\cdot8293 - (0\cdot9494 + 1\cdot0492)]$$
$$= \tfrac{1}{2}[1\cdot7859 - 1\cdot9986] = \tfrac{1}{2}[\bar{1}\cdot7873] = \bar{1}\cdot8936(5)$$

$$\therefore \qquad\qquad \tfrac{1}{2}X = 51°31\tfrac{1}{2}'$$
$$\therefore \qquad\qquad \underline{\text{angle } X = 103°3'}$$

(e) Solve the equation $\cos 3\theta + \cos \theta = 0$.
Express the left-hand side as a product

$$2 \cos \tfrac{1}{2}(3\theta + \theta) \cos \tfrac{1}{2}(3\theta - \theta) = 0$$
$$2 \cos 2\theta \cos \theta = 0$$

$\therefore \cos \theta = 0;$

$$\underline{\theta = 90°, 270°}$$

or $\cos 2\theta = 0;$

$$2\theta = 90°, 270°, 450°, 630°$$
$$\theta = 45°, 135°, 225°, 315°$$

Solutions are:

$$\underline{\theta = 45°, 90°, 135°, 225°, 270°, 315°}$$

Examples 9

1. Solve the following equations, giving solutions between $0°$ and $360°$:
   (i) $6 \sec^2 \theta + 7 \tan \theta - 8 = 0$, (ii) $\sec^2 \theta - \tan \theta = 4$,
   (iii) $2 \sec x - 3 \cos x = 5$, (iv) $3 \sec^2 \theta = 7 - 4 \tan \theta$,
   (v) $2 \tan x - 4 \cot x = 5$, (vi) $5 \cdot 1 \cos(2\theta - 35°) = 3 \cdot 7$.

2. Solve the following equations:
   (i) $2 \cos 2\theta + \cos \theta + 1 = 0$, (ii) $\cot 2x + \tan x = 2$,
   (iii) $6 \cos 2\theta + \cos \theta + 2 = 0$, (iv) $\cos 2x = 3 \sin x - 2$,
   (v) $5 \cos 2\theta + 6 \cos \theta + 2 = 0$, (vi) $\cos 2x = \sin x + 5$.

3. (i) Solve $\cos 2\theta - 5 \cos \theta = 2$; $0 \leqslant \theta \leqslant 360°$.
   (ii) Solve $\cos 2\theta + 5 \sin \theta = 1 \cdot 8$; $0 \leqslant \theta \leqslant 360°$.
   (iii) Solve $\cot 2x - 3 \tan x = 4$; $0 \leqslant x \leqslant 180°$
   (iv) Solve $\cot^2 \theta + 3 \csc \theta - 3 = 0$; $0 \leqslant \theta \leqslant 360°$.

4. Express $\sin 3\theta + \sin \theta$ as a product and hence solve the equation
   $\sin 3\theta + \sin \theta = 0$, $0 \leqslant \theta \leqslant 360°$.
   (ii) Solve the equation $\sin 4x + \sin x = 0$, $0 \leqslant x \leqslant 180°$.
   (iii) Solve the equation $\sin \theta + \sin 3\theta = \sin 2\theta$, $0 \leqslant \theta \leqslant 180°$.

5. Solve the following equations, giving, in each case solutions between $0°$ and
   $360°$:
   (i) $3 \cos x - 4 \sin x = 2$, (ii) $8 \cos \theta - 15 \sin \theta = 8 \cdot 5$,
   (iii) $4 \cos x - 3 \sin x = 2 \cdot 4$, (iv) $10 \sin \theta - 7 \cos \theta = 6 \cdot 3$,
   (v) $24 \sin \theta + 7 \cos \theta = 15$, (vi) $5 \cos x + 7 \sin x = 3 \cdot 85$.

6. (i) Express $30 \cos 100t + 40 \sin 100t$ in the form $R \cos(100t - \alpha)$ where α
   is in radians. Hence find the smallest positive value of t to satisfy the equation
   $30 \cos 100t + 40 \sin 100t = 23 \cdot 5$.
   (ii) Express $100 \cos \theta - 50 \sin \theta$ in the form $R \cos(\theta + \alpha)$ and hence find
   the solutions of $100 \cos \theta - 50 \sin \theta = 72 \cdot 8$ between $0°$ and $360°$.
   (iii) Find the solutions of $40 \sin t - 20 \cos t = 17$ for t between 0 and 2π
   radians.

7. Find the least values of t to satisfy the equations:
   (i) $20 \sin 100t + 40 \cos 100t = 30$,
   (ii) $25 \cos 40t + 38 \sin 40t = 23$,
   (iii) $8 \sin 2t - 5 \cos 2t = 6$.

8. Given $\sin \theta = 2t/(1 + t^2)$, $\cos \theta = (1 - t^2)/(1 + t^2)$ where $t = \tan \frac{1}{2} \theta$,
   express the equation $4 \sin \theta + 5 \cos \theta = 3 \cdot 5$ as a quadratic equation in t.
   Solve this equation and hence write down the solutions of the original equation between $0°$ and $360°$.

9. Prove the identity $\csc x - \cot x = \tan \frac{1}{2}x$ and use the identity to solve
   $\csc 3\theta - \cot 3\theta = 1 \cdot 2$, $0 \leqslant \theta \leqslant 360°$.

10. Use the identity $\sin 2\theta = 2 \sin \theta \cos \theta$ to factorize the expression
    $2 \sin 2\theta - \sin \theta - 12 \cos \theta + 3$ and hence solve the equation
    $2 \sin 2\theta - \sin \theta - 12 \cos \theta + 3 = 0$.

11. Using $\cos 2\theta = 1 - 2 \sin^2 \theta$ and $\cos A = (b^2 + c^2 - a^2)/(2bc)$ show that
    $\sin \frac{1}{2}A = \sqrt{\left[\dfrac{(s-b)(s-c)}{bc}\right]}$. Use the result to calculate angle A of the
    triangle ABC in which $a = 3 \cdot 7$ in, $b = 4 \cdot 5$ in and $c = 5 \cdot 1$ in.

12. For a triangle XYZ, using the method of 9.3.2 show that
    $\tan \frac{1}{2}(Y - Z) = \dfrac{y - z}{y + z} \cot(\frac{1}{2}X)$. Hence find the other angles of the triangle
    XYZ when $y = 13 \cdot 5$ cm, $z = 9 \cdot 8$ cm, angle $X = 73°12'$.

13. In the quadrilateral ABCD (lettered in order), AB = 7·8 in, BC = 5·5 in, CD = 8·7 in, DA = 7·1 in, and angle B = 85°. Calculate the length of AC, the angle D and the area of the quadrilateral.

14. Using appropriate half-angle formulae solve the triangles:
    (i) ABC, in which a = 15·6 ft, b = 11·9 ft, c = 8·7 ft,
    (ii) PQR, in which p = 11·9 in, q = 7·4 in, angle R = 67°28′.

15. ABCD is a quadrilateral in which AB = 1·3 ft, AD = 1·5 ft, CD = 1 ft, angle BCD = 98°8′ and angle BAD = 59°29′. Calculate the lengths of the diagonals, the angle ADC and the area of the quadrilateral.

16. From a ship two lighthouses X and Y are observed to bear E19°37′S and E59°43′S respectively. If the bearing of Y from X is W63°S and XY = 5 miles, find the distances of the ship from X and Y.

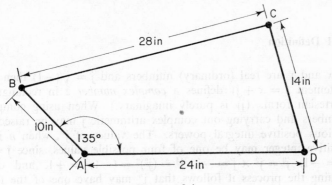

FIGURE 9.4

17. From two points A and B on the same side of a river, observations are made to two points P and Q on the opposite side of the river. The following observations are recorded: Distance AB = 750 ft, angle BAP = 33°5′, angle ABQ = 39°17′, angle BAQ = 79°15′, angle ABP = 69°28′. Calculate the distance PQ and the distance of P from B.

18. Two sides AB and BC of a quadrilateral are tangents touching a circle at A and C, while the other two sides CD and DA are chords of the circle of lengths 6 in and 10 in respectively meeting at an angle of 20°20′. Find the circumference of the circle and the perimeter of the quadrilateral.

19. The top of a tower 100 ft above the ground has an angle of elevation of 41° from a point A, S21°W of the foot of the tower. From a point B S.E. of A the bearing of the tower is N12°W. Find the distance between A and B and the angle of elevation of the top of the tower from B.

20. The diagram (Figure 9.4) represents a four pin-jointed link frame in which the link AD is kept fixed. If, from the position shown the link AB is rotated through 45° about A in a counter clockwise direction, through what angle will the link CD rotate about D?

10

Complex Numbers—Argand Diagram

10.1 Definition

If x and y are real (ordinary) numbers and $j = \sqrt{(-1)}$ then the statement $z = x + jy$ defines a *complex number* z in rectangular Cartesian form. (jy is purely imaginary.) When using complex numbers and carrying out complex arithmetic j may be raised to various positive integral powers. The values of j^n, when n is a positive integer may be one of four possible values, since $j = j$, $j^2 = -1$, $j^3 = j^2 \times j = -j$, $j^4 = (j^2)^2 = (-1)^2 = +1$, and continuing the process it follows that j^n may have one of the four values, j, $-j$, -1, or $+1$.

10.1.1 Algebraic rules

(Used to simplify complex number expressions.)
Consider the product of $z_1 = 4 + j3$ and $z_2 = 5 - j2$
(or $4 + 3j$ and $5 - 2j$).

$$z_1 z_2 = (4 + j3)(5 - j2) \text{ [using ordinary algebraic rules]}$$
$$= 4 \times 5 + (j3)5 - 4(j2) - (j3)(j2)$$
$$= 20 + 15j - 8j - 6j^2 = 20 + 7j - 6(-1) = \underline{26 + 7j}$$

(*Note:* jy and yj are equivalent and hence $26 + 7j$ and $26 + j7$ are identical.)

For other processes, normal algebraic processes are still used.

10.1.2 Conjugate complex numbers

Two complex numbers $z = x + jy$ and $\bar{z} = x - jy$ are referred to as *conjugate* complex numbers. Two such numbers possess important properties:

(i) *Sum:* $z + \bar{z} = (x + jy) + (x - jy) = 2x$ *(real)*
(ii) *Difference:* $z - \bar{z} = (x + jy) - (x - jy) = 2jy$ (imaginary)
(iii) *Product:* $z\bar{z} = (x + jy)(x - jy) = x^2 - (jy)^2 = x^2 - j^2y^2$
$= x^2 + y^2$

i.e. the product is *real*.

(*Note:* The rather special result of the product of two conjugate numbers is the *sum of two squares*.)

10.2 Modulus-argument Form

If x and y are expressed in *polar co-ordinate* form (see Chapter 7) then $x = r \cos \theta$, $y = r \sin \theta$, and $z = x + jy = r \cos \theta + jr \sin \theta = r(\cos \theta + j \sin \theta)$. This form is known as the *modulus-argument* form of a complex number.

The *modulus* of the number z is $r = \sqrt{(x^2 + y^2)}$.

The *argument* of the number z is $\theta = \tan^{-1}(y/x)$ where $0 \leqslant \theta \leqslant 2\pi$ [or $-\pi \leqslant \theta \leqslant \pi$]. The argument may be expressed in degrees and minutes, in which case $0 \leqslant \theta \leqslant 360°$ [or $-180° \leqslant \theta \leqslant 180°$].

The value of θ between 0 and 2π [or $0°$ and $360°$] *or* $-\pi$ and $+\pi$ [or $-180°$ and $180°$] is referred to as the *principal value of the argument*. θ may, for certain cases, be taken as a general angle.

(*Note:* In descriptive work, the terms *modulus* and *argument* may be abbreviated as follows:

$$\text{modulus of } z = \text{mod } z = |z|,$$
$$\text{argument of } z = \arg z.)$$

10.2.1 Alternative modulus-argument forms

(i) $r(\cos \theta + j \sin \theta) = r\underline{/\theta}$
(ii) $r(\cos \theta + j \sin \theta) = re^{j\theta}$ *or* $r \exp(j\theta)$

i.e. the various forms of representation of complex numbers may be abbreviated as follows:

$$z = x + jy = r(\cos \theta + j \sin \theta) = r\underline{/\theta} = r\,e^{j\theta} = r \exp(j\theta)$$

Example:

Express $z = 4 + 3j$ in the modulus-argument forms.

$$r = \sqrt{(4^2 + 3^2)} = 5; \quad \tan \theta = 3/4 = 0.75$$

$$\theta = \tan^{-1}(0.75) = 36° \ 52'$$

$$\therefore \qquad z = 4 + 3j = 5(\cos 36°52' + j \sin 36°52')$$
$$= 5\underline{/36°52'} = 5e^{j(36°52')} = 5 \exp(j36°52')$$

10.2.2 The Argand diagram

Complex numbers may be represented geometrically as points on an *Argand diagram* (Figure 10.1).

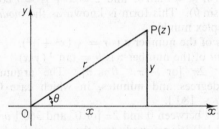

FIGURE 10.1

The *point* P represents the complex number z, $OP = \text{mod } z$, angle $POx = \arg z$. If P moves on a locus this locus may be described by a complex number *statement* or *equation*.

Example:

If P moves on a circle centre O then the *modulus* of the number representing P remains constant, i.e. the equation of the locus of P may be written as

$$|z| = \text{constant} \quad \text{(circle)}$$

Other loci may be described by complex number *statements*.

10.3 General Rules

(i) *Addition:*

If $z_1 = x_1 + jy_1$, $z_2 = x_2 + jy_2$, then
$$z_1 + z_2 = (x_1 + jy_1) + (x_2 + jy_2) = (x_1 + x_2) + j(y_1 + y_2)$$

(ii) *Subtraction:*

$$z_1 - z_2 = (x_1 + jy_1) - (x_2 + jy_2) = (x_1 - x_2) + j(y_1 - y_2)$$

(iii) *Multiplication:*

$$z_1z_2 = (x_1 + jy_1)(x_2 + jy_2) = x_1x_2 + j^2y_1y_2 + jy_1x_2 + jx_1y_2$$
$$= (x_1x_2 - y_1y_2) + j(x_1y_2 + y_1x_2)$$

(iv) *Division:*

Use is made of the very important property of the product of two *conjugate numbers*, i.e.

$$(x + jy)(x - jy) = x^2 + y^2$$

$$\frac{z_1}{z_2} = \frac{x_1 + jy_1}{x_2 + jy_2}$$

Multiply numerator and denominator by the conjugate $x_2 - jy_2$ in order to make the denominator *real*.

$$\frac{z_1}{z_2} = \frac{x_1 + jy_1}{x_2 + jy_2} \times \frac{x_2 - jy_2}{x_2 - jy_2} = \frac{x_1x_2 + y_1y_2 + j(y_1x_2 - y_2x_1)}{x_2{}^2 + y_2{}^2}$$

$$= \frac{x_1x_2 + y_1y_2}{x_2{}^2 + y_2{}^2} + j\frac{y_1x_2 - y_2x_1}{x_2{}^2 + y_2{}^2}$$

Generally results will be expressed in the form $A + jB$. It is not necessary to *memorize* these results, only the *methods* of obtaining them.

10.3.1 Products and quotients (using polar forms)

Generally addition and subtraction are better dealt with using Cartesian forms, but multiplication and division may be dealt with using polar forms.

(i) *Multiplication:*

Let

$$z_1 = r_1(\cos \theta_1 + j \sin \theta_1), \quad z_2 = r_2(\cos \theta_2 + j \sin \theta_2)$$

$$\begin{aligned}
z_1 z_2 &= r_1 r_2(\cos \theta_1 + j \sin \theta_1)(\cos \theta_2 + j \sin \theta_2) \\
&= r_1 r_2[(\cos \theta_1 \cos \theta_2 - \sin \theta_1 \sin \theta_2) \\
&\qquad\qquad\qquad + j(\sin \theta_1 \cos \theta_2 + \cos \theta_1 \sin \theta_2)] \\
&= r_1 r_2[\cos (\theta_1 + \theta_2) + j \sin (\theta_1 + \theta_2)] = r_1 r_2 \underline{/\theta_1 + \theta_2}
\end{aligned}$$

i.e. the *modulus* of the product = product of the separate moduli, the *argument* of the product = sum of the separate arguments.

(ii) *Division:*

$$\begin{aligned}
\frac{z_1}{z_2} &= \frac{r_1(\cos \theta_1 + j \sin \theta_1)}{r_2(\cos \theta_2 + j \sin \theta_2)} \\[2mm]
&= \frac{r_1}{r_2} \frac{(\cos \theta_1 + j \sin \theta_1)(\cos \theta_2 - j \sin \theta_2)}{(\cos \theta_2 + j \sin \theta_2)(\cos \theta_2 - j \sin \theta_2)} \\[2mm]
&= \frac{r_1}{r_2} \frac{[\cos (\theta_1 - \theta_2) + j \sin (\theta_1 - \theta_2)]}{\cos^2 \theta_2 + \sin^2 \theta_2} \\[2mm]
&= \frac{r_1}{r_2} [\cos (\theta_1 - \theta_2) + j \sin (\theta_1 - \theta_2)] = \frac{r_1}{r_2} \underline{/\theta_1 - \theta_2}
\end{aligned}$$

i.e. the *modulus* of the quotient = quotient of the separate moduli, the *argument* of the quotient = difference of the two arguments.

10.3.2 Geometrical interpretation of sum and difference

Sum:

(See Figure 10.2.)

Let P represent the number

$$z_1 = x_1 + jy_1, \quad OP = |z_1|$$

and Q represent the number

$$z_2 = x_2 + jy_2, \quad OQ = |z_2|$$

and O the origin of the Cartesian co-ordinates.

Complete the parallelogram OQRP, then R represents the sum of z_1 and z_2.

From geometrical considerations, the triangles ONP, QSR are congruent,

$$\therefore \qquad QS = ON = MT = x_1; \quad SR = NP = y_1;$$
$$OM = x_2; \quad MQ = ST = y_2;$$

at the point R,

$$z = OT + jTR = (OM + MT) + j(RS + ST)$$
or $$z = (x_1 + x_2) + j(y_1 + y_2) = z_1 + z_2$$

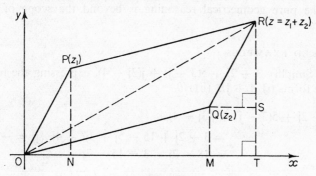

FIGURE 10.2

Difference:

(See Figure 10.3.)

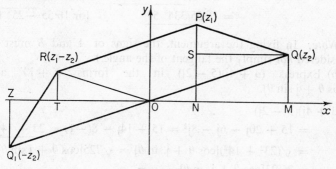

FIGURE 10.3

Let P represent z_1 and Q represent z_2. Q_1 represents $-z_2$. Complete the parallelogram OQPR, then R represents $z_1 - z_2$. From geometrical considerations, the triangles PSQ, RTO are congruent.

$$OT = -SQ = -(MO - NO) = -(x_2 - x_1) = x_1 - x_2$$
$$TR = SP = PN - SN = PN - QM = y_1 - y_2;$$

at the point R,

$$z = OT + jTR = (x_1 - x_2) + j(y_1 - y_2) = z_1 - z_2$$

Products and quotients may also be represented geometrically but the pure geometrical reasoning is beyond the scope of this volume.

WORKED EXAMPLES

(a) Simplify $3 + 2j + 5(3 - j) + j(3j - 4)$, expressing the result in the forms (i) $A + jB$, (ii) $r\underline{/\theta}$.

$3 + 2j + 5(3 - j) + j(3j - 4)$
$$= 3 + 2j + 15 - 5j + 3j^2 - 4j \quad [j^2 = -1]$$
$$= 18 - 7j - 3 = 15 - 7j$$
$$= \sqrt{(15^2 + 7^2)} \underline{/\tan^{-1}\left(\frac{-7}{+15}\right)}$$
$$= \sqrt{(225 + 49)}\underline{/\tan^{-1}(-0{\cdot}4667)}$$
$$= 16{\cdot}55\underline{/360° - 25° \ 1'}$$
$$= 16{\cdot}55\underline{/334° \ 59'} \qquad [\text{or } 16{\cdot}55\underline{\backslash-25° \ 1'}\,]$$

(*Note:* In fixing the argument, the signs of A and B must be considered *not* simply the tangent of the angle.)

(b) Express $(3 + 4j)(5 - 2j)$ in the forms $X + jY$ and $r(\cos\theta + j\sin\theta)$.

$(3 + 4j)(5 - 2j)$
$$= 15 + 20j - 6j - 8j^2 = 15 + 14j - 8(-1) = \underline{23 + 14j}$$
$$= \sqrt{(23^2 + 14^2)}[\cos\theta + j\sin\theta] = \sqrt{725}[\cos\theta + j\sin\theta]$$
$$= 26{\cdot}93[\cos\theta + j\sin\theta]$$

where $\tan \theta = +14/+23 = 0\cdot6087$,

$$\theta = 31° \ 20' \ \text{(1st quadrant angle)}$$

$\therefore \quad (3 + 4j)(5 - 2j) = 23 + 14j = 26\cdot93(\cos 31°20' + j \sin 31°20')$

(c) Simplify $(3 + 5j)/(4 + 3j)$ to the form $A + jB$.

This type of simplification is carried out by making the denominator real. This is done by multiplying numerator and denominator by the *conjugate* of the denominator.

$$\frac{3 + 5j}{4 + 3j} = \frac{(3 + 5j)(4 - 3j)}{(4 + 3j)(4 - 3j)} = \frac{12 + 20j - 9j - 15j^2}{4^2 - (3j)^2}$$

$$= \frac{12 + 11j + 15}{16 - 9j^2} = \frac{27 + 11j}{16 + 9} = \frac{27}{25} + \frac{11}{25}j = 1\cdot08 + 0\cdot44j$$

(d) Express $4 + 3j$ and $5 + 12j$ in the form $r/\underline{\theta}$ and hence find the product of the two numbers in $r/\underline{\theta}$ form.

$$z_1 = 4 + 3j = \sqrt{(4^2 + 3^2)}/\underline{\tan^{-1} (3/4)} = 5/\underline{36°52'}$$

$$z_2 = 5 + 12j = \sqrt{(5^2 + 12^2)}/\underline{\tan^{-1} (12/5)} = 13/\underline{67°23'}$$

$\therefore \quad z_1 z_2 = 5 \times 13/\underline{36°52' + 67°23'} = 65/\underline{104°15'}$

(e) Express $3 + 4j$ and $12 + 5j$ in $r \exp (j\theta)$ form and hence find $(3 + 4j)/(12 + 5j)$ in the form $A + jB$.

$3 + 4j = \sqrt{(3^2 + 4^2)} \exp [j \tan^{-1} (4/3)] = 5 \exp (j53°8')$
$12 + 5j = \sqrt{(12^2 + 5^2)} \exp [j \tan^{-1} (5/12)] = 13 \exp (j22°37')$

$\therefore \quad (3 + 4j)/(12 + 5j)$

$\quad = [5 \exp (j53° \ 8')]/[13 \exp (j22° \ 37')]$
$\quad = (5/13) \exp [j(53°8' - 22°37')] = (5/13) \exp (j30°31')$
$\quad = (5/13)[\cos 30°31' + j \sin 30° \ 31'] = (5/13)[0\cdot8615 + j0\cdot5078]$
$\quad = (1/13)[4\cdot3075 + j2\cdot5390] = 0\cdot3313 + j0\cdot1953$

(f) If $w = u + jv$ and $z = x + jy$ and $w = z^3$, express u and v in terms of x and y.

$u + jv = (x + jy)^3 = x^3 + 3x^2(jy) + 3x(jy)^2 + (jy)^3$
$\quad\quad = x^3 + 3jx^2y + 3xj^2y^2 + j^3y^3 \quad\quad\quad (j^2 = -1, \ j^3 = -j)$
$u + jv = x^3 - 3xy^2 + j(3x^2y - y^3)$

Two complex numbers can only be identically equal if they are represented by the same point on the Argand diagram. Hence the *real* parts must be *equal* and the *imaginary* parts must be *equal*. ∴ equating real and imaginary parts:

$$u = x^3 - 3xy^2; \quad v = 3x^2y - y^3$$

(g) If $z = 2 + j$, express z^2 and z^3 in $A + jB$ form. Represent the three numbers on an Argand diagram.

$$z = 2 + j; \quad z^2 = (2 + j)^2 = 4 + 4j + j^2 = 4 + 4j - 1 = \underline{3 + 4j}$$
$$z^3 = (2 + j)^3 = (2 + j)(3 + 4j) = 6 + 3j + 8j + 4j^2 = 6 + 11j - 4$$
$$= \underline{2 + 11j}$$

The points P, Q, R in Figure 10.4 represent z, z^2, z^3.

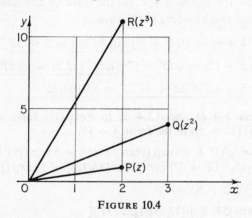

FIGURE 10.4

(*Note*: $|z| = OP = \sqrt{(2^2 + 1^2)} = \sqrt{5}$
$|z^2| = OQ = \sqrt{(3^2 + 4^2)} = 5,$
$|z^3| = OR = \sqrt{(2^2 + 11^2)} = \sqrt{125} = 5\sqrt{5}$

Clearly, $|z^2| = |z|^2; \quad |z^3| = |z|^3$

also if $\arg z = \theta, \quad \arg z^2 = 2\theta, \quad \arg z^3 = 3\theta.$)

(h) Show that if $z = \cos \theta + j \sin \theta$, then (i) $z^3 = \cos 3\theta + j \sin 3\theta$, (ii) $z^n = \cos n\theta + j \sin n\theta$ (n a positive integer).

As shown in 10.3.1 the argument of a product = sum of arguments and the modulus is the product of the moduli, now

$$|\cos\theta + j\sin\theta| = \sqrt{(\cos^2\theta + \sin^2\theta)} = 1$$
$$z^3 = zzz, \quad \arg z = \theta, \quad |z^3| = |z|^3 = 1$$
$$\arg z^3 = \arg z + \arg z + \arg z = 3\arg z = 3\theta$$

i.e.
$$z^3 = \cos 3\theta + j\sin 3\theta$$

Similarly $|z^n| = |z||z| \ldots (n \text{ times}) = |z|^n$,
$$\arg z^n = \arg z + \arg z + \ldots (n \text{ times}) = n\arg z = n\theta$$

i.e.
$$z^n = \cos n\theta + j\sin n\theta$$

Examples 10

1. Express $(5 + 4j)^2$ in the forms $A + jB$ and $r/\underline{\theta}$.
2. Simplify $(4 + 7j)(3 - 2j)$ by reducing to the form $A + jB$. What are the modulus and argument of the result?
3. Simplify $(7 - 2j)(4 + 5j) + (3 - j)(5 - 2j)$ to the form $A + jB$. Express the result in modulus-argument form. Represent the result on an Argand diagram.
4. Simplify the following to the form $x + jy$:
   (i) $(5 + 4j)(3 - 5j)$, (ii) $(3 + 2j)/(3 - 4j)$,
   (iii) $(7 + 3j)^2$, (iv) $\dfrac{2+j}{2-j} + \dfrac{2-j}{2+j}$.
5. Express $5 - 7j$ and $8 + 5j$ in the $r/\underline{\theta}$ form. Hence find the value of $(5 - 7j)(8 + 5j)$ and $(5 - 7j)/(8 + 5j)$ in $r/\underline{\theta}$ form. Represent the four numbers on an Argand diagram.
6. If $x + jy = 10\exp(j30°)$ find the values of x and y.
7. If $z = x + jy$ find z^2 and z^3 in $A + jB$ form. Represent z, z^2, z^3 by points on an Argand diagram.
8. Given $w = u + jv$ and $z = x + jy$ and $w = 3z^2$ express u and v in terms of x and y.
9. Express $-4 + 3j$ in $r/\underline{\theta}$ form and hence obtain the value of $(-4 + 3j)^4$ in the form $P + jQ$.
10. Express $(3 + 2j)(5 - 3j)/(3 - 4j)$ in the form $A + Bj$ and hence in the form $r/\underline{\theta}$. Represent the result on an Argand diagram.
11. (i) Express in exponential form:
    (a) $\cos 30° + j\sin 30°$, (b) $\cos 2\alpha + j\sin 2\alpha$, (c) $5(\cos 120° + j\sin 120°)$,
    (d) $5(\cos \pi/4 + j\sin \pi/4)$, (e) $b(\cos a/2 + j\sin a/2)$.

    (ii) Simplify the following:
    (a) $(\cos 50° + j\sin 50°)(\cos 45° + j\sin 45°)$,
    (b) $(\cos 25° + j\sin 25°)^2$, (c) $(\cos \pi/3 + j\sin \pi/3)(\cos \pi/5 + j\sin \pi/5)$.

5

12. (i) Express the following in $a + jb$ form and represent each number on an Argand diagram:

   (a) $5\exp(j\theta)$, (b) $10e^{-j60°}$, (c) $5e^{j\pi/3}$, (d) $10\exp(-2j\pi/3)$.

   (ii) Express the following complex numbers in modulus-argument form. Represent each number on an Argand diagram:

   (a) $0.5 + 0.8j$, (b) $-1.5 + 3.0j$, (c) $-2 - 4j$, (d) $3 - 4.5j$.

13. Solve the following equations giving your results in Cartesian and modulus-argument form.

   (i) $x^2 + x + 1 = 0$, (ii) $t^2 - 2t + 3 = 0$, (iii) $5y^2 + y + 2 = 0$.

14. If $w = u + jv$ and $z = x + jy$ express u and v in terms of x and y in the following cases:

   (i) $w = 2z + 3$, (ii) $w = 1/z$, (iii) $w = (z + 1)^2$,
   (iv) $w = 1/(z - 1)$, (v) $w = z + 1/z$.

15. If $w = u + jv$ and $z = x + jy$, find the expressions for $u^2 + v^2$ and $(u - v)^2$ in terms of x and y when $w = z + 2$.

16. (i) Using $e^{j\theta} = \cos\theta + j\sin\theta$ show that $e^{-j\theta} = \cos\theta - j\sin\theta$.
    (ii) If $z = e^{j\theta}$, simplify the expressions

$$\frac{1}{2}\left(z + \frac{1}{z}\right) \text{ and } \frac{1}{2j}\left(z - \frac{1}{z}\right).$$

17. Using $z = \cos\theta + j\sin\theta$ and $z^3 = \cos 3\theta + j\sin 3\theta$, by expanding $(\cos\theta + j\sin\theta)^3$ using the binomial expansion and equating real and imaginary parts show:

   (i) $\cos 3\theta = 4\cos^3\theta - 3\cos\theta$, (ii) $\sin 3\theta = 3\sin\theta - 4\sin^3\theta$.

18. In alternating current theory, the voltage V, current I and impedance Z may all be complex numbers and the basic relation between V, I, and Z is $I = V/Z$. Find in Cartesian form (i) I when $V = 10$, $Z = 4 + 3j$,
   (ii) V when $I = 3 + 8j$, $Z = 10 + 5j$,
   (iii) Z when $I = 8 + 5j$, $V = 20 - 8j$.

19. If a point P denoted by $z(= x + jy)$ moves in a circle of radius 5 units whose centre is denoted by $a = 2 + 3j$, write down the equation of the locus of P and express the equation in Cartesian form (*Note:* If P and Q are two points z_1 and z_2 in the Argand diagram, then $PQ = |z_1 - z_2|$). Show the centre and locus on an Argand diagram.

20. If two points A and B represent the numbers z_1 and z_2 then $AP = |z_1 - z_2|$ and the angle made by AB with the real axis $= \arg(z_1 - z_2)$. If A is represented by the number a and P by the number z describe the locus of P (for fixed A) if z satisfies the equations:

   (i) $|z - a| = r$ (constant), (ii) $\arg(z - a) = K$ (constant).

11

Some Applications of Complex Numbers—Vectors—A.C. Theory

11.1 Vectors

A vector quantity is a quantity having both magnitude and direction. Generally vectors may be compounded by additions and/or subtractions using parallelograms or triangles.

Consider two vectors **OP**, **OQ** (Figure 11.1).

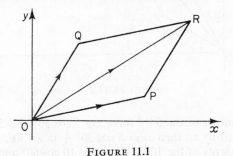

FIGURE 11.1

The resultant of **OP** and **OQ** is the vector represented by **OR**, i.e.

$$\mathbf{OP} + \mathbf{OQ} = \mathbf{OR}$$

Clearly, if P and Q are represented by the numbers $z_1 = x_1 + jy_1$ $z_2 = x_2 + jy_2$, then the complex number representing R is $z_1 + z_2$. Hence, if **OP** and **OQ** are taken as *equivalent* to z_1 and z_2 [with components (x_1, y_1), (x_2, y_2)] then the resultant of **OP** and **OQ** is given by **OR**, where the complex number of R, z, is given by $z = z_1 + z_2 = (x_1 + x_2) + j(y_1 + y_2)$. The *modulus* of the sum represents the *magnitude* of the resultant vector and the *argument* gives the *direction* of the resultant. This may be extended to obtain the resultant of any number of vectors (in two dimensions).

Example:

Two forces of magnitudes 5 lbf and 10 lbf act at angles of 30° and 60° to a direction Ox. Use complex numbers to find the magnitude and direction of the resultant force (Figure 11.2).

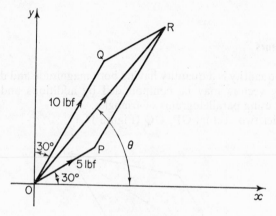

FIGURE 11.2

The components of the 5 lbf force are 5 cos 30° and 5 sin 30°. If P is represented by z_1, then $z_1 = 5 \cos 30° + j5 \sin 30°$.

The components of the 10 lbf force are 10 cos 60° and 10 sin 60°. If Q is represented by z_2, then $z_2 = 10 \cos 60° + j10 \sin 60°$.

Let z represent R, where **OR** represents the resultant force, then

$$
\begin{aligned}
z = z_1 + z_2 &= (5 \cos 30° + j5 \sin 30°) + (10 \cos 60° + j10 \sin 60°) \\
&= (5 \cos 30° + 10 \cos 60°) + j(5 \sin 30° + 10 \sin 60°) \\
&= (5 \times 0{\cdot}866 + 10 \times 0{\cdot}5) + j(5 \times 0{\cdot}5 + 10 \times 0{\cdot}866) \\
&= (4{\cdot}33 + 5{\cdot}0) + j(2{\cdot}5 + 8{\cdot}66) = \underline{9{\cdot}33 + j11{\cdot}16}
\end{aligned}
$$

The resultant is a force of magnitude

$$\sqrt{(9.33^2 + 11.16^2)} = \underline{14.89 \text{ lbf}}$$

The direction relative to Ox is given by

$$\theta = \arg z = \tan^{-1}(11.16/9.33) = \tan^{-1}(1.1930) = \underline{50°2'}$$

i.e.   the resultant is 14·89 lbf acting at 50°2' to the Ox axis

11.2 Further Applications of Complex Numbers

There are a number of *applications* of complex numbers. They may be used for the evaluation of powers and roots, *vectors*, and are used in *alternating current theory*. Other applications involve transformations (or complex projection), stress analysis, etc. The scope of this book does not allow of all applications to be dealt with here. The application to alternating current theory will, however, be dealt with here.

When an alternating voltage V is applied across a circuit the resultant opposition to current flow is called the *impedance* of the circuit. This is often denoted by Z and the current I in the circuit is given by $I = V/Z$.

If the alternating voltage has a frequency f and $\omega = 2\pi f$, then in complex notation the *impedance* of the circuit is provided by a combination of *resistance* and *reactance*. Reactance is provided by an *inductance* and/or *capacitance*. If an inductance is present, of magnitude L (henries) then the *reactance* due to the *inductance* is represented by jωL. If a capacitance of magnitude C (farads) is present then the *reactance* due to the *capacitance* is represented by $1/\text{j}\omega C$ or $-\text{j}/\omega C$. If jX is the resultant *reactance* of a circuit then the impedance Z (in complex form) is given by $Z = R + \text{j}X$, where R is the *resistance* present.

By considering the complex value in *modulus-argument* form the *phase* relationships between the various quantities may be established.

Examples:

(i) A voltage of magnitude 100 volt r.m.s. and frequency 50 c/s is applied across a coil equivalent to a resistance of 5 ohms in series with an inductance of 10 millihenries. Find the magnitude of the current and its phase angle with reference to the applied voltage.

The reactance of the inductance $= j\omega L = j2\pi f L$ (L in henries), i.e.

$$\text{Reactance} = j2\pi \times 50 \times 10 \times 10^{-3} = j\pi$$

The impedance $Z = R + j\omega L = 5 + j\pi$.

∴ The current I is given by $I = V/Z = 100/(5 + j\pi)$,

$$I = \frac{100(5 - j\pi)}{(5 + j\pi)(5 - j\pi)} = \frac{100(5 - j\pi)}{25 + \pi^2} = \frac{100(5 - j\pi)}{25 + 9 \cdot 87}$$
$$= \frac{100}{34 \cdot 87}(5 - j\pi) = 2 \cdot 867(5 - j\pi) = \underline{14 \cdot 335 - j9 \cdot 010}$$

Taking the voltage as real (see Figure 11.3), the magnitude of the current

$$= |I| = \sqrt{(14 \cdot 335^2 + 9 \cdot 01^2)} = \sqrt{(205 \cdot 4 + 81 \cdot 18)}$$
$$= \sqrt{286 \cdot 58} = \underline{16 \cdot 93 \text{ amp r.m.s.}}$$

The current *lags* the voltage by an angle θ where

$$\tan \theta = 9 \cdot 010/14 \cdot 335 = \pi/5 = 0 \cdot 6284$$
$$\theta = \tan^{-1}(0 \cdot 6284) = \underline{32°9'}$$

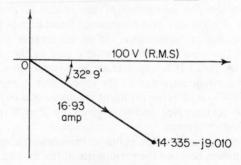

FIGURE 11.3

(ii) A resistance R ohms is in series with an inductance of L henries and a capacitance of C farads. A voltage of angular velocity ω is applied across the series circuit. Find the condition that the

current will be in *phase* with the applied voltage and what is the
current when this condition holds.

The impedance of the circuit

$$Z = R + j\omega L + 1/j\omega C = R + j\omega L - j/\omega C$$
$$= R + j(\omega L - 1/\omega C)$$

Consider the voltage E to be real, then the current I is given by

$$I = E/Z = E/[R + j(\omega L - 1/\omega C)]$$

The current will be *in phase* with the voltage if I is *real*, and this
will be so if $R + j(\omega L - 1/\omega C)$ is real, i.e. if $\omega L - 1/\omega C = 0$,

$$\omega^2 = 1/LC, \quad \underline{\omega = \sqrt{(1/LC)}}$$

When this condition holds, the current $I = E/R$, i.e. the circuit
behaves as a resistance only.

WORKED EXAMPLES

(*a*) Three forces of magnitudes 10, 15, 20 lbf act at angles of
30°, 70°, and 120° respectively, to the real axis Ox. Using complex
numbers find the magnitude and direction of the resultant force.

The forces are represented as in Figure 11.4.

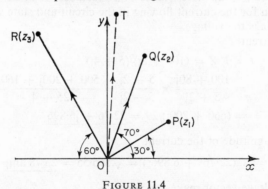

FIGURE 11.4

Let the forces be represented by the vectors **OP, OQ, OR**. The
points P, Q, and R are represented by

$$z_1 = 10(\cos 30° + j \sin 30°), \quad z_2 = 15(\cos 70° + j \sin 70°),$$
$$z_3 = 20(\cos 120° + j \sin 120°)$$

If the resultant force is represented by the vector **OT** and z is the complex number representing the point T, then

$$z = z_1 + z_2 + z_3$$
$$= (10 \cos 30° + 15 \cos 70° + 20 \cos 120°)$$
$$+ j(10 \sin 30° + 15 \sin 70° + 20 \sin 120°)$$
or $z = [10 \times 0\cdot8660 + 15 \times 0\cdot3420 + 20 \times (-0\cdot5)]$
$$+ j[10 \times 0\cdot5 + 15 \times 0\cdot9397 + 20 \times 0\cdot8660]$$
$$= (8\cdot66 + 5\cdot13 - 10) + j(5\cdot0 + 14\cdot0955 + 17\cdot32)$$
$$= 3\cdot79 + j36\cdot4155$$

The magnitude of the resultant force

$$= |z| = \sqrt{(3\cdot79^2 + 36\cdot4155^2)}$$
$$= \sqrt{(14\cdot36 + 1326\cdot5)} = \sqrt{(1340\cdot86)} = \underline{36\cdot62 \text{ lbf}}$$

The resultant makes an angle θ with Ox, where

$$\tan \theta = 36\cdot4155/3\cdot79 = 9\cdot610; \quad \theta = 84° \ 6'$$

i.e.   the resultant force is of magnitude 36·62 lbf at 84° 6′ to Ox

(*b*) A voltage represented by $100 + 80j$ is applied across an impedance of $5 + 2j$. (Resistance and reactance in ohm.) Find an expression for the current flowing in the circuit and state whether it leads or lags the voltage.

The current

$$I = V/Z = (100 + 8j)/(5 + 4j)$$
$$= \frac{100 + 80j}{5 + 2j} \times \frac{5 - 2j}{5 - 2j} = \frac{500 + 200j + 160}{25 + 4}$$
$$= (660 + 200j)/29 = \underline{22\cdot76 + j6\cdot896}$$

The magnitude of the current

$$= \sqrt{(22\cdot76^2 + 6\cdot896^2)} = \sqrt{565\cdot55} = \underline{23\cdot78 \text{ amp}}$$

The voltage vector makes an angle

$$\tan^{-1}(80/100) = \tan^{-1}(0\cdot8) = \underline{38° \ 40'}$$

with Ox and the current vector makes an angle

$$\tan^{-1}(6\cdot896/22\cdot76) = \tan^{-1} 0\cdot3030 = \underline{16° \ 51' \text{ with } Ox}$$

∴ The current lags the voltage by
$$38° \, 40' - 16°51' = \underline{21°49'}$$

Examples 11

1. Obtain the roots of the equation $Lm^2 + Rm + 1/C = 0$ when $R = 20$, $L = 0.01$, $C = 2 \times 10^{-6}$, expressing the roots in the form $a + jb$ and the polar form.

2. Show that $x = 3$ is a root of the equation $x^3 - 5x^2 + 8x - 6 = 0$. Hence obtain the quadratic equation to give the other two roots. Show that the two roots are complex and express them in Cartesian and modulus-argument form.

3. Two forces of magnitudes 10 lbf and 20 lbf act at angles 30° and 70° to the x-axis. Use complex numbers to find the resultant force in magnitude and direction.

4. Three forces of magnitudes 20, 30, 15 lbf act respectively at angles of 40°, 110°, and 210° with the real axis Ox. Find the magnitude and direction of the resultant force using complex numbers.

5. The impedance of a circuit is given by $Z = 8 + 6j$ and the current is represented by $10 + 8j$. Find, in complex form the voltage across the impedance. (*Note:* The units involved are ohms, amps and volts.) Hence find the magnitude of the voltage and the phase angle between the current and voltage, indicating whether the current leads or lags on the voltage.

6. An electrical circuit consists of a resistance of 10 ohms in series with an inductance of 10 millihenries. An alternating voltage of frequency 100 cycles per second and of magnitude 75 volts R.M.S. is applied across the circuit. Find the magnitude and phase angle of the resulting current.

7. In an electrical network problem the following condition holds: $R_2 - j/\omega C_2$
$= 1 / \left(\dfrac{1}{R_1} + j\omega C_1 \right)$ where ω is the angular velocity, R_1 and R_2 resistances, C_1 and C_2 capacitances. By simplifying the complex numbers find R_2 and C_2 in terms of ω, R_1 and C_1.

8. The characteristic impedance Z of a transmission line is given by $Z = \sqrt{\left(\dfrac{R + j\omega L}{G + j\omega C} \right)}$. Show that Z is real when $RC = LG$. What is the value of Z in this case?

   If $R = 2.2$ (ohms), $L = 0.16 \times 10^{-3}$ (henries), $G = 4.2 \times 10^{-6}$ (mhos), $C = 1.2 \times 10^{-9}$ (farads), and $\omega = 10^4$ obtain the modulus and argument of Z.

9. An alternating voltage of frequency $f = 100$ cycles/second and magnitude 250 volts is applied across a series circuit consisting of a resistance of 200 ohms and capacitance of 5 microfarads. Write down the expression for the complex impedance of the circuit and obtain the current in complex form. Express the current in modulus-argument form and state whether the curtren leads or lags the voltage.

10. The resultant impedance Z of two parallel circuits of impedances Z_1 and Z_2 is given by the formula $\dfrac{1}{Z} = \dfrac{1}{Z_1} + \dfrac{1}{Z_2}$. Find the resultant impedance when $Z_1 = 3 + 4j$, $Z_2 = 4 - 2j$ expressing it in Cartesian and modulus-argument form.

12

Finite Differences—Interpolation—
Newton's Formula

12.1 Finite Differences

An introduction to the theory of finite differences was given in Volume I. The theory dealt with will be revised and extended in this chapter.

In many problems, a set of corresponding numerical values of a function $f(x)$ and the argument x are obtained. From these tabulated values it is possible, by making use of *finite difference* theory to estimate intermediate values of the function or values of the function outside the range of values of x given. Estimation of intermediate values *within* the range of values given is known as *interpolation*, and estimation of values outside the range is known as *extrapolation*.

Generally numerical values of a function $f(x)$ are tabulated for *equal* increments of the argument x. These increments may be fractional or integral values. Such increments are called *finite differences* and the corresponding *finite differences* for the function may be written down.

12.1.1 Notation

If x increases by Δx, then the corresponding first order finite difference (increment) in y is Δy, the second order finite difference in y is $\Delta^2 y$ and so on. The nth order finite difference in y is denoted by $\Delta^n y$.

Example:

The following corresponding values of argument x and function $f(x)$ are obtained. Tabulate the finite differences for $y = f(x)$ up to the *fourth order*.

x	0	1	2	3	4	5	6	7	8	9	10
y	0	1	8	27	64	125	216	343	512	729	1000
Δy		1	7	19	37	61	91	127	169	217	271
$\Delta^2 y$			6	12	18	24	30	36	42	48	54
$\Delta^3 y$				6	6	6	6	6	6	6	6
$\Delta^4 y$					0	0	0	0	0	0	0

It will be noted that the *third order* differences are constant and the *fourth order* differences zero. This is a special property of *polynomial functions* (in this case a cubic).

12.2 Polynomial Functions

If the values of a polynomial function of degree n for equal increments in x are tabulated and the finite differences written down, the differences of order $(n + 1)$ vanish. Hence, if the $(n + 1)$th order differences of a tabulated function (for equal intervals of the argument) vanish, it can be deduced that the function is a polynomial of degree n. In practical problems, if the differences of order $(n + 1)$ are very small, then the function may be taken to be approximately a polynomial of degree n.

Example:

Show that the following tabulated numerical values of y obey a *quadratic* polynomial and hence find y as a function of x.

x	0	1	2	3	4	5	6
y	4	3	6	13	24	39	58
Δy	−1	3	7	11	15	19	
$\Delta^2 y$		4	4	4	4	4	
$\Delta^3 y$			0	0	0	0	

Since the third order differences vanish, y is a quadratic function of x. Let $y = ax^2 + bx + c$,

when $x = 0$ $4 = c$

when $x = 1$ $3 = a + b + c$, $\underline{a + b = -1}$

when $x = 2$ $6 = 4a + 2b + c$, $\underline{4a + 2b = 2}$

$2a + b = 1$, hence $a = 2$, $b = -3$,

$$\therefore \qquad\qquad y = 2x^2 - 3x + 4$$

12.3 Interpolation

This may be carried out with different degrees of accuracy depending on the orders of the finite differences used. An illustration of simple interpolation using *proportional parts* (first order differences) is given here.

Example:

Given $(2{\cdot}6)^3 = 17{\cdot}576$, $(2{\cdot}7)^3 = 19{\cdot}683$, use proportional parts (first order differences) to estimate $(2{\cdot}64)^3$.

The argument x increases from $2{\cdot}6$ to $2{\cdot}7$, i.e. $\Delta x = 0{\cdot}1$,

$$\Delta y = 19{\cdot}683 - 17{\cdot}576 = 2{\cdot}107$$

To obtain $(2{\cdot}64)^3$ use $\Delta x = 2{\cdot}64 - 2{\cdot}60 = 0{\cdot}04$.
For this increment in x, the corresponding increment in y

$$\Delta y = \frac{0{\cdot}04}{0{\cdot}10} \times 2{\cdot}107 \quad \text{(using proportion)}$$

$$= \frac{2}{5} \times 2{\cdot}107 = \frac{4{\cdot}214}{5} = 0{\cdot}8428$$

Hence

$$(2{\cdot}64)^3 \simeq 17{\cdot}576 + 0{\cdot}8428 = 18{\cdot}4188 \simeq 18{\cdot}419$$

(Using four figure logarithms $(2{\cdot}64)^3 = 18{\cdot}40$.)

    (*Note:* Different degrees of accuracy are obtained by using *significant* figures.)

For instance, in the above example, if 17·58 and 19·68 had been used for $(2·6)^3$ and $(2·7)^3$ a slightly different value would have been obtained for $(2·64)^3$.

In order to obtain greater accuracy of estimation it is necessary to use extended interpolation formulae. There are a number of such formulae. The formulae used are *comparable* with applying binomial approximations of different orders.

12.4 Numerical Computations

There are many aids available in order to carry out repetitive calculations and computation. Such aids include *tabulated values* of certain functions, e.g. trigonometric functions, logarithms (common), logarithms (natural or Napierian), exponential functions, cubes, squares, square roots, etc. In addition to tabulated values of functions there are available a number of *mechanical methods for calculation*, including slide rules, calculating machines, digital computers. Depending on the type of instrument or machine used, different *degrees of accuracy* may be obtained.

Even the use of tables will produce more significant figures as there are available 4, 5, 6, 7, and sometimes 9 figure tables. Calculations using slide rules are limited in accuracy, but, even so are extremely valuable to provide approximate checking of computations. It cannot be stressed too highly that, with any method of calculation, some form of *checking* results should be employed, even if these are only *rough checks*.

Where possible it is advisable to familiarize oneself with some form of desk calculator (whether mechanical hand-operated or electro-mechanically operated). In modern computational mathematics it may become essential at a later stage to become acquainted with a form of *digital computing system*. These, like any system for calculation require to be *programmed*. Such programming may be initially dealt with in a simple way by the use of desk calculators.

12.4.1 Extended Interpolation

When a set of tabulated values of a function is given (normally for equal steps of the argument x) it is possible to make better estimates of the values of the function for intermediate values of x by using extended interpolation formulae. There are a number of these available employing *forward differences*, *central differences*, or *backward differences*.

The techniques employed with these formulae are similar but it is impossible to deal with many of these formulae at this level. One formula only will be discussed. This formula is known as *Newton's forward difference interpolation formula*. Examples will follow illustrating the use of the formula.

12.4.2 Newton's forward difference interpolation formula

Notation: Let numerical values of a function $u(x)$ be given for values of $x = 0, h, 2h, 3h, \ldots$, i.e. in equal steps of x. Let u_0 be the value of $u(x)$ when $x = 0$ and the various orders of finite differences be denoted by $\Delta u_0, \Delta^2 u_0, \Delta^3 u_0, \ldots$, originating at $x = 0$, then

$$\Delta u_0 = u(h) - u(0)$$
$$\Delta^2 u_0 = \Delta[u(h) - u(0)] = \Delta u(h) - \Delta u(0)$$
$$= u(2h) - u(h) - [u(h) - u(0)] = u(2h) - 2u(h) + u(0)$$
$$\Delta^3 u_0 = \Delta[u(2h) - 2u(h) + u(0)] = \Delta u(2h) - 2\Delta u(h) + \Delta u(0)$$
$$= u(3h) - u(2h) - 2[u(2h) - u(h)] + u(h) - u(0)$$
$$= u(3h) - 3u(2h) + 3u(h) - u(0)$$

i.e. the different orders of differences associated with $u(0)$ can all be expressed in terms of the numerical values of the function.

Let $E = 1 + \Delta$ (an *operator*), let $u(0) = u_0$, then

$$E[u(0)] = (1 + \Delta)u_0 = u_0 + \Delta u_0 = u_0 + u(h) - u_0 = u(h)$$
$$E^2[u(0)] = (1 + \Delta)^2 u_0 = (1 + 2\Delta + \Delta^2)u_0$$
$$= u_0 + 2\Delta u_0 + \Delta^2 u_0$$
$$= u_0 + 2[u(h) - u(0)] + u(2h) - 2u(h) + u(0)$$
$$= u_0 + 2u(h) - 2u(0) + u(2h) - 2u(h) + u(0)$$
$$= u(2h)$$

and so on. Similarly

$$E^3[u(0)] = (1 + \Delta)^3 u_0 = u(3h)$$

and, more generally, if there are sufficient terms

$$E^n[u(0)] = (1 + \Delta)^n u(0) = u(nh)$$

or

$$u(x) = 1 + n\Delta u_0 + \frac{n(n-1)}{2!}\Delta^2 u_0 + \frac{n(n-1)(n-2)}{3!}\Delta^3 u_0 + \cdots$$

where $x = nh$, h being the uniform interval of x.

This is the *Gregory-Newton* forward (or advancing) difference interpolation formula.

In the above formula n was an *integer*, but it can be shown that, even if n is *not an integer*, and if $x = ph$, where p may be fractional, then

$$u(x) = (1 + \Delta)^p u(0) = 1 + p\Delta u_0 + \frac{p(p-1)}{2!}\Delta^2 u_0 + \cdots$$

This is a generalisation of the *Gregory-Newton* formula and may be used to estimate intermediate values of the function $u(x)$.

If the function is an exact *polynomial*, then at some stage, all the differences are zero. In approximate polynomial functions the differences do not vanish, but at some stage they become very small and the formula may be used to estimate $u(x)$ approximately.

Examples:

(i) The following values of a function $u(x)$ are given for the argument x from $0\cdot1(0\cdot1)0\cdot6$. Use the Newton formula to estimate $u(1\cdot5)$, $u(0\cdot45)$, and $u(0\cdot27)$. Show that the value of $u(0\cdot45)$ is the same when computed from $u(0\cdot1)$ and $u(0\cdot4)$.

x	0·1	0·2	0·3	0·4	0·5	0·6
$u(x)$	0·03	0·12	0·27	0·48	0·75	1·08
Δu		0·09	0·15	0·21	0·27	0·33
$\Delta^2 u$			0·06	0·06	0·06	0·06
$\Delta^3 u$				0	0	0

[*Note:* If it is assumed that the values continue in the same form, it follows that, since $\Delta^3 u = 0$ then $u(x)$ is a quadratic function of x.]

Take $u(0)$ as the function value when $x = 0.1$, $h = 0.1$.

(1) <u>To obtain $u(0.15)$</u>, take $ph = 0.15 - 0.10 = 0.05$, i.e.

$$p = 0.05/0.10 = \underline{0.5} \text{ (or } \tfrac{1}{2})$$

$$u(x) = (1 + \Delta)^p u(0) = u_0 + p\Delta u_0 + \frac{p(p-1)}{2!}\Delta^2 u_0 + \ldots$$

$$u_0 = 0.03, \quad \Delta u_0 = 0.09, \quad \Delta^2 u_0 = 0.06, \quad \Delta^3 u_0 = 0, \ldots$$

$$\therefore \quad u(0.15) = 0.03 + 0.5(0.09) + \frac{0.5(-0.5)}{1 \times 2}(0.06)$$

$$+ \frac{(0.5)(-0.5)(-1.5)}{1 \times 2 \times 3}(0) + \ldots$$

$$= 0.03 + 0.045 - 0.0075 = 0.075 - 0.0075 = \underline{0.0675}$$

(2) <u>To obtain $u(0.45)$</u>, use u_0, Δu_0 as above and $ph = 0.45 - 0.1$ $= 0.35$, i.e. for this function value

$$p = 0.35/0.1 = \underline{3.5} \text{ (or } \tfrac{7}{2})$$

$$u(0.45) = 0.03 + 3.5(0.09) + \frac{(3.5)(2.5)}{1 \times 2}(0.06)$$

$$+ \frac{(3.5)(2.5)(1.5)}{1 \times 2 \times 3}(0) + \ldots$$

$$= 0.03 + 0.315 + 0.2625 = \underline{0.6075}$$

(3) <u>To obtain $u(0.27)$</u>, use $u_0 = u(0.2) = 0.12$ and the corresponding differences; $ph = 0.27 - 0.20 = 0.07$,

$$p = 0.07/0.1 = \underline{0.7}$$

$$u(0.27) = 0.12 + (0.7)(0.15) + \frac{(0.7)(-0.3)}{1 \times 2}(0.06) + \ldots$$

$$= 0.12 + 0.105 - 0.7 \times 0.009 = 0.2250 - 0.0063$$

$$= \underline{0.2187}$$

(4) To obtain $u(0·45)$ using $u(0·4)$,

$$ph = 0·45 - 0·40 = 0·05, \quad p = 0·05/0·1 = 0·5 \text{ (or } \tfrac{1}{2})$$

$$u(0·45) = 0·48 + 0·5(0·27) + \frac{(0·5)(-0·5)}{1 \times 2}(0·06) + \ldots$$

$$= 0·48 + 0·135 - 0·0075 = 0·6150 - 0·0075$$

$$= 0·6075$$

Notice that with proper choice of *initial value* the computation may be simplified.

(ii) The following values of $f(x)$ are taken from a table of the normal integral. Using interpolation obtain estimates of $f(0·37)$, $f(0·55)$, and extrapolate the value of $f(0·9)$.

x	0·20	0·30	0·40	0·50	0·60	0·70	0·80
$f(x)$	0·0793	0·1179	0·1554	0·1915	0·2257	0·2580	0·2881
Δf		0·0386	0·0375	0·0361	0·0342	0·0323	0·0301
$\Delta^2 f$			−0·0009	−0·0014	−0·0019	−0·0019	−0·0022
$\Delta^3 f$				−0·0005	−0·0005	0·0000	−0·0003
$\Delta^4 f$					0·0000	0·0005	−0·0003

The fourth order differences are small and irregular so they will be ignored in the approximation.

(1) Let $u_0 = f(0·30)$, $h = 0·10$ (the constant interval).

To evaluate $f(0·37)$ take $ph = 0·07$, $p = 0·07/0·10 = 0·7$.

$$f(0·37) = u_0 + p\Delta u_0 + \frac{p(p-1)}{1 \times 2}\Delta^2 u_0 + \frac{p(p-1)(p-2)}{1 \times 2 \times 3}\Delta^3 u_0$$

$$= 0·1179 + 0·7(0·0375) + \frac{(0·7)(-0·3)}{1 \times 2}(-0·0014)$$

$$+ \frac{(0·7)(-0·3)(-1·3)}{1 \times 2 \times 3}(-0·0005)$$

$$= 0·1179 + 0·02625 + 0·000147 - 0·0000273$$

$$= 0·1442970 - 0·0000273 = 0·1442697$$

$$= 0·1443 \text{ (4 figures)}$$

(this agrees with the table from which the extract is taken).

(2) Let $u_0 = f(0.5)$, $h = 0.10$.

To evaluate $f(0.55)$, take $ph = 0.05$, $p = 0.05/0.10 = 0.5$.

$$f(0.55) = 0.1915 + (0.5)(0.0342) + \frac{(0.5)(-0.5)}{1 \times 2}(-0.0019)$$

$$+ \frac{(0.5)(-0.5)(-1.5)}{1 \times 2 \times 3}(-0.0003)$$

$$= 0.1915 + 0.0171 + 0.000238 - 0.000019$$

$$= 0.208838 - 0.000019 = 0.208819 = \underline{0.2088(2)}$$

(3) To extrapolate $f(0.9)$.

Let $u_0 = f(0.6)$, take $ph = 0.3$, $p = 3$.

$$f(0.9) = 0.2257 + 3(0.0323) + \frac{(3)(2)}{1 \times 2}(-0.0022)$$

$$= 0.2257 + 0.0969 - 0.0066 = 0.3226 - 0.0066$$

$$= \underline{0.3160}$$

(the value from the original table is 0.3160).

WORKED EXAMPLES

(a) For the given tabulated values of y and argument x, tabulate the differences for y. Deduce that y is a cubic polynomial and find the polynomial.

x	0	1	2	3	4	5	6	7	8	9	10
y	2	6	28	80	174	322	536	828	1210	1694	2292
Δy		4	22	52	94	148	214	292	382	484	598
$\Delta^2 y$			18	30	42	54	66	78	90	102	114
$\Delta^3 y$				12	12	12	12	12	12	12	12
$\Delta^4 y$					0	0	0	0	0	0	0

Since all the fourth order differences vanish, the function $f(x)$ is a polynomial of third degree, i.e.

$$y = ax^3 + bx^2 + cx + d$$

When $x = 0$; $y = 2$,

$\therefore$ $$2 = d$$

When $x = 1$; $y = 6$,

$\therefore$ $\qquad a + b + c + d = 6$; $\quad a + b + c = 4$

When $x = 2$; $y = 28$,

$\therefore$ $\qquad 8a + 4b + 2c + d = 28$; $\quad 8a + 4b + 2c = 26$

When $x = 3$; $y = 80$,

$\therefore$ $\qquad 27a + 9b + 3c + d = 80$; $\quad 27a + 9b + 3c = 78$

$\therefore$ $\quad a + b + c = 4$, $\quad 4a + 2b + c = 13$, $\quad 9a + 3b + c = 26$

Hence $3a + b = 9$, $5a + b = 13$, $2a = 4$, $a = 2$,

$\therefore$ $\qquad\qquad a = 2$, $\quad b = 3$, $\quad c = -1$, $\quad d = 2$

Hence

$$y = 2x^3 + 3x^2 - x + 2$$

(b) The following values of square roots are read off for values of x at intervals of 0·2. Use interpolation to obtain $\sqrt{4·15}$ and $\sqrt{4·86}$.

x	4·0	4·2	4·4	4·6	4·8	5·0	5·2	5·4
$u = \sqrt{x}$	2·0	2·049	2·098	2·145	2·191	2·236	2·280	2·324
Δu		0·049	0·049	0·047	0·046	0·045	0·044	0·044
$\Delta^2 u$			0	−0·002	−0·001	−0·001	−0·001	0·000
$\Delta^3 u$				−0·002	0·001	0	0	0·001

$$u(x) = u_0 + p\Delta u_0 + \frac{p(p-1)}{1 \times 2} \Delta^2 u_0 + \frac{p(p-1)(p-2)}{1 \times 2 \times 3} \Delta^3 u_0 + \ldots$$

(1) Take $u_0 = 2·0$, $h = 0·2$, $ph = 4·15 - 4·0 = 0·15$, $p = 0·75$.

$$u(4·15) = 2·0 + (0·75)(0·049) + \frac{(0·75)(-0·25)}{1 \times 2}(0) + \ldots$$

$$\text{ignoring } \Delta^3 u$$

$$= 2·0000 + 0·03675 = 2·03675 = \underline{2·037} \text{ (3 d.p.)}$$

(2) Take $u_0 = 2.191$ $(x = 4.8)$, $ph = 4.86 - 4.8 = 0.06$, $p = 0.3$.

$$u(4.86) = 2.191 + (0.3)(0.045) + \frac{(0.3)(-0.7)}{1 \times 2}(-0.001) + \cdots$$
$$= 2.1910 + 0.0135 + 0.0001 = 2.2046 = \underline{2.205} \text{ (3 d.p.)}$$

(c) If $y = n^3 + n^2 + n + 1$, find expressions for Δy, $\Delta^2 y$, $\Delta^3 y$, $\Delta^4 y$ for an increment of 1 in n.

$$\Delta y_n = y_{n+1} - y_n = (n+1)^3 + (n+1)^2 + (n+1) + 1$$
$$- [n^3 + n^2 + n + 1]$$
$$= n^3 + 3n^2 + 3n + 1 + n^2 + 2n + 1 + n + 1 + 1$$
$$- n^3 - n^2 - n - 1$$
$$= 3n^2 + 5n + 3$$
$$\Delta^2 y_n = \Delta y_{n+1} - \Delta y_n = 3(n+1)^2 + 5(n+1) + 3 - (3n^2 + 5n + 3)$$
$$= 3n^2 + 6n + 3 + 5n + 5 + 3 - 3n^2 - 5n - 3 = \underline{6n + 8}$$
$$\Delta^3 y_n = \Delta^2 y_{n+1} - \Delta^2 y_n = 6(n+1) + 8 - (6n + 8) = \underline{6}$$
$$\Delta^4 y_n = 0 \text{ since } \Delta^3 y_n \text{ is a constant.}$$

Note: This is the way in which it is deduced that, for a polynomial of degree r the $(r+1)$th order differences vanish for equal steps of the argument.

Examples 12

1. The following values of $f(x)$ for increasing values of x are given. Show that the values obey a polynomial in x, state the order of the polynomial and find the polynomial:

x	0	1	2	3	4	5	6	7	8	9	10
$f(x)$	3	5	23	93	275	653	1335	2453	4163	6645	10103

2. Given $e^{0.30} = 1.3499$ and $e^{0.35} = 1.4191$ use first order interpolation to estimate the values of $e^{0.32}$ and $e^{0.335}$.
3. The following values are taken from a table of tangents, $\tan 30° = 0.5774$, $\tan 30°30' = 0.5890$.
    Use first order interpolation (i.e. use first order differences only) to estimate the values of $\tan 30°6'$, $\tan 30°15'$, $\tan 30°24'$.
4. Given $\log_5 (7.5) = 1.2510$ and $\log_5 (8.5) = 1.3290$ use first order interpolation to estimate $\log_5 (8.0)$ and $\log_5 (7.85)$.

140 O.N.C. MATHEMATICS II

5. The following values of a function $f(x)$ are given for intervals of 0·2 in the argument x.

x	0·2	0·4	0·6	0·8	1·0	1·2	1·4
$f(x)$	1·221	1·492	1·822	2·225	2·718	3·320	4·055

Tabulate the differences for $f(x)$ up to the fourth order. Hence, using the Newton interpolation forward difference formula estimate the values of $f(0·5), f(0·9)$. Also extrapolate the value of $f(1·6)$. (*Note*: Use any suitable initial value for u_0.)

6. The following tabulated values of $f(x)$ are given for equal intervals in the argument x. Show that $f(x)$ is a polynomial in x and find the polynomial function.

x	0	1	2	3	4	5	6	7	8
$f(x)$	7	10	33	88	187	342	565	868	1263

7. The following values of acceleration in ft/s² after t seconds are recorded. Tabulate the finite differences for f and use Newton's formula to estimate the acceleration at times $t = 2·5$ and 9·5 seconds.

t	0	2	4	6	8	10	12	14	16
$f(t)$[acc.]	11·2	10·5	9·9	9·1	8·5	8·0	7·5	7·0	6·4

8. The following values are given of an exact polynomial function of the variable x. Write down the differences up to the order where all are zero. Hence, working back progressively find the values of $f(x)$ when $x = 7, 8, 9$:

x	0	1	2	3	4	5	6
$f(x)$	3	7	29	81	175	323	537

9. The following table gives y as a function of x for equal steps of the argument x. Use finite differences and the Newton formula to find y when $x = 0·3$ and y when $x = 0·9$.

x	0·2	0·4	0·6	0·8	1·0	1·2	1·4
y	0·90703	0·82646	0·75621	0·69462	0·64034	0·59229	0·54960

10. The following values are extracted from a table of Bessel Functions $J_0(x)$. Use finite difference interpolation to estimate $J_0(0·84)$ and $J_0(0·93)$.

x	0·80	0·85	0·90	0·95	1·00
$J_0(x)$	0·8463	0·8274	0·8075	0·7868	0·7652

11. The following table shows the cubes of integers from $x = 8$ to 12 in steps of
1 unit. Use Newton's formula to calculate $8 \cdot 9^3$ and $9 \cdot 7^3$. Also (since this is a
polynomial function) use Newton's formula to extrapolate the values of 15^3
and 20^3.

x	8	9	10	11	12
$f(x) = x^3$	512	729	1000	1331	1728

12. Write down the table of differences for the function $f(x)$ for values of x from
5 to 11 up to the fourth order of differences:

x	5	6	7	8	9	10	11
$f(x)$	16·25	18·43	19·84	20·70	21·15	21·24	20·98

Neglecting the fourth order differences use Newton's formula to interpolate to find $f(x)$ when $x = 5 \cdot 5$ and $x = 7 \cdot 3$. Also, using the initial value
when $x = 8$ extrapolate the value of $f(x)$ for $x = 12$.

13. Taking $f(x) = ax^2 + bx + c$, show that for an increment of 1 unit in x:
$\Delta f(x) = 2ax + a + b$, $\Delta^2 f(x) = 2a$, $\Delta^3 f(x) = 0$.
    Show that the following functional values obey a quadratic function and
hence, using the above results find the coefficients a, b and c and hence the
function:

x	0	1	2	3	4	5	6
$f(x)$	5	4	9	20	37	60	89

14. Taking $f(x) = ax^3 + bx^2 + cx + d$, show that, for an increment of 1 unit
in x, $\Delta f(x) = 3ax^2 + x(3a + 2b) + a + b + c$, $\Delta^2 f(x) = 6ax + 6a + 2b$,
$\Delta^3 f(x) = 6a$, $\Delta^4 f(x) = 0$.
    Show that the following values of $f(x)$ obey a cubic polynomial function
and hence, using the above results find the values of a, b, c and d and hence
the function

x	0	1	2	3	4	5	6
$f(x)$	7	10	15	28	55	102	175

15. The tabulated values of $\tan \theta$ have been taken from four figure tables:

θ	65°	66°	67°	68°	69°
$\tan \theta$	2·1445	2·2460	2·3559	2·4751	2·6051

Use finite differences to estimate $\tan 65°30'$ and $\tan 66°45'$.

13

Approximate Rates of Change—Derivatives from First Principles—Derivative of x^n

13.1 Tabulation of Functions

In many practical problems the results of observations are usually numerical values, relating some variable y with a variable x. It is often assumed that y is some function of x although the exact functional relation is not known. From tabulated values, however, it is possible to estimate the *approximate rates of change* of the function for different values of x. Proceeding from approximate rates of change it is possible, using *limiting processes*, to obtain values for *instantaneous rates of change* of the function at specified values of x.

Example:

Tabulate the values of $y = 3x^2$ for steps of 0·1 in x from $x = 1$ to 2. Calculate the value of

$$\frac{\Delta y}{\Delta x} = \frac{\text{increment of } y}{\text{increment of } x}$$

for each of the intervals 0·1 in x.

x	1·0	1·1	1·2	1·3	1·4	1·5	1·6	1·7	1·8	1·9	2·0
$y = 3x^2$	3·0	3·63	4·32	5·07	5·88	6·75	7·68	8·67	9·72	10·83	12·0
Δy		0·63	0·69	0·75	0·81	0·87	0·93	0·99	1·05	1·11	1·17
$\Delta y/\Delta x$		6·3	6·9	7·5	8·1	8·7	9·3	9·9	10·5	11·1	11·7

It will be seen that the ratios calculated are simply related as follows:

(i) The ratios increase at a constant rate *or* the values of the ratios are a linear function of *x*.

(ii) In each case the ratio of the increments is exactly six times the mid-value of *x*.

13.2 Approximate Rates of Change (Functional Notation)

A general increment in the variable *x* will be represented by δx (delta *x*) and the *corresponding* increment in *y* (where *y* is a function of *x*) will be represented by δy (delta *y*). During the interval δx *the average rate of change of y with respect to x is* given by $\delta y / \delta x$. This ratio gives the average increase of *y for unit increase in x* during the interval δx.

If $y = f(x)$ (i.e. *y* is a function of *x*) then, if *x* increases by δx, $f(x)$ increases to $f(x + \delta x)$, hence

$$\delta y = f(x + \delta x) - f(x)$$

$$\therefore \quad \frac{\delta y}{\delta x} = \frac{f(x + \delta x) - f(x)}{\delta x}$$

$$= \text{average rate of change of } y \text{ with respect to } x.$$

Examples:

(i) If $y = x^2 + 5x - 4$ obtain an expression for the average rate of change of *y* with respect to *x*.

Let $y = f(x) = x^2 + 5x - 4$.

If *x* increases by δx and *y* increases correspondingly by δy, then the value of *x* increases to $x + \delta x$ and *y* to $y + \delta y$, then

$$\begin{aligned}
\delta y &= f(x + \delta x) - f(x) \\
&= (x + \delta x)^2 + 5(x + \delta x) - 4 - [x^2 + 5x - 4] \\
&= x^2 + 2x\delta x + (\delta x)^2 + 5x + 5\delta x - 4 - x^2 - 5x + 4 \\
\delta y &= 2x\delta x + 5\delta x + (\delta x)^2
\end{aligned}$$

dividing by δx:

$$\frac{\delta y}{\delta x} = 2x + 5 + \delta x = \text{average rate of change of } y$$

(ii) For the function $y = x^3$ tabulate the values of y for $x = 2 \cdot 0$, $2 \cdot 10$, $2 \cdot 08$, $2 \cdot 06$, $2 \cdot 04$, $2 \cdot 02$, $2 \cdot 01$, $2 \cdot 001$. Obtain the values of $\delta y/\delta x$ to two decimal places for an initial value of $x = 2$ and increments $0 \cdot 10$, $0 \cdot 08$, $0 \cdot 06$, $0 \cdot 04$, $0 \cdot 02$, $0 \cdot 01$, $0 \cdot 001$ in x. What is the likely value of the rate of change when $x = 2$?

x	$f(x) = x^3$	δx	δy	$\delta y/\delta x$
2·00	8·0000			
2·10	9·2610	0·10	1·2610	12·61
2·08	8·9989	0·08	0·9989	12·48
2·06	8·7416	0·06	0·7416	12·34
2·04	8·4896	0·04	0·4896	12·24
2·02	8·2424	0·02	0·2424	12·12
2·01	8·1206	0·01	0·1206	12·06
2·001	8·0120	0·001	0·0120	12·00

From the manner in which $\delta y/\delta x$ is changing it would seem that the most likely value of $\delta y/\delta x$ when $x = 2$ would be $12 \cdot 00$. It will be shown in 13·3 that this is the exact value.

13.3 Instantaneous Rates of Change—Derivatives from First Principles

Consider the graph of the function $y = f(x)$ [see Figure 13.1].

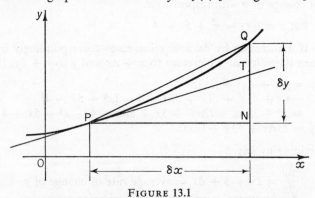

FIGURE 13.1

Let P be the point (x, y) and Q the point $(x + \delta x, y + \delta y)$, then $\delta y/\delta x$ = gradient (or slope) of the chord PQ. As δx tends to zero (P fixed), i.e. as $\delta x \to 0$, the slope of the chord PQ tends to the slope of the tangent PT at the point P, i.e. the limiting value of $\delta y/\delta x$ as $\delta x \to 0$ will be the gradient of the tangent to the curve at P. This is denoted by dy/dx (= limit of $\delta y/\delta x$ as $\delta x \to 0$),

$$or \quad \frac{dy}{dx} = \lim_{\delta x \to 0} \left(\frac{\delta y}{\delta x} \right)$$

This expression is a measure of the *instantaneous rate of change* of y with respect to x.

Depending on the type of problem, dy/dx may be described as:

(i) the differential coefficient of y with respect to x,
(ii) the derivative of y with respect to x,
(iii) the derived function,
(iv) the gradient of the function,
(v) the gradient (slope) of the graph of $y = f(x)$,
(vi) the gradient (slope) of the tangent to the graph of $y = f(x)$.

Example:

Find, from first principles the derivative of y with respect to x when $y = x^3$.

Let x increase by δx and y increase correspondingly by δy, then

$$y + \delta y = (x + \delta x)^3; \quad y = x^3$$

$$\therefore \quad \delta y = (x + \delta x)^3 - x^3 = x^3 + 3x^2\delta x + 3x(\delta x)^2 + (\delta x)^3 - x^3$$
$$\therefore \quad \delta y = 3x^2\delta x + 3x(\delta x)^2 + (\delta x)^3$$

divide by δx:

$$\frac{\delta y}{\delta x} = 3x^2 + 3x\delta x + (\delta x)^2$$

as $\delta x \to 0$, this approaches $3x^2$, i.e.

$$\frac{dy}{dx} = \lim_{\delta x \to 0} \left(\frac{\delta y}{\delta x} \right) = 3x^2$$

[*Note:* Referring to 13.2, Example (ii), when $x = 2$, $dy/dx = 12$.]

13.4 Derivative of a General Function from First Principles

Let $y = f(x)$, x increase by δx and y increase correspondingly by δy, then

$$\delta y = f(x + \delta x) - f(x)$$

$$\frac{\delta y}{\delta x} = \frac{f(x + \delta x) - f(x)}{\delta x}$$

$$\therefore \qquad \frac{dy}{dx} = \lim_{\delta x \to 0} \frac{\delta y}{\delta x} = \lim_{\delta x \to 0} \left[\frac{f(x + \delta x) - f(x)}{\delta x} \right]$$

Generally this limit is readily obtained for simple algebraic functions. With more difficult functions (e.g. trigonometric) special limits need to be used.

The method described here, of obtaining dy/dx is general for finding dy/dx from first principles.

13.4.1 Derivative of x^n

Let $y = x^n$, x increase by δx and y correspondingly by δy,

$$y + \delta y = (x + \delta x)^n$$

$$\delta y = (x + \delta x)^n - x^n = x^n[1 + \delta x/x]^n - x^n$$

$$= x^n \left[1 + n \frac{\delta x}{x} + \frac{n(n-1)}{1 \times 2} \left(\frac{\delta x}{x} \right)^2 + \ldots \right] - x^n$$

$$\therefore \qquad \delta y = nx^{n-1}\delta x + \frac{n(n-1)}{2} x^{n-2}(\delta x)^2 + \text{terms containing } (\delta x)^3$$

$$\therefore \qquad \frac{\delta y}{\delta x} = nx^{n-1} + \text{terms containing at least } \delta x$$

as $\delta x \to 0$ all the terms on the right-hand side $\to 0$ (except the first),

$$\therefore \qquad \underline{\frac{dy}{dx} = \lim_{\delta x \to 0} \left(\frac{\delta y}{\delta x} \right) = nx^{n-1}}$$

This is the general rule for *writing down* the derivative of x^n.

13.4.2 Derivative of ax^n (a is a constant)

Since dy/dx is a *rate of change*, the rate of change of ax^n will be $a \times$ rate of change of x^n, i.e. if

$$y = ax^n, \quad \frac{dy}{dx} = nax^{n-1}$$

13.4.3 Derivative of the sum of a number of functions

If $y = u + v + w + \ldots$, where u, v, w are separate functions of x, since dy/dx = rate of change of y with respect to x it follows that

$$\frac{dy}{dx} = \text{sum of the rates of change of } u, v, w, \ldots$$

i.e.
$$\frac{dy}{dx} = \frac{du}{dx} + \frac{dv}{dx} + \frac{dw}{dx} + \ldots$$

WORKED EXAMPLES

(*a*) The equation of a graph is $y = x^2 + 3x - 1$. Estimate the approximate rates of change of y with respect to x when $x = 2$ and $x = 5$. Compare these values with those found by writing down the derivative of the function.

(*Note:* Any small increment for δx may be used. The smaller the increment, the better the estimate.)

Tabulation:

x	1·8	2·2	4·8	5·2
y	7·64	10·44	36·44	41·64

(*Note* that values of y have been found at equal small intervals of x from $x = 2$ and $x = 5$.)

Near $x = 2$,

$$\delta x = 2\cdot2 - 1\cdot8 = 0\cdot4, \quad \delta y = 10\cdot44 - 7\cdot64 = 2\cdot80$$

$$\therefore \quad \frac{\delta y}{\delta x} = \frac{2\cdot80}{0\cdot40} = \underline{7\cdot0}$$

Near x = 5,

$$\delta x = 5 \cdot 2 - 4 \cdot 8 = 0 \cdot 4, \quad \delta y = 41 \cdot 64 - 36 \cdot 44 = 5 \cdot 20$$

$$\therefore \quad \frac{\delta y}{\delta x} = \frac{5 \cdot 20}{0 \cdot 40} = \underline{13 \cdot 0}$$

Now if $y = x^2 + 3x - 1$, $dy/dx = 2x + 3$

when $x = 2$,

$$\frac{dy}{dx} = 2 \times 2 + 3 = 7$$

when $x = 5$,

$$\frac{dy}{dx} = 2 \times 5 + 3 = 13$$

It will be seen that the estimates and exact values agree.

(*b*) The height h feet of a body, projected vertically upwards, is given at time t seconds by $h = 80t - 16t^2$ feet. Estimate the velocity of the body at $t = 1 \cdot 5$ seconds and write down the exact velocity when $t = 1 \cdot 5$ seconds.

When $t = 1 \cdot 6$ sec,

$$h = 80(1 \cdot 6) - 16(1 \cdot 6)^2 = 87 \cdot 04 \, \text{ft}$$

When $t = 1 \cdot 4$ sec,

$$h = 80(1 \cdot 4) - 16(1 \cdot 4)^2 = 80 \cdot 64 \, \text{ft}$$

Let the increase in height during the time δt be δh, then

$$\delta h = 87 \cdot 04 - 80 \cdot 64 = 6 \cdot 40 \, \text{ft}, \quad \delta t = 1 \cdot 6 - 1 \cdot 4 = 0 \cdot 2 \, \text{sec}$$

$$\therefore \quad \delta h / \delta t = 6 \cdot 40 / 0 \cdot 2 = \underline{32 \, \text{ft/s} = \text{velocity}}$$

Using differentiation

$$\text{velocity} = \frac{dh}{dt} = 80 - 32t$$

when $t = 1 \cdot 5$,

$$\frac{dh}{dt} = 80 - 32 \times 1 \cdot 5 = 80 - 48 = \underline{32 \, \text{ft/s}}$$

(c) Using the general rule, write down the derivatives of (i) $x^3 + 4x^2 - x + 3$, (ii) $t^2 + 5t - 1$, (iii) $\sqrt{x}$, (iv) $z^2 + 3/z - \sqrt{z^3}$.

(i) Let $y = x^3 + 4x^2 - x + 3$, $\dfrac{dy}{dx} = \underline{3x^2 + 8x - 1}$

(ii) Let $s = t^2 + 5t - 1$, $\dfrac{ds}{dt} = \underline{2t + 5}$

(iii) Let $y = \sqrt{x} = x^{\frac{1}{2}}$, $\dfrac{dy}{dx} = \frac{1}{2}x^{\frac{1}{2}-1} = \frac{1}{2}x^{-\frac{1}{2}} = \underline{1/(2\sqrt{x})}$

(iv) Let $y = z^2 + 3/z - \sqrt{z^3} = z^2 + 3z^{-1} - z^{\frac{3}{2}}$

$$\frac{dy}{dz} = 2z + 3(-1)z^{-2} - \frac{3}{2}z^{\frac{3}{2}-1} = \underline{2z - 3/z^2 - \frac{3}{2}\sqrt{z}}$$

(d) Determine the values of x and y when $dy/dx = 0$ if $y = 4x^3 - 3x^2 - 36x + 11$. [*Note:* such values of y are called *stationary values*.]

$dy/dx = 12x^2 - 6x - 36 = 6(2x^2 - x - 6) = 6(2x + 3)(x - 2)$

$\therefore dy/dx = 0$, when $(2x + 3)(x - 2) = 0$, i.e. when

$$\underline{x = 2 \text{ or } -3/2}$$

when $x = 2$,

$$y = 4(2)^3 - 3(2)^2 - 36(2) + 11 = \underline{-41}$$

when $x = -3/2$,

$$y = 4(-3/2)^3 - 3(-3/2)^2 - 36(-3/2) + 11 = \underline{44\tfrac{3}{4}}$$

i.e. the function has *stationary values* of -41 and $+44\tfrac{3}{4}$

(e) Determine the equations of the tangents and normals to the curve $y = 2x^2 - 3x + 2$ at the points where $x = 2$ and $x = -1$. Find the stationary value of y.

$$\frac{dy}{dx} = 4x - 3 = 0 \text{ for stationary value of } y \text{ when } \underline{x = \tfrac{3}{4}},$$

when $x = \tfrac{3}{4}$,

$$y = 2(\tfrac{3}{4})^2 - 3(\tfrac{3}{4}) + 2 = \underline{7/8}$$

which is the stationary value of y.

The slope of the tangent to the graph $= dy/dx$.
The slope of the normal $= -1/$slope of tangent $= -1/(dy/dx)$.
At the point where $x = 2$, $dy/dx = 4(2) - 3 = +5$.
At the point where $x = -1$, $dy/dx = 4(-1) - 3 = -7$.
Equation of tangent at $x = 2$ is $y - y_1 = m(x - x_1)$, where $x_1 = 2$, $y_1 = 4$, $m = 5$, i.e. tangent at $(2, 4)$ is

$$y - 4 = 5(x - 2) \quad \text{or} \quad \underline{y = 5x - 6}$$

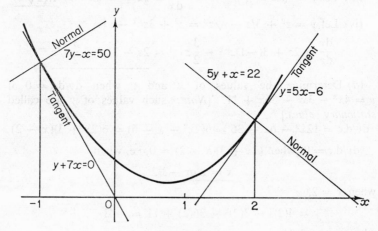

FIGURE 13.2

Equation of normal at $(2, 4)$, slope $= -1/5$, is

$$y - 4 = -\tfrac{1}{5}(x - 2) \quad \text{or} \quad \underline{5y + x = 22}$$

Equation of tangent at the point where $x = -1$, $y = 7$ is

$$y - 7 = m[x - (-1)], \quad (m = -7)$$
$$y - 7 = -7[x + 1], \quad \underline{y + 7x = 0}$$

Equation of normal at $(-1, 7)$, slope $= +\tfrac{1}{7}$,

$$y - 7 = +\tfrac{1}{7}(x + 1), \quad \underline{7y - x = 50}$$

Figure 13.2 shows the various lines.

Examples 13

1. For the function $y = 2x^2 + 3x$, tabulate the values of y for $x = 1(0\cdot1)\ 1\cdot6$. Tabulate the first order differences for y and hence estimate the average rates of change of y with respect to x in each of the intervals of $0\cdot1$ in x.

2. For the function $y = 3x^2 - 2x + 1$ tabulate the values of y for $x = 2\cdot0$, $2\cdot10$, $2\cdot08$, $2\cdot06$, $2\cdot04$, $2\cdot02$, $2\cdot01$, $2\cdot001$ (to 4 decimal places). Tabulate the values of δy for the increments of x from 2 and also the values of $\delta y/\delta x$ for these increments. What is the likely value of dy/dx when $x = 2$?

3. The following values of the function e^x are given for steps of $0\cdot1$ in x:

x	0·5	0·6	0·7	0·8	0·9	1·0
e^x	1·649	1·822	2·014	2·226	2·460	2·718

Estimate the rates of change of e^x in the five equal intervals of x. Compare these values with the values of e^x for the mid-values of x in each interval.

4. Find, from first principles, the derivatives of the following functions:
(i) $y = 3x^2 + 2x - 1$ (ii) $y = 1/x$ (iii) $y = \sqrt{x}$.

5. For the function $y = x^2 + 3x - 1$ find the values of $\delta y/\delta x$ for intervals of $\delta x = 0\cdot1,\ 0\cdot2,\ 0\cdot3,\ 0\cdot4$ measured from the initial value $x = 1$.

6. Estimate the approximate rate of change of the function $p = 5/v$ when $v = 2$ using a suitable small interval for v. Compare the result with the value of dp/dv when $v = 2$.

7. The distance s feet travelled by a body in time t seconds is given by $s = 0\cdot5t^2 + 1\cdot5t - 2$. Estimate approximately the speed of the body in feet per second, when $t = 1, 2, 3$ seconds. Compare the results with the velocity $= ds/dt$ at these times.

8. Use the general rule for differentiating x^n to write down the derivatives of the following functions with respect to the appropriate independent variable:
(i) $4x^3 + 2x - 5$, (ii) $(t + 3)^2$, (iii) $x^{3/2}$, (iv) $(z - 3)^2$, (v) $\theta^4 + \theta^2 + 1$, (vi) $1/u^2$, (vii) $15/\sqrt{t}$, (viii) v^{-n}, (ix) $(\sqrt{x} + 1)^2$, (x) $7y^4 - 6y^3 + 5y^2 + 3^2$, (xi) $v^{1\cdot4}$.

9. Write down the rates of change of p (pressure) with respect to v (volume) when the functional relationships between p and v are:
(i) $pv = c$ (constant), (ii) $pv^n = c$ (constant).

10. Determine the stationary values of the function $y = 4x^3 + 3x^2 - 18x + 15$.

11. If the distance s feet, travelled by a body in time t seconds is given by $s = 2t^3 - 15t^2 + 24t + 6$, write down the expression for the velocity $v = ds/dt$.

Hence (i) find the velocity when $t = 3$ seconds,
(ii) find the times when the velocity is zero,
(iii) find the distance s at the times when the velocity is zero.

12. The angle θ (in radians) turned through by a rotating shaft in time t seconds is given by $\theta = 0\cdot1t^2 - 1\cdot5t + 4$. Find the angular velocity ($d\theta/dt$) in radians per second when $t = 1, 5$ seconds. At what time will the shaft come to rest?

13. The equation to a rectangular hyperbola is given by $xy = 5$. Find the equations of the tangents to the curve at the points where (i) $x = 2$, (ii) $x = -1$. Find also the equations of the normals to the curve at these points.

6

14. The equation to a parabola is given by $y^2 = 4x$. Obtain the equations of the tangent and normal to the curve at the point (1, 2).

15. The current i amp in an electrical circuit is given at time t by the expression $i = 10\,e^{-kt}$, where $k = 2$. Tabulate the values of i for $t = 0(0\cdot1)0\cdot6$ and using the tabulated values estimate the approximate rates of change of i with respect to t when $t = 0\cdot05, 0\cdot15, 0\cdot25, 0\cdot35, 0\cdot45, 0\cdot55$ seconds.

16. Draw the graph whose equation is $y = x^3 + x^2$ for values of x from 0 to 2 in steps of $0\cdot2$ for x. By drawing suitable chords, estimate the slopes of the tangents to the curve at the points where $x = 0\cdot5, 1\cdot0, 1\cdot5$. Verify the results by writing down dy/dx and evaluating when x has these three values.

14

Differentiation of Trigonometric Functions, Exponential and Logarithmic Functions

14.1 Small Angles

If θ is measured in radians and θ is a small angle, approximations for the trigonometric ratios $\sin \theta$, $\cos \theta$, $\tan \theta$ can be found in terms of θ.

Consider the numerical values of $\sin \theta$, $\cos \theta$, $\tan \theta$ tabulated below (these values are read off from four-figure tables).

θ (degrees)	5°	4°	3°	2°	1°	30′	18′	6′	0
θ (radians)	0·0873	0·0698	0·0524	0·0349	0·0175	0·0087	0·0052	0·0017	0
$\sin \theta$	0·0872	0·0698	0·0523	0·0349	0·0175	0·0087	0·0052	0·0017	0
$\cos \theta$	0·9962	0·9976	0·9986	0·9994	0·9998	1·0000	1·0000	1·0000	1·00
$\tan \theta$	0·0875	0·0699	0·0524	0·0349	0·0175	0·0087	0·0052	0·0017	0

The following points will be noted, as θ tends to zero:

  (i) $\sin \theta$ and θ approach equality, i.e. as $\theta \to 0$, $\sin \theta \to \theta$,
 (ii) $\tan \theta$ and θ approach equality,
(iii) $\cos \theta \to 1$.

In particular $\sin \theta / \theta \to 1$ *or* when θ is in radians as θ tends to zero,

$$\lim_{\theta \to 0} \frac{\sin \theta}{\theta} = 1$$

153

14.1.1 Expansions of sin θ and cos θ

When θ is in radian measure, series can be obtained for $\sin \theta$ and $\cos \theta$ as follows:

$$\sin \theta = \theta - \frac{\theta^3}{3!} + \frac{\theta^5}{5!} - \frac{\theta^7}{7!} + \ldots \text{ to infinity}$$

$$\cos \theta = 1 - \frac{\theta^2}{2!} + \frac{\theta^4}{4!} - \frac{\theta^6}{6!} + \frac{\theta^8}{8!} + \ldots \text{ to infinity}$$

Using these series, various orders of approximation may be used for $\sin \theta$ and $\cos \theta$ when θ is small. These series may also be used to calculate the values of $\sin \theta$ and $\cos \theta$.

Example:

Using the series for $\sin \theta$ and $\cos \theta$, evaluate (i) $\sin 0 \cdot 1$, (ii) $\cos 0 \cdot 1$ to four significant figures.

(i) $\sin 0 \cdot 1 = 0 \cdot 1 - \dfrac{(0 \cdot 1)^3}{3!} + \dfrac{(0 \cdot 1)^5}{5!} - \dfrac{(0 \cdot 1)^7}{7!} + \ldots$

$\qquad = 0 \cdot 1 - \dfrac{0 \cdot 001}{6} + \dfrac{0 \cdot 00001}{120} - \dfrac{0 \cdot 0000001}{5040} + \ldots$

$\qquad = 0 \cdot 1 - 0 \cdot 00016667 + 0 \cdot 000000083 + \ldots$

$\qquad = 0 \cdot 100000083 - 0 \cdot 00016667$

$\qquad = 0 \cdot 099833413 \simeq 0 \cdot 0998$

(*Note:* From four figure tables $0 \cdot 1$ rad $= 5°44'$, $\sin 5°44' = 0 \cdot 0999$.)

(ii) $\cos (0 \cdot 1) = 1 - \dfrac{(0 \cdot 1)^2}{2!} + \dfrac{(0 \cdot 1)^4}{4!} - \dfrac{(0 \cdot 1)^6}{6!} + \ldots$

$\qquad = 1 - \dfrac{0 \cdot 01}{2} + \dfrac{0 \cdot 0001}{24} - \dfrac{0 \cdot 000001}{720} + \ldots$

$\qquad = 1 \cdot 0 - 0 \cdot 005 + 0 \cdot 000004$

$\qquad = 0 \cdot 995 + 0 \cdot 000004 = 0 \cdot 995004 \simeq 0 \cdot 9950$

(*Note:* From four figure tables $\cos 0 \cdot 1$ rad $= \cos 5°44' = 0 \cdot 9950$.)

14.1.2 Geometrical illustration

Consider the Figure 14.1.

PQ is the arc of a circle of radius r. Let QN and RP be perpendicular to OP, then when θ is in radians, arc PQ $= r\theta$, QN $= r \sin \theta$,

$ON = r \cos \theta$, $RP = r \tan \theta$. As θ tends to zero, arc $QP \to$ chord QP,
$\therefore$ for small θ;

$$QN < QP < PR$$

or
$$r \sin \theta < r\theta < r \tan \theta$$

divide by r:

$$\sin \theta < \theta < \tan \theta$$

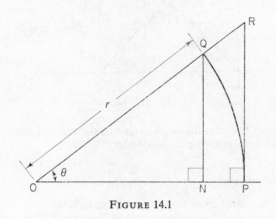

FIGURE 14.1

divide by $\sin \theta$:

$$1 < \frac{\theta}{\sin \theta} < \frac{1}{\cos \theta}$$

as $\theta \to 0$, $\cos \theta \to 1$,

$\therefore$
$$\frac{\theta}{\sin \theta} \to 1 \text{ as } \theta \to 0,$$

i.e.
$$\lim_{\theta \to 0} \frac{\theta}{\sin \theta} = 1 \quad \text{and} \quad \lim_{\theta \to 0} \frac{\sin \theta}{\theta} = 1$$

14.2 Differentiation of Trigonometric Functions

By consideration of the manner in which $\sin x$ and $\cos x$ vary as x
increases from 0 to $\pi/2$ radians (see Figure 14.2) it can be seen that

as *x increases*, sin *x increases*, hence the rate of increase of sin *x* with respect to *x will be positive*. Similarly, as *x increases*, cos *x decreases*, hence the rate of change of cos *x* with respect to *x will be negative*.

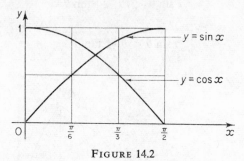

FIGURE 14.2

From such considerations, the *signs* of the rates of change of *any* trigonometric functions may be deduced.

14.2.1 Derivative of sin *x*

This may be found from first principles as follows.

Let $y = \sin x$ (x in radians) and x increase by δx, y increase correspondingly by δy, then

$$y + \delta y = \sin (x + \delta x)$$

$$\therefore \qquad \delta y = \sin (x + \delta x) - \sin x \qquad \text{[express as product]}$$

$$= 2 \cos \tfrac{1}{2}(x + \delta x + x) \sin \tfrac{1}{2}(x + \delta x - x)$$

$$= 2 \cos (x + \tfrac{1}{2}\delta x) \sin (\tfrac{1}{2}\delta x)$$

Divide by δx

$$\frac{\delta y}{\delta x} = 2 \cos (x + \tfrac{1}{2}\delta x) \frac{\sin (\tfrac{1}{2}\delta x)}{\delta x}$$

$$= \cos (x + \tfrac{1}{2}\delta x) \left[\frac{\sin (\tfrac{1}{2}\delta x)}{\tfrac{1}{2}\delta x} \right]$$

as $\delta x \to 0$, $\tfrac{1}{2}\delta x \to 0$ and hence

$$\lim_{\delta x \to 0} \left[\frac{\sin \tfrac{1}{2}\delta x}{\tfrac{1}{2}\delta x} \right] = 1$$

and also
$$\lim_{\delta x \to 0} \cos (x + \tfrac{1}{2}\delta x) = \cos x$$

Hence
$$\frac{dy}{dx} = \lim_{\delta x \to 0} \left(\frac{\delta y}{\delta x} \right) = \cos x$$

or
$$\frac{d}{dx} (\sin x) = \cos x \ [x \text{ in radians}]$$

14.2.2 Derivative of cos x

Let $y = \cos x$ (x in radians), and x increase by δx, y increase correspondingly by δy, then

$$y + \delta y = \cos (x + \delta x)$$
$$\therefore \quad \delta y = \cos (x + \delta x) - \cos x \qquad \text{[express as product]}$$
$$= -2 \sin \tfrac{1}{2}(x + \delta x + x) \sin \tfrac{1}{2}(x + \delta x - x)$$
$$= -2 \sin (x + \tfrac{1}{2}\delta x) \sin (\tfrac{1}{2}\delta x)$$

Divide by δx

$$\frac{\delta y}{\delta x} = -2 \sin (x + \tfrac{1}{2}\delta x) \frac{\sin (\tfrac{1}{2}\delta x)}{\delta x}$$

$$= -\sin (x + \tfrac{1}{2}\delta x) \left[\frac{\sin \tfrac{1}{2}\delta x}{\tfrac{1}{2}\delta x} \right]$$

as $\delta x \to 0$, $\tfrac{1}{2}\delta x \to 0$ and hence

$$\lim_{\delta x \to 0} \left[\frac{\sin \tfrac{1}{2}\delta x}{\tfrac{1}{2}\delta x} \right] = 1$$

and also
$$\lim_{\delta x \to 0} \sin (x + \tfrac{1}{2}\delta x) = \sin x$$

Hence
$$\frac{dy}{dx} = \lim_{\delta x \to 0} \left(\frac{\delta y}{\delta x} \right) = - \sin x$$

or
$$\frac{d}{dx} (\cos x) = -\sin x \ [x \text{ in radians}]$$

14.2.3 Derivative of tan x

Let $y = \tan x = \dfrac{\sin x}{\cos x}$ and x increase by δx and y increase correspondingly by δy, then

$$y + \delta y = \frac{\sin (x + \delta x)}{\cos (x + \delta x)}$$

$$\therefore \qquad \delta y = \frac{\sin (x + \delta x)}{\cos (x + \delta x)} - \frac{\sin x}{\cos x}$$

$$= \frac{\sin (x + \delta x) \cos x - \cos (x + \delta x) \sin x}{\cos (x + \delta x) \cos x}$$

The numerator is the form $\sin A \cos B - \cos A \sin B = \sin (A - B)$,

$$\delta y = \frac{\sin (x + \delta x - x)}{\cos (x + \delta x) \cos x} = \frac{\sin \delta x}{\cos (x + \delta x) \cos x}$$

$$\therefore \qquad \frac{\delta y}{\delta x} = \frac{[\sin \delta x / \delta x]}{\cos (x + \delta x) \cos x}$$

$$\lim_{\delta x \to 0} \left[\frac{\sin \delta x}{\delta x} \right] = 1, \quad \lim_{\delta x \to 0} \cos (x + \delta x) = \cos x$$

$$\therefore \qquad \frac{dy}{dx} = \lim_{\delta x \to 0} \left(\frac{\delta y}{\delta x} \right) = \frac{1}{\cos x \cos x} = \frac{1}{\cos^2 x} = \underline{\sec^2 x}$$

or

$$\frac{d}{dx} (\tan x) = \sec^2 x$$

The derivatives of other trigonometric functions may be similarly obtained from *first principles*.

14.3 Derivatives of e^x and $\log_e x$

These will be obtained by considering first the series for e^x which was used previously, i.e.

$$e^x = 1 + \frac{x}{1!} + \frac{x^2}{2!} + \frac{x^3}{3!} + \frac{x^4}{4!} + \ldots \, ad\infty$$

14.3.1 Derivative of e^x

This may be obtained as follows:

Let $y = e^x$, x increase by δx, y by δy, then

$$y + \delta y = e^{x+\delta x}$$

$$\therefore \quad \delta y = e^{x+\delta x} - e^x = e^x e^{\delta x} - e^x = e^x[e^{\delta x} - 1]$$
$$= e^x[1 + \delta x + \tfrac{1}{2}(\delta x)^2 + \text{terms in } (\delta x)^3 + \ldots - 1]$$
$$= e^x[\delta x + \tfrac{1}{2}(\delta x)^2 + \text{terms in } (\delta x)^3 \text{ at least}]$$

Divide by δx

$$\frac{\delta y}{\delta x} = e^x[1 + \tfrac{1}{2}(\delta x) + \text{terms in } (\delta x)^2 \text{ at least}]$$

$$\therefore \quad \frac{dy}{dx} = \lim_{\delta x \to 0}\left(\frac{\delta y}{\delta x}\right) = e^x[1 + 0 + \ldots] = e^x$$

or

$$\frac{d}{dx}(e^x) = e^x$$

14.3.2 Derivative of $\log_e x$

This is most readily obtained using the following result.

For finite relative increments δx and δy,

$$\frac{\delta y}{\delta x} = \frac{1}{\delta x/\delta y}$$

So long as δx is finite this result is exact, hence it may be assumed that as δx and δy both tend to zero

$$\lim_{\delta x \to 0}\frac{\delta y}{\delta x} = \frac{1}{\lim\limits_{\delta y \to 0}\delta x/\delta y} \quad \text{or} \quad \frac{dy}{dx} = \frac{1}{dx/dy}$$

Let $y = \log_e x$, then $x = e^y$,

$$\frac{dx}{dy} = e^y = x,$$

$$\therefore \quad \frac{dy}{dx} = \frac{1}{dx/dy} = \frac{1}{x} \quad \text{or} \quad \frac{d}{dx}(\log_e x) = \frac{1}{x}$$

WORKED EXAMPLES

(a) Obtain, from first principles the derivative of $\cot x$.

Let $y = \cot x = \dfrac{\cos x}{\sin x}$, x increase by δx, y by δy, then

$$y + \delta y = \frac{\cos(x + \delta x)}{\sin(x + \delta x)}$$

$$\therefore \qquad \delta y = \frac{\cos(x + \delta x)}{\sin(x + \delta x)} - \frac{\cos x}{\sin x}$$

$$= \frac{\cos(x + \delta x)\sin x - \cos x \sin(x + \delta x)}{\sin(x + \delta x)\sin x}$$

$$= \frac{-\sin(x + \delta x - x)}{\sin(x + \delta x)\sin x} = \frac{-\sin \delta x}{\sin(x + \delta x)\sin x}$$

Divide by δx

$$\frac{\delta y}{\delta x} = \frac{-[\sin \delta x / \delta x]}{\sin(x + \delta x)\sin x}$$

$$\lim_{\delta x \to 0}\left[\frac{\sin \delta x}{\delta x}\right] = 1, \quad \lim_{\delta x \to 0}\sin(x + \delta x) = \sin x$$

$$\frac{dy}{dx} = \lim_{\delta x \to 0}\left(\frac{\delta y}{\delta x}\right) = -\frac{1}{\sin x \sin x} = -\operatorname{cosec}^2 x$$

or

$$\frac{d}{dx}(\cot x) = -\operatorname{cosec}^2 x$$

(b) Obtain the equation of the tangent to the graph of $y = \sin x$ at the point where $x = \pi/4$.

The slope of the tangent to the graph is dy/dx,

$$\frac{dy}{dx} = \cos x$$

when $x = \pi/4$,

$$\frac{dy}{dx} = \cos\frac{\pi}{4} = \frac{1}{\sqrt{2}}$$

The equation of the tangent to the curve at (x_1, y_1) is given by:

$$y - y_1 = m(x - x_1)$$

where m = slope of tangent = $1/\sqrt{2}$.

When $x = \pi/4$,

$$y = \cos \pi/4 = 1/\sqrt{2}$$

$\therefore$ the equation of the tangent is

$$y - 1/\sqrt{2} = (1/\sqrt{2})(x - \pi/4) \quad \text{or} \quad \underline{y\sqrt{2} - x - 1 + \pi/4 = 0}$$

(c) If $y = e^x - \log_e x + x^2$ obtain the approximate increase in y when x increases from 1·0 to 1·1.

$$\frac{dy}{dx} = e^x - \frac{1}{x} + 2x$$

But when δx and δy are small,

$$\frac{\delta y}{\delta x} \simeq \frac{dy}{dx}$$

i.e.

$$\delta y \simeq \frac{dy}{dx}(\delta x)$$

when $x = 1$,

$$\frac{dy}{dx} = e^1 - 1 + 2 = 2·718 + 1 = 3·718$$

$\therefore$ the approximate increase in y is given by

$$\delta y \simeq 3·718 \times 0·1 = 0·3718 \simeq \underline{0·372}$$

(d) Find a maximum and minimum value of the function $4 \sin t + 3 \cos t$.

Let $s = 4 \sin t + 3 \cos t$,

$$\frac{ds}{dt} = 4 \cos t - 3 \sin t; \quad \frac{d^2s}{dt^2} = -4 \sin t - 3 \cos t$$

For stationary values of s (maximum or minimum),

$$\frac{ds}{dt} = 0,$$

i.e. when $\qquad 4 \cos t - 3 \sin t = 0,$

$\qquad \tan t = 4/3$ for stationary values of s

(i) Let $t = t_1$ (acute), $\tan t_1 = 4/3$, $\sin t_1 = 4/5$, $\cos t_1 = 3/5$, then

$$\frac{d^2 s}{dt^2} = -4 \times \frac{4}{5} - 3 \times \frac{3}{5} = -5$$

$\therefore$ s is a maximum when

$$t = t_1 = \tan^{-1}\left(\frac{4}{3}\right) \text{ [acute]}$$

Maximum value of $s = 4 \times 4/5 + 3 \times 3/5 = +5$

(ii) Let $t = t_2$ (reflex), $\tan t_2 = 4/3$, $\sin t_2 = -4/5$, $\cos t_2 = -3/5$, then

$$\frac{d^2 s}{dt^2} = -4 \times \left(-\frac{4}{5}\right) - 3 \times \left(-\frac{3}{5}\right) = +5$$

$\therefore$ s is a minimum when

$$t = t_2 = \tan^{-1}(4/3) \text{ [reflex]}$$
$$t_2 = \pi + t_1$$

Minimum value of $s = +4 \times (-4/5) + 3 \times (-3/5) = -5$

Examples 14

1. Using the series for $\sin \theta$ estimate the approximate value of $\sin 0.2$ rad to four significant figures. Verify the result by converting the angle to degrees and minutes and finding the sine from four figure tables.
2. Evaluate $\cos 0.2$ rad using the series for $\cos \theta$ to four significant figures. Verify the result by reference to four figure tables.
3. Tabulate the values of $\sin x$ for x in radians from $x = 0$ to 0.5 in steps of 0.1 (radians). Estimate the average rates of change of $\sin x$ in the five intervals of 0.1. Compare these with the values of $\frac{d}{dx}(\sin x)$ when $x = 0.05$, 0.15, 0.25, 0.35 and 0.45.
4. Tabulate the values of e^x for $x = 0(0.2)1.0$. Estimate the average rates of change of e^x in the five intervals and compare the results with the values of $\frac{d}{dx}(e^x)$ at the mid-points of the five intervals.

5. Sketch the graph of $y = e^x$. Determine the equations of the tangent and normal to the graph at the point where $x = 1$.

6. If $s = 20 \sin t + 40 \cos t$, show that $\dfrac{d^2s}{dt^2} + s = 0$.

7. If $y = 10\,e^x$, show that $\dfrac{d^2y}{dx^2} - y = 0$.

8. Show that the function $y = e^x - 2x + 3$ has a minimum value when $x = \log_e 2$ and find the minimum value of y to 3 significant figures.

9. Show, from first principles that the derivative of e^{2x} is $2\,e^{2x}$.

10. Show, from first principles that the derivative of $\sin 2x$ is $2 \cos 2x$.

11. Find, from first principles the derivative of $\tan 3x$.

12. Using first principles show that, if n is a constant then (i) $\dfrac{d}{dx}(\sin nx) = n \cos nx$ (ii) $\dfrac{d}{dx}(\cos nx) = -n \sin nx$ when x is in radian measure.

13. If K is a constant, using first principles show that the derivative of e^{Kx} is $K\,e^{Kx}$.

14. Write down the derivatives of the following functions using the results in examples 12 and 13:

   (i) $\cos 3x$ (ii) $5 \sin 4x$ (iii) $\cos(t/2)$ (iv) $\sin(\tfrac{3}{2}\theta)$ (angle in radians)
   (v) $10\,e^{3x}$ (vi) $3\,e^{2t}$ (vii) e^{-4x} (viii) $e^{0.5t}$.

15. The displacement s of a body at time t seconds is given by $s = 3 \sin 2t + 5 \cos 2t$. Show that s satisfies the equation $\dfrac{d^2s}{dt^2} + 4s = 0$.

16. If $y = 5e^{3x} + 3e^{-2x}$ show that y satisfies the (differential) equation $\dfrac{d^2y}{dx^2} - \dfrac{dy}{dx} - 6y = 0$.

17. Find a maximum and minimum value of the function $3 \sin x - 4 \cos x$ stating the values of x at which the maximum and minimum values occur.

18. Obtain the equations of the tangent and normal to the graph of $y = \cos 2x$ at the point where $x = \pi/2$. Sketch the graph and the tangent and normal.

19. If $y = e^{2x} + 2 \log_e x + x$ write down dy/dx. Hence find the approximate increase (δy) in y when x increases from 0·5 to 0·6.

20. Given $s = 5 \sin 2t - 3 \cos 2t$, write down ds/dt. Hence find the approximate increase in s when t increases from 0·6 to 0·7 (*Note:* the angles involved are in radian measure).

15

Function of a Function—Products and Quotients

15.1 Derivatives by Substitution

Using some form of substitution and the standard rules for the differentiation of simple functions it is possible to write down the derivatives of more complicated functions. The basic rule which is applied to simplify differentiation is called the *function of a function* rule.

Additional rules, which allow the derivatives of *products* and *quotients* to be written down, may also be used. The three rules necessary to carry out the differentiation of functions by *substitution* and the differentiation of *products* and *quotients* will be established and examples worked to show the applications of the rules.

15.1.1 Function of a function

Let $y = f(z)$ be a general function of z. If z be a function of a third variable x then, if $z = z(x)$, when the substitution is effected y becomes a function of x, i.e.

$$y \text{ is a } function \text{ } of \text{ } z \text{ which is a } function \text{ } of \text{ } x$$

Illustrations:

 (i) If $y = (x^2 + 1)^{3/2}$, let $z = x^2 + 1$ [i.e. $x = \sqrt{(1 - z)}$], then $y = z^{3/2}$ or y is a standard function of z.

164

(ii) If $y = \log_e (\sqrt{x} + 1)$, let $z = \sqrt{x} + 1$ [i.e. $x = (z - 1)^2$], then $y = \log_e z$ or y is a standard function of z.

15.1.2 Function of a function rule

Let $y = f(z)$ and $z = z(x)$. If x increases by δx, z increases correspondingly by δz and y by δy, then, so long as δx, δz, and δy are finite (or measurable), the identity

$$\frac{\delta y}{\delta x} = \frac{\delta y}{\delta z} \times \frac{\delta z}{\delta x}$$

holds, since δz may be cancelled for finite values.

As δx, δz, and δy all tend to zero, the identity still holds, hence it is assumed that the result will hold in the limit as δx, δz, and δy tend to zero, i.e.

$$\lim_{\delta x \to 0} \frac{\delta y}{\delta x} = \lim_{\delta z \to 0} \frac{\delta y}{\delta z} \times \lim_{\delta x \to 0} \frac{\delta z}{\delta x}$$

or

$$\underline{\frac{dy}{dx} = \frac{dy}{dz} \times \frac{dz}{dx}}$$

(*Note:* Any letter may be used in place of z.)

Examples:

(i) If $y = (x^2 + 1)^4$ write down the derivative of y with respect to x.

Let $z = x^2 + 1$, then $y = z^4$

$$\frac{dy}{dz} = 4z^3, \quad \frac{dz}{dx} = 2x$$

$\therefore \quad \dfrac{dy}{dx} = \dfrac{dy}{dz} \times \dfrac{dz}{dx} = 4z^3 \times 2x = 8 \; x \; z^3 = \underline{8x(x^2 + 1)^3}$

(ii) If $s = (t^2 - 1)^{3/2}$ write down ds/dt.

Let $u = t^2 - 1$, then $s = u^{3/2}$,

$$\frac{ds}{du} = \frac{3}{2} u^{\frac{1}{2}}, \quad \frac{du}{dt} = 2t$$

$\therefore \quad \dfrac{ds}{dt} = \dfrac{ds}{du} \times \dfrac{du}{dt} = \dfrac{3}{2} u^{\frac{1}{2}} \times 2t = 3tu^{\frac{1}{2}} = \underline{3t(t^2 - 1)^{\frac{1}{2}}}$

(iii) Find dV/dx given $V = 5 \log_e (x^3 + 1)$.

Let $z = x^3 + 1$, $V = 5 \log_e z$,

$$\frac{dV}{dz} = 5 \times \frac{1}{z}, \quad \frac{dz}{dx} = 3x^2$$

$$\therefore \quad \frac{dV}{dx} = \frac{dV}{dz} \times \frac{dz}{dx} = 5 \times \frac{1}{z} \times 3x^2 = \frac{15x^2}{z} = \frac{15x^2}{x^3 + 1}$$

15.1.3 Extension of the function of a function rule

The rule established above may be extended, e.g. if $y = f(z)$, $z = z(u)$, $u = u(x)$, then

$$\frac{dy}{dx} = \frac{dy}{dz} \times \frac{dz}{du} \times \frac{du}{dx}$$

Example:

If $y = \log_e (\sin x^3)$ write down dy/dx.

Let $u = x^3$, $z = \sin u$, then $y = \log_e z$,

$$\frac{du}{dx} = 3x^2, \quad \frac{dz}{du} = \cos u, \quad \frac{dy}{dz} = \frac{1}{z}$$

$$\therefore \quad \frac{dy}{dx} = \frac{dy}{dz} \times \frac{dz}{du} \times \frac{du}{dx} = \frac{1}{z} \times \cos u \times 3x^2$$

$$= \frac{3x^2 \cos u}{z} = \frac{3x^2 \cos x^3}{\sin x^3} = 3x^2 \cot x^3$$

15.2 Products of Functions

Consider the function $y = x^3 \sin x$. This may be treated as the *product* of two separate functions, i.e. if $u = x^3$, $v = \sin x$, then $y = uv$.

15.2.1 Product rule

Let $y = uv$, where u and v are functions of x. Let x increase by δx, u increase correspondingly by δu, v by δv, and y by δy, then

$$y + \delta y = (u + \delta u)(v + \delta v)$$
$$y = uv$$
$$\therefore \quad \delta y = (u + \delta u)(v + \delta v) - uv$$
$$= uv + v\delta u + u\delta v + \delta u \delta v - uv$$
$$= v\delta u + u\delta v + \delta u \delta v$$

Divide by δx

$$\frac{\delta y}{\delta x} = v\frac{\delta u}{\delta x} + u\frac{\delta v}{\delta x} + \delta u\frac{\delta v}{\delta x} \qquad (1)$$

Now as $\delta x \to 0$, $\delta u \to 0$, $\delta v \to 0$, and $\delta y \to 0$

$$\lim_{\delta x \to 0}\frac{\delta y}{\delta x} = \frac{dy}{dx}, \quad \lim_{\delta x \to 0}\frac{\delta u}{\delta x} = \frac{du}{dx}, \quad \lim_{\delta x \to 0}\frac{\delta v}{\delta x} = \frac{dv}{dx}$$

Hence the limiting value of (1) becomes

$$\frac{dy}{dx} = v\frac{du}{dx} + u\frac{dv}{dx}$$

or

$$\frac{d}{dx}(uv) = v\frac{du}{dx} + u\frac{dv}{dx} \qquad (2)$$

Examples:

(i) Write down dy/dx when $y = x^2(x - 1)^3$.

Let $u = x^2$, $v = (x - 1)^3$, $y = uv$, then

$$\frac{du}{dx} = 2x, \quad \frac{dv}{dx} = 3(x - 1)^2$$

$$\therefore \quad \frac{dy}{dx} = v\frac{du}{dx} + u\frac{dv}{dx} = (x - 1)^3 \times 2x + x^2 \times 3(x - 1)^2$$

$$= x(x - 1)^2[2(x - 1) + 3x] = \underline{x(x - 1)^2(5x - 2)}$$

(ii) Without using substitutions write down the derivative of $x^3 \sin x$.

Let $y = x^3 \sin x$, then

$$\frac{dy}{dx} = \sin x \frac{d}{dx}(x^3) + x^3 \frac{d}{dx}(\sin x)$$

$$= \sin x \times 3x^2 + x^3 \times \cos x = \underline{x^2(3 \sin x + x \cos x)}$$

15.3 Quotients of Functions

Consider the function $y = (x + 1)^3/(\cos 2x)$. This may be treated as the *quotient* of two separate functions, i.e. if $u = (x + 1)^3$, $v = \cos 2x$, then $y = u/v$.

15.3.1 Quotient rule

Let $y = u/v$, where u and v are functions of x.

Let x, u, v, and y increase by corresponding amounts δx, δu, δv, and δy, then

$$y + \delta y = \frac{u + \delta u}{v + \delta v}; \quad y = \frac{u}{v}$$

$\therefore \qquad \delta y = \frac{u + \delta u}{v + \delta v} - \frac{u}{v} = \frac{v(u + \delta u) - u(v + \delta v)}{v(v + \delta v)}$

$$= \frac{vu + v\delta u - uv - u\delta v}{v(v + \delta v)} = \frac{v\delta u - u\delta v}{v(v + \delta v)}$$

Divide by δx,

$$\frac{\delta y}{\delta x} = \frac{v(\delta u/\delta x) - u(\delta v/\delta x)}{v(v + \delta v)} \tag{3}$$

Now as $\delta x \to 0$, $\delta u \to 0$, $\delta v \to 0$, $\delta y \to 0$,

$$\lim_{\delta x \to 0} \frac{\delta y}{\delta x} = \frac{dy}{dx}, \quad \lim_{\delta x \to 0} \frac{\delta u}{\delta x} = \frac{du}{dx}, \quad \lim_{\delta x \to 0} \frac{\delta v}{\delta x} = \frac{dv}{dx}$$

and

$$\lim_{\delta v \to 0} v(v + \delta v) = v^2$$

Hence, the limiting value of (3) becomes

$$\frac{dy}{dx} = \frac{v\,(du/dx) - u\,(dv/dx)}{v^2}$$

or

$$\frac{d}{dx}\left(\frac{u}{v}\right) = \frac{v\,(du/dx) - u\,(dv/dx)}{v^2} \tag{4}$$

Examples:

(i) Write down dy/dx, when $y = x^3/(\sin 2x)$.

Let $u = x^3$, $v = \sin 2x$, then $y = u/v$,

$$\frac{du}{dx} = 3x^2; \quad \frac{dv}{dx} = 2\cos 2x$$

$$\frac{dy}{dx} = \frac{v(du/dx) - u(dv/dx)}{v^2}$$

$$= \frac{\sin 2x(3x^2) - x^3(2\cos 2x)}{(\sin 2x)^2}$$

$$= \frac{3x^2 \sin 2x - 2x^3 \cos 2x}{\sin^2 2x}$$

$$= \frac{x^2(3\sin 2x - 2x\cos 2x)}{\sin^2 2x}$$

(ii) Obtain ds/dt if $s = e^{3t}/t^2$ without substitution.

$$\frac{ds}{dt} = \frac{t^2(d/dt)\,(e^{3t}) - e^{3t}\,(d/dt)(t^2)}{(t^2)^2}$$

$$= \frac{t^2(3e^{3t}) - e^{3t}(2t)}{t^4} = \frac{te^{3t}(3t - 2)}{t^4} = \frac{e^{3t}(3t - 2)}{t^3}$$

[*Notes:* (i) the rules should be memorized in derivative form, (ii) quotients *may* be replaced by products since $y = u/v$ $= uv^{-1}$, hence the quotient could be considered as the product of the functions u and v^{-1}.]

WORKED EXAMPLES

(a) Write down dy/dx, when $y = 5/(x^2 - 1)^3$.
Let $u = x^2 - 1$, then $y = 5/u^3 = 5u^{-3}$,

$$\frac{dy}{du} = 5(-3)u^{-4}, \quad \frac{du}{dx} = 2x$$

$$\therefore \frac{dy}{dx} = \frac{dy}{du} \times \frac{du}{dx} = -15u^{-4} \times 2x = -30x/u^4 = \underline{-30x/(x^2 - 1)^4}$$

(b) If $s = e^{-t}(\cos t + \sin t)$, show that $\dfrac{d^2s}{dt^2} + 2\dfrac{ds}{dt} + 2s = 0$.

$$\frac{ds}{dt} = e^{-t}\frac{d}{dt}(\cos t + \sin t) + (\cos t + \sin t)\frac{d}{dt}(e^{-t})$$

$$= e^{-t}(-\sin t + \cos t) + (\cos t + \sin t)(-e^{-t})$$
$$= e^{-t}(-\sin t + \cos t - \cos t - \sin t) = -2\sin t\, e^{-t}$$

$$\frac{d^2s}{dt^2} = -2\frac{d}{dt}[\sin t\, e^{-t}] = -2\left[e^{-t}\frac{d}{dt}(\sin t) + \sin t\frac{d}{dt}(e^{-t})\right]$$

$$= -2[e^{-t}\cos t + \sin t(-e^{-t})] = -2e^{-t}[\cos t - \sin t]$$

$$\therefore \frac{d^2s}{dt^2} + 2\frac{ds}{dt} + 2s$$

$$= e^{-t}[-2(\cos t - \sin t) + 2(-2\sin t) + 2(\cos t + \sin t)]$$
$$= e^{-t}[-2\cos t + 2\sin t - 4\sin t + 2\cos t + 2\sin t] \underline{= 0}$$

(c) If $y = x^2 \log_e x$, find a value of x, not zero when dy/d$x = 0$.
What is the value of d^2y/dx^2 when x has this value?

$$\frac{dy}{dx} = x^2 \frac{d}{dx}(\log_e x) + \log_e x \frac{d}{dx}(x^2)$$
$$= x^2 \times 1/x + \log_e x \times (2x)$$
$$= x(1 + 2\log_e x) = 0$$

when

$$x = 0 \quad \text{or} \quad 1 + 2\log_e x = 0$$

i.e. other than $x = 0$, $2\log_e x = -1$, $\log_e x = -0.5$, $\underline{x = e^{-0.5}}$.

$$\frac{d^2y}{dx^2} = x\frac{d}{dx}(1 + 2\log_e x) + (1 + 2\log_e x)\frac{d}{dx}(x)$$

$$= x \times (2/x) + (1 + 2\log_e x)1 = 3 + 2\log_e x$$

When $x = e^{-0.5}$

$$\frac{d^2y}{dx^2} = 3 + 2(-0.5) = 2$$

(d) Write down the derivatives of (i) e^{2x}/x^3, (ii) $(3t - 1)/(3t + 1)$.

(i) If $y = e^{2x}/x^3$,

$$\frac{dy}{dx} = \left[x^3 \frac{d}{dx}(e^{2x}) - e^{2x}\frac{d}{dx}(x^3) \right] / (x^3)^2$$

$$= [x^3(2e^{2x}) - e^{2x}(3x^2)]/x^6$$

$$= [x^2 e^{2x}(2x - 3)]/x^6 = e^{2x}(2x - 3)/x^4$$

(ii) If $y = (3t - 1)/(3t + 1)$,

$$\frac{dy}{dt} = \left[(3t + 1)\frac{d}{dt}(3t - 1) - (3t - 1)\frac{d}{dt}(3t + 1) \right] / (3t + 1)^2$$

$$= [(3t + 1)3 - (3t - 1)3]/(3t + 1)^2 = 6/(3t + 1)^2$$

(e) Using the *product rule*, write down the derivative of $e^{2x}/(x + 1)^3$.

Let $y = e^{2x}/(x + 1)^3 = e^{2x}(x + 1)^{-3}$.

$$\frac{dy}{dx} = (x + 1)^{-3}\frac{d}{dx}(e^{2x}) + e^{2x}\frac{d}{dx}(x + 1)^{-3}$$

$$= (x + 1)^{-3} \times 2e^{2x} + e^{2x}[-3(x + 1)^{-4}]$$

$$= e^{2x}(x + 1)^{-4}[2(x + 1) - 3]$$

$$= e^{2x}(x + 1)^{-4}(2x - 1) = e^{2x}(2x - 1)/(x + 1)^4$$

Examples 15

1. Using the function of a function rule write down the derivatives of:

   (i) $(3x + 1)^5$, (ii) $(x^2 + 3x)^4$, (iii) $(3x^4 + 6)^{1/2}$, (iv) $\sqrt{(x^2 + 2x)}$.

2. Write down the differential coefficients of:

   (i) $3/(1 - 2x)^2$, (ii) $1/(4x + 3)^3$, (iii) $(3x^2 + 1)^{3/2}$, (iv) $\sqrt{(a^2 + 2x^2)}$.

3. Differentiate the following functions with respect to x:

   (i) e^x, (ii) e^{2x+3}, (iii) e^{x^2+3x}, (iv) $\frac{1}{2}(e^{3x} + e^{-3x})$.

4. Write down the derivatives of the following functions with respect to the appropriate variables:

   (i) $\log_e(1 + 2x)$, (ii) $\log_e(3t^2 + t - 1)$, (iii) $\log_e(2u + 1)^3$,
   (iv) $\log_e(z^4 + z^3)$.

5. Using an appropriate rule write down dy/dx when:

(i) $y = \log_e (\sin x)$, (ii) $y = \log_e (\cos x)$, (iii) $y = e^{\tan x}$, (iv) $y = 5e^{x^3}$.

6. Write down the differential coefficients of the following functions with respect to the appropriate variables:

(i) $4 \sin (t^3 + 1)$, (ii) $\sin^2 (3x)$, (iii) $e^{\sin t}$, (iv) $(2x + 3/x)^3$, (v) $e^{\sin 2t}$, (vi) $\log_e (\tan \theta)$, (vii) $\sqrt{(\sin 2x)}$, (viii) $3e^{1/t}$, (ix) $\cos^4 u$, (x) $\log_e (e^x + e^{-x})$.

7. Using the product rule differentiate the following functions with respect to x:

(i) $x(2 + x^2)$, (ii) $x(12 - 2x)^2$, (iii) $(3x + 5)(x - 2)^2$, (iv) $x\sqrt{(x + 1)}$.

8. Write down the derivatives of the functions:

(i) $e^{3x}(5x + 3)$, (ii) $(2x + 1)^3 e^{2x}$, (iii) $x^3 e^{2x}$, (iv) $(x + 1)^3 e^{4x}$.

9. Differentiate the following functions with respect to the appropriate variables:

(i) $t^3 \sin 2t$, (ii) $\sqrt{x} \log_e x$, (iii) $5 e^x \sin 3x$, (iv) $e^u \sin u$.

10. Write down the derivatives of the functions:

(i) $x^2 \cos 2x$, (ii) $e^{-x} \cos 3x$, (iii) $\sin t \cos t$, (iv) $t^3 \sin^2 t$.

11. Differentiate the following functions with respect to the appropriate variables:

(i) $\sin 2t \cos 3t$, (ii) $x^2 \sin (x/2)$, (iii) $\log_e (u \sin u)$, (iv) $\sqrt{x} \tan x$, (v) $(e^t + e^{-t})(\sin t + \cos t)$, (vi) $x^2 e^{1/x}$.

12. Using the quotient rule differentiate the functions:

(i) $(x - 1)/(x + 1)$, (ii) $(2x + 3)/(3x - 4)$, (iii) $(x^2 - 1)/(x^2 + 1)$, (iv) $(x^2 + 1)/(x + 1)$.

13. Write down the derivatives of the functions:

(i) e^{2x}/x^3, (ii) $(e^x - 1)/(e^x + 1)$, (iii) $\sin x/(2x)$, (iv) $(\log_e x)/x$.

14. Differentiate the following functions with respect to the appropriate variables:

(i) $x^2/(\cos x)$, (ii) $(\log_e t)/t^3$, (iii) $\sin t/t^2$, (iv) $\sin x/\cos^3 x$.

15. Express the following quotients as products and hence write down the derivatives:

(i) $\tan x/x$, (ii) $x/\log_e x$, (iii) $\sin x/(1 + \cos x)$, (iv) $(\sin x)/x$.

16. (i) Show that $y = e^{-x} \cos 2x$ satisfies the differential equation

$$\frac{d^2y}{dx^2} + 2\frac{dy}{dx} + 5y = 0.$$

(ii) Show that $s = e^{-2t} \cos 3t$ satisfies the differential equation

$$\frac{d^2s}{dt^2} + 4\frac{ds}{dt} + 13s = 0.$$

17. Verify that $y = (A + Bx) e^{-2x}$ satisfies the equation $\dfrac{d^2y}{dx^2} + 4\dfrac{dy}{dx} + 4y = 0$

(A and B are constants).

18. Show that $s = e^t \sin t$ satisfies the equation $\dfrac{d^2s}{dt^2} - 2\dfrac{ds}{dt} + 2s = 0$.

19. Find the possible values of m if $y = e^{mx}$ satisfies the equation

$$\frac{d^2y}{dx^2} + \frac{dy}{dx} - 20y = 0.$$

20. (i) If $s = t e^{2t}$, show that $\dfrac{d^2s}{dt^2} - \dfrac{ds}{dt} - 2s = 3e^{2t}$.

(ii) Verify that $s = e^{-Kt} \sin nt$ satisfies the equation

$$\frac{d^2s}{dt^2} + 2K\frac{ds}{dt} + (K^2 + n^2)s = 0.$$

16

Applications of the Differential Calculus—Kinematics—Rates of Change—Maxima and Minima

16.1 Applications of the Differential Calculus

Since a derivative basically measures *rate of change* or the *slope* (*gradient*) of a curve there are many applications of the differential calculus. Some of these are:

   (i) *Rates of change:* rates of cooling, rates of increase of pressure, rates of change of currents, rates of expansion.
  (ii) *Kinematics:* involving relationships between distance and time, velocity and time, acceleration and time (both linear and angular).
 (iii) *Geometry of curves:* involving the determination of equations of tangents and normals to curves.
 (iv) *Turning* (*stationary*) *values:* determination of maxima and minima of functions and the practical determination of maxima and minima of surfaces, volumes, etc.

16.1.1 Rates of change

The function of a function rule may be applied in a number of problems. The method is illustrated by means of the following example.

173

Example:

A metal sphere expands when heated such that the radius increases at a rate of 0·005 in./s. Obtain an expression for the rate at which the volume is increasing when the radius is r in. What is the rate of increase when the radius is 1·5 in.?

Let the volume of the sphere at time t seconds be $V = (4/3)\pi r^3$, where r is a function of t, i.e. V is a function of r, which is a function of t, then

$$\frac{dV}{dt} = \frac{dV}{dr} \times \frac{dr}{dt} \text{ (function of a function rule)}$$

$$\frac{dV}{dr} = 4\pi r^2; \quad \frac{dV}{dt} = 4\pi r^2 \frac{dr}{dt}$$

but $$\frac{dr}{dt} = 0·005 \text{ in/s}$$

$$\frac{dV}{dt} = 4\pi r^2 \times 0·005 = 0·02\pi r^2 \text{ in}^3/\text{s}$$

when $r = 1·5$ in.,

$$\frac{dV}{dt} = 0·02\pi \times (1·5)^2 = 0·045\pi = \underline{0·1414 \text{ in}^3/\text{s}}$$

16.1.2 Kinematics

The elements of kinematics (motion ignoring forces) have been dealt with in Volume I. If $s = f(t)$ represents the distance of a particle from a fixed point on a line, then

velocity $= v =$ rate of change of distance $= \dfrac{ds}{dt}$,

acceleration $= f$ (or a) $=$ rate of change of velocity

$$= \frac{dv}{dt} = \frac{d}{dt}\left(\frac{ds}{dt}\right) = \frac{d^2s}{dt^2}$$

Example:

The distance s feet of a particle from a fixed point O on a line is given at time t seconds by the formula $s = 4t^3 - 21t^2 + 18t + 32$.

Find (i) when the velocity is zero, (ii) the accelerations at these times, (iii) the distances from O at these times.

$$velocity = v = \frac{ds}{dt} = 12t^2 - 42t + 18 = 6(2t^2 - 7t + 3)$$
$$= \underline{6(2t - 1)(t - 3)\ \text{ft/s}}$$

$$acceleration = f = \frac{dv}{dt} = \frac{d^2s}{dt^2} = \underline{24t - 42\ \text{ft/s}^2}$$

(i) the velocity is zero when $v = 0$, i.e.

$$(2t - 1)(t - 3) = 0, \quad \text{or} \quad \underline{t = \tfrac{1}{2}\ \text{or}\ 3\ \text{seconds}}$$

(ii) when $t = \tfrac{1}{2}, f = 24 \times \tfrac{1}{2} - 42 = \underline{-30\ \text{ft/s}^2}$ (retardation)

when $t = 3, f = 24 \times 3 - 42 = \underline{+30\ \text{ft/s}^2}$ (acceleration)

(iii) when $t = \tfrac{1}{2}, s = 4 \times (\tfrac{1}{2})^3 - 21 \times (\tfrac{1}{2})^2 + 18 \times \tfrac{1}{2} + 32$
$$= \underline{36\tfrac{1}{4}\ \text{ft from O}}$$

when $t = 3, s = 4 \times 3^3 - 21 \times 3^2 + 18 \times 3 + 32$
$$= \underline{5\ \text{ft from O}}$$

16.1.3 Parametric differentiation

This is an application involving rates of change when the co-ordinates of a point are given in parametric form (as shown in Chapter 7).

If
$$x = \text{a function of } t = x(t)$$
$$y = \text{a function of } t = y(t)$$

then for a small finite increment δt in t

$$\frac{\delta y}{\delta x} = \frac{\delta y / \delta t}{\delta x / \delta t}$$

hence

$$\lim_{\delta x \to 0} \frac{\delta y}{\delta x} = \lim_{\delta t \to 0}\left(\frac{\delta y}{\delta t}\right) \div \lim_{\delta t \to 0}\left(\frac{\delta x}{\delta t}\right)$$

or

$$\frac{dy}{dx} = \frac{dy}{dt} \div \frac{dx}{dt} = \frac{dy}{dt} \Big/ \frac{dx}{dt}$$

Example:

Find the gradient of the tangent to the curve whose parametric equations are $x = a \cos t$, $y = b \sin t$.

$$\frac{dx}{dt} = -a \sin t, \quad \frac{dy}{dt} = b \cos t$$

$$\therefore \qquad \frac{dy}{dx} = \frac{dy}{dt} \Big/ \frac{dx}{dt} = \frac{b \cos t}{-a \sin t} = -\frac{b}{a} \cot t$$

(generally the result is in terms of the parameter).

16.2 Tangents and Normals to Plane Curves

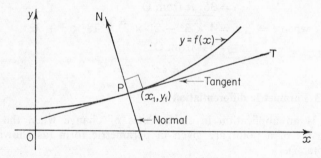

FIGURE 16.1

Consider the sketch in Figure 16.1. PT is the tangent to the curve $y = f(x)$ at the point $P(x_1, y_1)$. The line PN which is *perpendicular* to the tangent is called the *normal* to the curve at the point P.

16.2.1 Equations of tangent and normal

The slope of the tangent is $m = dy/dx$ [where dy/dx is evaluated at the point (x_1, y_1)].

Since the normal is defined as being perpendicular to the tangent, therefore the slope of the normal $m_1 = -1/m = -1/(\mathrm{d}y/\mathrm{d}x)$ (the condition of perpendicularity was derived in Chapter 7).

Hence the *equation of the tangent* at $P(x_1, y_1)$ is

$$y - y_1 = m(x - x_1) \quad \text{or} \quad y - y_1 = \left(\frac{\mathrm{d}y}{\mathrm{d}x}\right)_1 (x - x_1)$$

and the *equation of the normal* at $P(x_1, y_1)$ is

$$y - y_1 = m_1(x - x_1) = -\frac{1}{m}(x - x_1)$$

or

$$y - y_1 = -\left(\frac{\mathrm{d}x}{\mathrm{d}y}\right)_1 (x - x_1)$$

Example:

Determine the equations of the tangent and normal to the curve $y = x^2 - 2x - 3$ at the point where $x = 2$. Find the co-ordinates of the points where these lines meet the two co-ordinate axes (Figure 16.2).

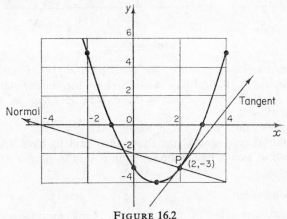

FIGURE 16.2

If $y = x^2 - 2x - 3$, $\mathrm{d}y/\mathrm{d}x = 2x - 2 =$ slope of tangent,

slope of tangent at $P(2, -3)$ is $m = 2 \times 2 - 2 = \underline{+2}$

slope of normal at $P(2, -3)$ is $m_1 = -1/m = \underline{-\tfrac{1}{2}}$

Equation of the tangent at P is

$$y - (-3) = +2(x - 2) \quad \text{or} \quad \underline{2x - y - 7 = 0} \qquad (1)$$

Equation of the normal at P is

$$y - (-3) = -\tfrac{1}{2}(x - 2) \quad \text{or} \quad \underline{x + 2y + 4 = 0} \qquad (2)$$

(1) meets $y = 0$ (x-axis), where $x_{\cdot} = 7/2$, i.e. at $(7/2, 0)$ and meets $x = 0$ (y-axis), where $y = -7$, i.e. at $(0, -7)$.

(2) meets $y = 0$ (x-axis), where $x = -4$, i.e. at $(-4, 0)$, and meets $x = 0$ (y-axis), where $y = -2$, i.e. at $(0, -2)$.

16.3 Turning Values of a Function

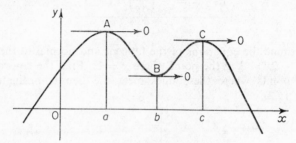

FIGURE 16.3

Consider the sketch graph of the function $y = f(x)$ [Figure 16.3]. At points like A, B, and C, y is said to have turning (or stationary) values. At such points $dy/dx = 0$, i.e. turning values of y occur when $dy/dx = 0$.

This gives values of $x = a$, b, and c.

At points like A and C the function is said to have *Maximum* values and at points like B the function is said to have a *Minimum* value.

16.3.1 Maximum

At a *maximum* value of y, $dy/dx = 0$. Considering Figure 16.4 it is seen that, as x *increases* through the value at a maximum y, dy/dx *decreases* from positive through zero to negative, i.e.

$$\frac{d}{dx}\left(\frac{dy}{dx}\right) \quad \text{or} \quad \frac{d^2y}{dx^2} \text{ is } Negative \text{ at a } maximum$$

16.3.2 Minimum

At a *minimum* value of y, $dy/dx = 0$. Considering Figure 16.5 it is seen that, as x *increases* through the value at a minimum y, dy/dx *increases* from negative through zero to positive, i.e.

$$\frac{d}{dx}\left(\frac{dy}{dx}\right) \quad \text{or} \quad \frac{d^2y}{dx^2} \text{ is } \textit{Positive} \text{ at a } \textit{minimum}$$

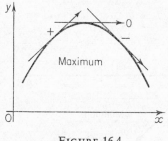

FIGURE 16.4

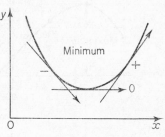

FIGURE 16.5

Example:

Determine the stationary values of the function $2x^3 - 3x^2 - 36x + 15$. Sketch the graph of the function.

Let $y = 2x^3 - 3x^2 - 36x + 15$,

$$\frac{dy}{dx} = 6x^2 - 6x - 36 = 6(x^2 - x - 6) = 6(x - 3)(x + 2)$$

For stationary values of y, $dy/dx = 0$, i.e. stationary values occur when

$$(x - 3)(x + 2) = 0$$

i.e.   when $x = 3$ or -2

$$\frac{d^2y}{dx^2} = 12x - 6.$$

When $x = 3$, $d^2y/dx^2 = 12 \times 3 - 6 = +30$, i.e. *positive*,

$\therefore$ y is a minimum when $x = 3$

Minimum value of $y = 2(3)^3 - 3(3)^2 - 36(3) + 15 = \underline{-66}$.

When $x = -2$, $d^2y/dx^2 = 12 \times (-2) - 6 = -30$, i.e. *negative*,

∴ y is a maximum when $x = -2$

Maximum value of $y = 2(-2)^3 - 3(-2)^2 - 36(-2) + 15 = \underline{+59}$.
For sketch see Figure 16.6.

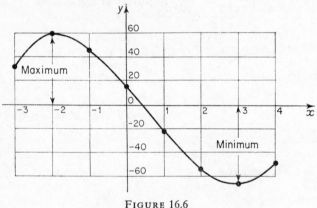

FIGURE 16.6

16.3.3 Problems involving stationary values

In order to determine stationary values involving, say, dimensions in practical problems it is first necessary to establish an expression for the quantity whose stationary value is required in terms of *one variable quantity*. The methods outlined above may then be used.

Example:

A cylindrical can is to be made from thin sheet metal in order to contain 100 in.³ of fluid. Find the dimensions so that the least (minimum) amount of metal shall be used.

Let r in. be the radius of the cylinder and h in. the height, then

$$\text{Volume } V = \pi r^2 h = 100$$

and

$$\text{Surface } S = 2\pi r^2 + 2\pi rh$$

(S is equivalent to the amount of metal.)

But $h = V/(\pi r^2) = 100/(\pi r^2)$,

$$\therefore \quad S = 2\pi r^2 + 2\pi r \times \frac{100}{\pi r^2} = 2\pi r^2 + \frac{200}{r}$$

For maximum or minimum value of S, $\mathrm{d}S/\mathrm{d}r = 0$,

$$\frac{\mathrm{d}S}{\mathrm{d}r} = 4\pi r - \frac{200}{r^2} = 0, \quad \text{when } 4\pi r^3 = 200$$

$$r^3 = 50/\pi, \quad r = \sqrt[3]{(50/\pi)}$$

$$\frac{\mathrm{d}^2 S}{\mathrm{d}r^2} = 4\pi + \frac{400}{r^3}$$

when $r = \sqrt[3]{(50/\pi)}$,

$$\frac{\mathrm{d}^2 S}{\mathrm{d}r^2} = 4\pi + 8\pi = 12\pi$$

Since $\mathrm{d}^2 S/\mathrm{d}r^2$ is positive therefore S is a minimum when the radius $r = \sqrt[3]{(50/\pi)} = 2\cdot516$ in. Then the height

$$h = \frac{100}{\pi r^2} = \frac{100}{\pi (50/\pi)^{2/3}} = \underline{2 \times \sqrt[3]{(50/\pi)}}$$

i.e. for minimum metal, height = $2 \times$ radius, where $r = \underline{2\cdot516 \text{ in.}}$

WORKED EXAMPLES

(a) Water runs at the rate of 5 in.^{3} per minute into an inverted conical can of depth 12 in. and base radius 5 in. Calculate the rate of rise of the water in the can when the depth is 8 in. (Figure 16.7).

Let the depth at time t minutes be x in. By proportion, if r is the radius of the water surface then $r/x = 5/12$,

$$\therefore \quad r = \frac{5x}{12} \text{ in.} \qquad (1)$$

If V is the volume of water at time t minutes,

$$V = \tfrac{1}{3}\pi r^2 x = \tfrac{1}{3}\pi \left(\frac{5x}{12}\right)^2 x = \frac{25\pi}{432} x^3 \qquad (2)$$

$$\frac{dV}{dt} = \frac{dV}{dx} \times \frac{dx}{dt} = \frac{25\pi}{432} \times 3x^2 \times \frac{dx}{dt} = \frac{25\pi}{144} x^2 \frac{dx}{dt} \qquad (3)$$

$\dfrac{dV}{dt}$ = rate of increase of volume = 5 in.^{3} per minute

$\dfrac{dx}{dt}$ = rate of rise of the water surface at time t.

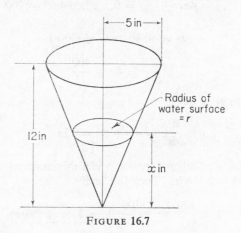

FIGURE 16.7

$\therefore$ from (3),

$$\frac{dx}{dt} = \frac{144}{25\pi x^2} \frac{dV}{dt} = \frac{144}{25\pi x^2} \times 5 = \frac{144}{5\pi x^2} \text{ in./min}$$

Hence, when the depth $x = 8$ in.,

$$\text{Rate of rise} = \frac{dx}{dt} = \frac{144}{5\pi (8)^2} = \frac{9}{20\pi} = \underline{0\cdot143(2) \text{ in./min}}$$

(b) The parametric equations of a parabola are $x = at^2$, $y = 2at$. Determine the equations of the tangent and normal to the curve at the point t (Figure 16.8).

The slope of the tangent PT at 't' is dy/dx,

$$\frac{dy}{dx} = \frac{dy}{dt} \Big/ \frac{dx}{dt} = 2a/2at = 1/t$$

∴ the slope of the normal PN at the point 't'

$$= -1/(dy/dx) = -t$$

The equation of the *tangent* at P is

$$y - 2at = \frac{1}{t}(x - at^2) \quad \text{or} \quad y = \frac{x}{t} + at$$

The equation of the *normal* at P is

$$y - 2at = -t(x - at^2) \quad \text{or} \quad tx + y = 2at + at^3$$

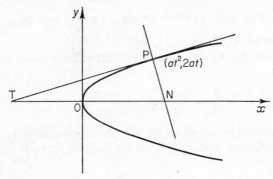

FIGURE 16.8

(c) If the distance s inches of a point in a mechanism at time t seconds from a fixed point O is given by $s = a\,e^{-Kt}\sin nt$, find values of t when s is a maximum or minimum. Show that the magnitudes of the successive maxima and minima form a geometric progression and find the common ratio of the progression.

$$s = a\,e^{-Kt}\sin nt$$

$$\frac{ds}{dt} = a[e^{-Kt}n\cos nt - K\,e^{-Kt}\sin nt] = a\,e^{-Kt}[n\cos nt - K\sin nt]$$

For stationary values of s (maxima and minima), $ds/dt = 0$, i.e.

$$n\cos nt - K\sin nt = 0, \quad \tan nt = n/K$$

$$nt = \tan^{-1}(n/K), \quad \therefore t = \frac{1}{n}\tan^{-1}(n/K)$$

Let the smallest value of $\tan^{-1}(n/K) = \alpha$ (radians, acute), then for stationary values

$$nt = s\pi + \alpha \quad (s = 0, 1, 2, 3, \ldots, \text{etc.})$$
$$t = \frac{s\pi}{n} + \frac{\alpha}{n}$$

Stationary values of s are $s = a\,e^{-K(s\pi+\alpha)/n}\sin(s\pi+\alpha)$ but for s an integer, $\sin(s\pi+\alpha) = \sin\alpha$ (numerically), therefore the magnitudes of the successive stationary values are

$$a\,e^{-K(s\pi+\alpha)/n}\sin\alpha$$

i.e. $a\,e^{-K\alpha/n}\sin\alpha,\ a\,e^{-K(\pi+\alpha)/n}\sin\alpha,\ a\,e^{-K(2\pi+\alpha)/n}\sin\alpha,\ \ldots$

Let these be $A_1, A_2, A_3, \ldots$, etc.

Clearly $A_1, A_2, A_3, \ldots$ form a Geometric Progression with common ratio

$$= \frac{A_2}{A_1} = \frac{a\,e^{-K(\pi+\alpha)/n}\sin\alpha}{a\,e^{-K\alpha/n}\sin\alpha} = \underline{e^{-K\pi/n}}$$

[Such a motion described by the above expression is called a *damped oscillatory motion*, K being the *damping factor*.]

Examples 16

1. The radius of a metal sphere subject to heat is increasing at the rate of 0·01 in./min. What is the rate of increase of (i) the surface area (ii) the volume, when the radius is 5 in.?

2. The volume of a sphere is increasing at the rate of 10 in³/s. Find:
   (i)   the rate of increase of the radius,
   (ii) the rate of increase of the surface area, at the instant when the radius is $2\frac{1}{2}$ in.

3. The torque in the crank shaft of a certain engine in lbf ft is given by $T = 1000 + 400\sin\theta - 150\sin 2\theta$ where θ is the angle turned through by the crank. If the crankshaft is rotating at a speed of 150 rad/s find, in lbf ft/min the rate at which the torque is changing when $\theta = \pi/3$.

4. A solution is poured into a conical filter of base radius 2 in. and vertical height 8 in at a uniform rate of 4 in³/s and filters out at 1·5 in³/s. What is the rate of rise of the level of the solution when the depth is 5 in.?

5. If $pv = 1000$ is the formula connecting pressure p in lbf/in² and the volume v in in³ and v is increasing at the rate of 5 in³/s, find the rate of decrease of p in lbf/in² per second when $v = 20$ in³.

6. At time t seconds, the distance s feet of a particle moving in a straight line, from a fixed point O on the line is given by $s = t^3 - 5t^2 + 3t + 9$.
   Find (i) the times at which the particle comes to rest,
        (ii) the accelerations at these times,
        (iii) the distance from O when the acceleration is zero.

7. The angle θ radians turned through by a wheel in t seconds is given by $\theta = 75 + 16t - t^2$. Find (i) the time when the wheel comes to rest, (ii) the retardation of the wheel.

8. The displacement equation of a vibrating mass is $s = 12 \sin (4\pi t + \pi/4)$ where s is the distance in inches from a fixed point and t is the time in seconds from a fixed instant. Find the velocity and acceleration of the mass when $t = 1/48$th second.

9. Find the equations of the tangent and normal to the curve $y^2 = 4x$ at the point $(1, 2)$. Find the co-ordinates of the points where these lines meet the axis of x.

10. Sketch the curve $xy = 5$. Determine the equations of the tangent and normal to the curve at the point where $x = 2$. Find the co-ordinates of the points where these lines meet the axes.

11. Find the slope of the tangent to the curve $y = x\,e^x + 1$ at the point $(0, 1)$. Hence obtain the equation of the tangent to the curve at this point.

12. Sketch the curve $y = x + 1/x$. Determine the equations of the tangent and normal to the curve at the point where $x = 2$.

13. The parametric equations of a hyperbola are $x = 4t$, $y = 4/t$. Find the equations of the tangent and normal to the curve at the point where $t = 2$.

14. Find the equations of the tangent and normal to the ellipse $x = 2 \cos t$, $y = \sin t$ at the point 't'.

15. Find the equations of the tangent and normal to the parabola $x = 4t^2$, $y = 8t$ at the point $t = 2$.

16. Find the maximum and minimum values of $4x^3 - 15x^2 - 18x + 7$.

17. Determine the maximum and minimum ordinates of the curve $y = 3x - 2 + 3/x$. Sketch the curve.

18. Find two successive stationary values of $5 \cos x - 8 \sin x$.

19. The current i in an electrical circuit is given at time t by
    $$i = 30 \cos 100t + 40 \sin 100t.$$
    Find the first two turning values of i stating the values of t at which these occur. Distinguish between the maximum and minimum.

20. The bending moment M of a uniform beam of length l at a distance x from a fixed end of the beam is given by $M = \frac{1}{2}(wlx - wx^2)$. Show that the bending moment is a maximum at the mid-point and find its value.

21. The power supplied to an external circuit by a generator of internal resistance r and E.M.F. e when the current is I is given by $W = eI - I^2 r$. Find for what current this is a maximum when $e = 20$ volts, $r = 1\cdot 8$ ohms.

22. The vertical cross-section of a tunnel is a rectangle surmounted by a semi-circle. If the area of the section is to be 350 ft², find the dimensions of the tunnel so that the perimeter of the section shall be a minimum.

23. An open channel is to have a cross-section in the form of a trapezium with sides sloping at 45°. It is running full and the cross section is 20 ft². Determine the dimensions so that the wetted perimeter shall be a minimum.

24. A cylindrical tin is to have a volume of 30 in³. It is to be soldered once up the curved surface (along a generator) and around both the top and bottom edges. If the line of solder is uniform calculate the dimensions so that a minimum amount of solder will be used.

25. A cylindrical container, made of thin material has a close fitting lid which overlaps the curved surface by a distance equal to one third of the radius of the container. The total amount of material to be used in the container is 3π ft². Show that the volume V of the container is given by $V = \pi r(9 - 8r^2)/6$ where r is the radius of the container. Hence find the maximum value of V.

17

Partial Differentiation—Applications

17.1 Functions of More Than One Variable

In many practical problems, functions are used which involve more than one variable quantity. In such problems it may be necessary to find how a quantity varies when more than one of the variables changes. In an elementary manner such variations have been dealt with by using binomial approximations to find percentage changes and errors.

More generally such changes may be dealt with by the use of *partial differentiation* giving rise to *partial* and *total* changes.

17.1.1 Illustration of partial and total changes

Consider the changes involved in the area of a rectangle whose initial dimensions are x and y, when x increases by δx and y increases by δy (see Figure 17.1).

The initial area $A = xy$ (ABCD).

If x increases by δx and y does not increase, then the *partial increase* of the area is represented by the rectangle BCHK $= y\,\delta x$. If y increases by δy and x does not increase, then the *partial increase* of the area is represented by the rectangle DEFC $= x\,\delta y$. When x and y increase simultaneously the final area is represented by the rectangle AEGK $= (x + \delta x)(y + \delta y)$, i.e. if the total increase in area is δA then

$$\delta A = (x + \delta x)(y + \delta y) - xy = y\,\delta x + x\,\delta y + \delta x\,\delta y$$

187

Total change = partial change due to an increase in x

+ partial change due to an increase in y + $\delta x \, \delta y$.

$\delta x \, \delta y$ (when δx and δy are small) is called a change of the *second order* of small quantities and may be neglected, i.e.

$$\underline{\delta A = y \, \delta x + x \, \delta y} \text{ (approximately)}$$

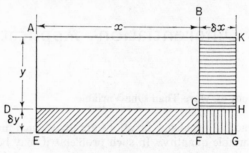

FIGURE 17.1

17.1.2 Partial derivatives

If $u = u(x, y)$ is a function of the two independent variable quantities x and y, the *rate of change* of u with respect to x *treating y as constant* is called the *partial derivative* of u with respect to x and is written as $\partial u/\partial x$ (read as partial dee u by dee x). Similarly the *rate of change* of u with respect to y *treating x as constant* is called the *partial derivative of u* with respect to y and is written as $\partial u/\partial y$. These may be readily written down (using the ordinary rules of differentiation) for a given function.

Examples:

(i) If $u = xy$, then $\partial u/\partial x = y$, $\partial u/\partial y = x$.

(ii) If $u = x^2 + 3xy + y^2$, show that $x(\partial u/\partial x) + y(\partial u/\partial y) = 2u$.

$$\frac{\partial u}{\partial x} = 2x + 3y, \quad \frac{\partial u}{\partial y} = 3x + 2y$$

$\therefore \qquad x\dfrac{\partial u}{\partial x} + y\dfrac{\partial u}{\partial y} = x(2x + 3y) + y(3x + 2y)$

$$= 2x^2 + 6xy + 2y^2 = \underline{2u}$$

17.1.3 Partial derivatives from first principles

Let $u = u(x, y)$ be a function of the two independent variables x and y, then

$$\frac{\partial u}{\partial x} = \lim_{\delta x \to 0} \left(\frac{\delta u}{\delta x} \right) \text{ (keeping } y \text{ constant)}$$

If x increases by δx and y remains constant, then

$$\delta u = u(x + \delta x, y) - u(x, y)$$

$$\frac{\delta u}{\delta x} = \frac{u(x + \delta x, y) - u(x, y)}{\delta x}$$

$$\therefore \qquad \frac{\partial u}{\partial x} = \lim_{\delta x \to 0} \left(\frac{\delta u}{\delta x} \right) = \lim_{\delta x \to 0} \left[\frac{u(x + \delta x, y) - u(x, y)}{\delta x} \right]$$

similarly

$$\frac{\partial u}{\partial y} = \lim_{\delta y \to 0} \left(\frac{\delta u}{\delta y} \right) = \lim_{\delta y \to 0} \left[\frac{u(x, y + \delta y) - u(x, y)}{\delta y} \right]$$

Example:

If $u = yx^2 + 2xy^2$, find $\partial u / \partial x$ from first principles.
Let x increase by δx, y remain constant, and u increase by δu, then

$$u + \delta u = y(x + \delta x)^2 + 2(x + \delta x)y^2$$
$$= y[x^2 + 2x\delta x + (\delta x)^2] + 2y^2 x + 2y^2 \, \delta x$$
$$u = yx^2 + 2xy^2$$
$$\therefore \qquad \delta u = 2xy \, \delta x + 2y^2 \, \delta x + y(\delta x)^2$$
$$\therefore \qquad \frac{\delta u}{\delta x} = 2xy + 2y^2 + y \, \delta x$$
$$\therefore \qquad \frac{\partial u}{\partial x} = \lim_{\delta x \to 0} \left(\frac{\delta u}{\delta x} \right) = 2xy + 2y^2$$

(*Note:* It is seen that the partial derivative can be written down in the ordinary way treating y as constant.)

17.2 Total First Order Change of a Function of Two Variables

Let $u = u(x, y)$, x increase by δx, y increase by δy, and u increase by δu, then

$$u + \delta u = u(x + \delta x, y + \delta y)$$
$$\delta u = u(x + \delta x, y + \delta y) - u(x, y)$$
$$= u(x + \delta x, y + \delta y) - u(x, y + \delta y) + u(x, y + \delta y) - u(x, y)$$

i.e. $\delta u = \dfrac{u(x + \delta x, y + \delta y) - u(x, y + \delta y)}{\delta x} \cdot \delta x$

$$+ \frac{u(x, y + \delta y) - u(x, y)}{\delta y} \cdot \delta y$$

as $\delta x \to 0$, $\dfrac{u(x + \delta x, y + \delta y) - u(x, y + \delta y)}{\delta x} \to \dfrac{\partial u}{\partial x}$

as $\delta y \to 0$, $\dfrac{u(x, y + \delta y) - u(x, y)}{\delta y} \to \dfrac{\partial u}{\partial y}$

Hence, for small changes δx, δy, δu, to the *first order* of small quantities,

$$\delta u = \frac{\partial u}{\partial x} \delta x + \frac{\partial u}{\partial y} \delta y \text{ (approximately)}$$

In 'differential' form this may be written as

$$du = \frac{\partial u}{\partial x} dx + \frac{\partial u}{\partial y} dy$$

These results may be extended to functions of more than two variables, e.g. if

$$u = u(x, y, z)$$

then the *total change* of u for small increments δx, δy, δz, is

$$\delta u = \frac{\partial u}{\partial x} \delta x + \frac{\partial u}{\partial y} \delta y + \frac{\partial u}{\partial z} \delta z \text{ (approximately)}$$

17.2.1 Applications of the results to small 'errors'

This will be illustrated by the following examples.

Examples:

(i) Find the approximate error in calculating the area of a rectangle of sides 3·1 and 4·7 in., if small errors of $+0·02$ in. and $+0·03$ in. are made in the measurements.

Let the sides of the rectangle be x (3·1) and y (4·7), then the area $A = xy$,

$$\text{error } \delta A = \frac{\partial A}{\partial x}\,\delta x + \frac{\partial A}{\partial y}\,\delta y \text{ (to the first order)}$$

$$\frac{\partial A}{\partial x} = y, \quad \frac{\partial A}{\partial y} = x, \quad \delta x = 0·02, \quad \delta y = 0·03$$

$$\therefore \quad \delta A = y\,\delta x + x\,\delta y = 4·7 \times 0·02 + 3·1 \times 0·03 \text{ in}^2$$
$$= 0·094 + 0·093 = \underline{0·187 \text{ in}^2}$$

(ii) Find the approximate *percentage* increase in the volume of a cylinder if the radius increases from 3 in. to 3·01 in. and the height increases from 5 in. to 5·02 in.

Let the volume of the cylinder be V, the radius r, height h, then

$$V = \pi r^2 h$$
$$\delta V = \frac{\partial V}{\partial r}\,\delta r + \frac{\partial V}{\partial h}\,\delta h \text{ (approximately)}$$
$$\frac{\partial V}{\partial r} = 2\pi rh, \quad \frac{\partial V}{\partial h} = \pi r^2$$
$$\therefore \quad \delta V = 2\pi rh\,\delta r + \pi r^2\,\delta h$$

Proportional increase:

$$\frac{\delta V}{V} = \frac{2\pi rh\,\delta r + \pi r^2\,\delta h}{\pi r^2 h} = \frac{2\delta r}{r} + \frac{\delta h}{h}$$

Percentage increase

$$= \frac{\delta V}{V} \times 100\% = \left(2\frac{\delta r}{r} + \frac{\delta h}{h}\right) \times 100\%$$

where $r = 3$ in., $\delta r = 0·01$ in., $h = 5$ in., $\delta h = 0·02$ in.

Percentage increase

$$= \left(2 \times \frac{0 \cdot 01}{3} + \frac{0 \cdot 02}{5}\right) \times 100\%$$

$$= (0 \cdot 0067 + 0 \cdot 004) \times 100 = 0 \cdot 0107 \times 100 = \underline{1 \cdot 07\%}$$

17.2.2 Higher partial derivatives

Any first order partial derivative may still be a function of all the variables, consequently the partial derivatives may be differentiated partially again. Such higher partial derivatives are written as $\partial^2 u/\partial x^2$, $\partial^2 u/\partial y^2$. In addition *mixed* partial derivatives may be written down, e.g.

$$\frac{\partial}{\partial y}\left(\frac{\partial u}{\partial x}\right) \text{ written as } \frac{\partial^2 u}{\partial y\,\partial x} \text{ and } \frac{\partial}{\partial x}\left(\frac{\partial u}{\partial y}\right) \text{ as } \frac{\partial^2 u}{\partial x\,\partial y}$$

Generally

$$\frac{\partial^2 u}{\partial y\,\partial x} = \frac{\partial^2 u}{\partial x\,\partial y}$$

Higher partial derivatives may be written down, but, at this stage will not be used.

Examples:

(i) Write down $\partial u/\partial x$, $\partial u/\partial y$, $\partial^2 u/\partial x^2$, $\partial^2 u/\partial y^2$ and show that

$$\frac{\partial^2 u}{\partial y\,\partial x} = \frac{\partial^2 u}{\partial x\,\partial y} \text{ when } u = x^3 y + y^3 x - 3x^2 y^2.$$

$$\frac{\partial u}{\partial x} = 3x^2 y + y^3 - 6xy^2, \quad \frac{\partial u}{\partial y} = x^3 + 3y^2 x - 6x^2 y$$

$$\frac{\partial^2 u}{\partial x^2} = 6xy - 6y^2, \quad \frac{\partial^2 u}{\partial y^2} = 6yx - 6x^2$$

$$\frac{\partial^2 u}{\partial y\,\partial x} = \frac{\partial}{\partial y}\left(\frac{\partial u}{\partial x}\right) = 3x^2 + 3y^2 - 12xy$$

$$\frac{\partial^2 u}{\partial x\,\partial y} = \frac{\partial}{\partial x}\left(\frac{\partial u}{\partial y}\right) = 3x^2 + 3y^2 - 12xy$$

Hence

$$\frac{\partial^2 u}{\partial y\, \partial x} = \frac{\partial^2 u}{\partial x\, \partial y}$$

(ii) If $u = \sin x \cos y$, find the value of $\partial^2 u/\partial x^2 + \partial^2 u/\partial y^2$.

$$\frac{\partial u}{\partial x} = \cos x \cos y, \quad \frac{\partial^2 u}{\partial x^2} = -\sin x \cos y$$

$$\frac{\partial u}{\partial y} = -\sin x \sin y, \quad \frac{\partial^2 u}{\partial y^2} = -\sin x \cos y$$

$$\therefore \quad \frac{\partial^2 u}{\partial x^2} + \frac{\partial^2 u}{\partial y^2} = -\sin x \cos y - \sin x \cos y = \underline{-2 \sin x \cos y}$$

WORKED EXAMPLES

(a) The relationship between pressure (p), volume (v), and temperature (T) is $pv = RT$, where R is a constant. Express p in terms of v and T. Hence obtain an expression for the percentage increase in pressure when v increases by δv and T increases by δT.

$$p = \frac{RT}{v}, \quad \frac{\partial p}{\partial v} = -\frac{RT}{v^2}, \quad \frac{\partial p}{\partial T} = \frac{R}{v}$$

$$\delta p = \frac{\partial p}{\partial v}\, \delta v + \frac{\partial p}{\partial T}\, \delta T$$

$$\therefore \quad \delta p = -\frac{RT}{v^2}\, \delta v + \frac{R}{v}\, \delta T$$

$$\frac{\delta p}{p} = \left(-\frac{RT}{v^2}\, \delta v + \frac{R}{v}\, \delta T\right)\Big/\left(\frac{RT}{v}\right)$$

$$= -\frac{RT}{v^2}\, \frac{v}{RT}\, \delta v + \frac{R}{v}\, \frac{v}{RT}\, \delta T$$

$$= -\frac{\delta v}{v} + \frac{\delta T}{T}$$

$$\therefore \quad \text{percentage increase in } p = 100\,\frac{\delta p}{p} = \underline{100\left(\frac{\delta T}{T} - \frac{\delta v}{v}\right)}$$

(b) The current i in an electrical circuit is given by $i = V/R$. If V increases from 50 to 52 volt and R increases from 20 to 21 ohm calculate the approximate change in i.

$$i = \frac{V}{R}, \quad \delta i = \frac{\partial i}{\partial V}\delta V + \frac{\partial i}{\partial R}\delta R = \frac{1}{R}\delta V - \frac{V}{R^2}\delta R \text{ amp}$$

But $V = 50$, $R = 20$, $\delta V = 2$, $\delta R = 1$,

$$\therefore \quad \delta i = \frac{1}{20} \times 2 - \frac{50}{20^2} \times 1 = \frac{1}{10} - \frac{50}{400} = \frac{1}{10} - \frac{1}{8}$$

$$\delta i = \frac{4-5}{40} = -\frac{1}{40} = \underline{-0.025 \text{ amp}}$$

i.e. the current *decreases* by approximately 0·025 amp.

(c) If $u = \sin(y/x)$, show that $x(\partial u/\partial x) + y(\partial u/\partial y) = 0$.

$$\frac{\partial u}{\partial x} = \cos\left(\frac{y}{x}\right) \times -\frac{y}{x^2} = -\frac{y}{x^2}\cos\left(\frac{y}{x}\right)$$

$$\frac{\partial u}{\partial y} = \cos\left(\frac{y}{x}\right) \times \frac{1}{x} = \frac{1}{x}\cos\left(\frac{y}{x}\right)$$

$$\therefore \quad x\frac{\partial u}{\partial x} + y\frac{\partial u}{\partial y} = -x \times \frac{y}{x^2}\cos\left(\frac{y}{x}\right) + \frac{y}{x}\cos\left(\frac{y}{x}\right)$$

$$= -\frac{y}{x}\cos\left(\frac{y}{x}\right) + \frac{y}{x}\cos\left(\frac{y}{x}\right) = 0$$

(d) If $z = (x^2 + y^2)/(x + y)$, find $\partial^2 z/\partial x^2$ and $\partial^2 z/\partial y\partial x$. (The rule for differentiating a *quotient* is used).

$$\frac{\partial z}{\partial x} = \frac{(x+y)2x - (x^2+y^2)1}{(x+y)^2} = \frac{2x^2 + 2xy - x^2 - y^2}{(x+y)^2}$$

$$= \frac{x^2 + 2xy - y^2}{(x+y)^2}$$

$$\frac{\partial^2 z}{\partial x^2} = \frac{(x+y)^2(2x+2y) - (x^2+2xy-y^2)2(x+y)1}{(x+y)^4}$$

$$= \frac{2(x+y)[(x+y)^2 - (x^2+2xy-y^2)]}{(x+y)^4}$$

$$= \frac{2[x^2+2xy+y^2-x^2-2xy+y^2]}{(x+y)^3} = \frac{4y^2}{(x+y)^3}$$

$$\frac{\partial^2 z}{\partial y \, \partial x} = \frac{\partial}{\partial y}\left(\frac{\partial z}{\partial x}\right) = \frac{\partial}{\partial y}\left[\frac{x^2 + 2xy - y^2}{(x + y)^2}\right]$$

$$= \frac{(x + y)^2(2x - 2y) - (x^2 + 2xy - y^2)2(x + y)1}{(x + y)^4}$$

$$= \frac{2(x + y)[(x + y)(x - y) - (x^2 + 2xy - y^2)]}{(x + y)^4}$$

$$= \frac{2[x^2 - y^2 - x^2 - 2xy + y^2]}{(x + y)^3} = \frac{-4xy}{(x + y)^3}$$

Examples 17

1. If $u = x^2 + xy$, show, from first principles that $\frac{\partial u}{\partial x} = 2x + y$.

2. If $u = 3x^2 + 2xy$, show that $x\frac{\partial u}{\partial x} + y\frac{\partial u}{\partial y} = 2u$ and write down expressions for $\frac{\partial^2 u}{\partial x^2}$, $\frac{\partial^2 u}{\partial y^2}$ and $\frac{\partial^2 u}{\partial y \, \partial x}$ in terms of x and y.

3. Given $z = \cos\left(\frac{y}{x}\right)$, show that $x\frac{\partial z}{\partial x} + y\frac{\partial z}{\partial y} = 0$.

4. For the function $z = \cos 2x \sin y$ show that $\frac{\partial^2 z}{\partial x^2} - 4\frac{\partial^2 z}{\partial y^2} = 0$.

5. For the function $u = \frac{x - y}{x + y}$ verify that $\frac{\partial^2 u}{\partial y \, \partial x} = \frac{\partial^2 u}{\partial x \, \partial y}$.

6. Show that, if $u = \log_e\left(\frac{y}{x}\right)$, then

   (i) $x\frac{\partial u}{\partial x} + y\frac{\partial u}{\partial y} = 0$,　(ii) $x^2\frac{\partial^2 u}{\partial x^2} + y^2\frac{\partial^2 u}{\partial y^2} = 0$.

7. Write down the following partial derivatives:

   (i) $\frac{\partial v}{\partial d}$ and $\frac{\partial v}{\partial P}$ when $V = 2d^2\sqrt{P}$,

   (ii) $\frac{\partial C}{\partial A}$ and $\frac{\partial C}{\partial t}$ when $C = \frac{KA}{4\pi t}$,

   (iii) $\frac{\partial v}{\partial u}$ and $\frac{\partial v}{\partial t}$ when $v = u\,e^{-Kt}$.

8. (i) If $T = 2\pi \sqrt{\left(\dfrac{l}{g}\right)}$ write down $\dfrac{\partial T}{\partial l}$ and $\dfrac{\partial T}{\partial g}$.

   (ii) If $n = \dfrac{K\sqrt{T}}{m}$ write down $\dfrac{\partial n}{\partial m}$ and $\dfrac{\partial n}{\partial T}$.

   (iii) Write down $\dfrac{\partial T}{\partial L}$ and $\dfrac{\partial T}{\partial C}$ when $T = \dfrac{1}{2\pi\sqrt{(LC)}}$.

   (iv) Write down $\dfrac{\partial T}{\partial H}$ given $T = 2\pi \sqrt{\left(\dfrac{I}{MH}\right)}$.

9. Write down the expression for the area A of a minor segment of a circle of radius r and angle θ radians. Hence write down $\dfrac{\partial A}{\partial r}$ and $\dfrac{\partial A}{\partial \theta}$.

10. Write down the formula for the area A of one face of a washer of internal radius r and external radius R. Hence write down expressions for $\dfrac{\partial A}{\partial R}$, $\dfrac{\partial^2 A}{\partial R^2}$, $\dfrac{\partial A}{\partial r}$ and $\dfrac{\partial^2 A}{\partial r^2}$.

11. Using the formula for the volume V of a cone of base radius r and altitude h, find the percentage change in the volume of a cone (to the first order) when the radius increases by 0·8 per cent and the altitude increases by 1·2 per cent.

12. The dimensions of a rectangular block are x, y and z in. Write down expressions for V (the volume) and S (the total surface area) in terms of x, y and z.

    If x, y and z increase respectively by small amounts δx, δy and δz obtain expressions for the increments δV and δS.

    What is the percentage increase in V when x increases by 1 per cent, y increases by 1·5 per cent and z increases by 2 per cent?

13. The Kinetic energy E of a moving body of weight W is given by $E = WV^2/(2g)$. If V increases by 2 per cent and g decreases by 1 per cent find, using partial differentiation the percentage change in E to the first order.

14. The voltage V volt in an electrical circuit is given by $V = iR$ where i is the current in amp and R is the resistance in ohms. If i increases from 4 amp to 4·05 amp and R increases from 25 ohms to 25·1 ohms use partial differentiation to calculate the change of V to the first order.

15. The field strength H on the axis of a small bar magnet of pole strength m, length $2l$ at a distance r from the centre of the magnet is given by $H = (4ml)/r^3$. If m remains constant, find an expression for the proportional change in H when l increases by δl and r increases by δr. (*Note*: the proportional change $= \delta H/H$.)

18

Integration—Standard Forms

18.1 Integration

This process may be treated in one of two ways: (i) as the reverse of the process of differentiation, or (ii) as the process of determining areas.

Both processes may be employed in the solution of practical problems. An elementary treatment of *integration* was dealt with in the final chapter of Volume I. However, for completeness some of the work is repeated.

18.1.1 Indefinite integral

Consider the solution of the following: if $dy/dx = 3x^2 + 2x - 1$, express y as a function of x.

This implies that, given the derivative of a function, in terms of the independent variable, what is the function?

The process of finding this function is called *Integration*. The solution is written as

$$y = \int (3x^2 + 2x - 1)\, dx$$

[read as the integral of $(3x^2 + 2x - 1)$ with respect to x *or* as y is the integral of $(3x^2 + 2x - 1)$ dee x].

The result depends on the results which were obtained for differentiation (in this case of powers of x), i.e.

$$y = x^3 + x^2 - x + C \tag{1}$$

where C is an *arbitrary* constant.

The result is called the *Indefinite Integral*.

18.1.2 Definite integral

If the result obtained in 18.1.1 can be made definite, then the result becomes a *Definite Integral*, i.e. if the curve represented by the equation (1) is to pass through the point (2, 5), i.e. if $y = 5$ when $x = 2$, then

$$5 = 2^3 + 2^2 - 2 + C = 10 + C$$

$$\therefore \qquad C = -5$$

Hence the *Definite integral* is

$$y = x^3 + x^2 - x - 5 \qquad (2)$$

18.1.3 Geometrical interpretation

(See Figure 18.1)

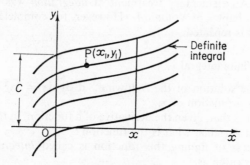

FIGURE 18.1

The indefinite integral represents an *infinite* set of *'parallel'* curves, i.e. parallel in the sense that, at a given value of x, the slope of all the curves is the same, or all the curves cross a vertical line *at the same angle*. The constant C is the *intercept* on the vertical axis. The particular curve of the set of *integral curves* which passes through a particular point $P(x_1, y_1)$ is the *definite integral curve*.

18.2 Definite Integral Between Limits

In many applications the independent variable changes between two definite values called the *limits of integration*. Such a definite integral is written as $\int_a^b f(x)\,dx$. If the indefinite integral of $f(x)$ be denoted as $F(x)$ [without the arbitrary constant], then

$$\int_a^b f(x)\,dx = \left[F(x)\right]_a^b = F(b) - F(a)$$

a is called the *lower limit* of integration, b is called the *upper limit* of integration (a and b may be negative or positive).

Examples:

(i) $\int_1^3 (x^2 + x)\,dx = \left[\dfrac{x^3}{3} + \dfrac{x^2}{2}\right]_1^3 = \left(\dfrac{3^3}{3} + \dfrac{3^2}{2}\right) - \left(\dfrac{1^3}{3} + \dfrac{1^2}{2}\right)$

$\qquad = 9 + \dfrac{9}{2} - \dfrac{1}{3} - \dfrac{1}{2} = 8\tfrac{2}{3} + 4 = \underline{12\tfrac{2}{3}\text{ unit}^2}$

(ii) $\int_1^4 \sqrt{x}\,dx = \int_1^4 x^{\frac{1}{2}}\,dx = \left[\dfrac{2}{3} x^{\frac{3}{2}}\right]_1^4$

$\qquad = \dfrac{2}{3}\left[4^{\frac{3}{2}} - 1^{\frac{3}{2}}\right] = \dfrac{2}{3}\left[8 - 1\right]$

$\qquad = \dfrac{2}{3} \times 7 = \dfrac{14}{3} = \underline{4\tfrac{2}{3}\text{ unit}^2}$

18.3 Integration of Standard Functions

From the results obtained for the derivatives of some of the simpler functions, the integrals of a number of simple standard functions may be deduced. In general these are quoted in order to carry out integrals. More difficult functions may be integrated by using a *substitution* process to reduce integrals to *standard forms*. Substitution will be dealt with in Chapter 19.

18.3.1 Integration of standard algebraic functions

These will be tabulated (they can be deduced from corresponding derivatives).

(1) $\int x^n \mathrm{d}x = \dfrac{x^{n+1}}{n+1} + C$ [except when $n = -1$]

(2) $\int \dfrac{1}{x} \, \mathrm{d}x = \log_e x + C$

(3) $\int e^x \, \mathrm{d}x = e^x + C$

This list may be extended for simple functions as follows:

(4) $\int (ax + b)^n \, \mathrm{d}x = \dfrac{(ax + b)^{n+1}}{a(n+1)} + C$ [except when $n = -1$]

(5) $\int \dfrac{1}{ax+b} \, \mathrm{d}x = \dfrac{1}{a} \log_e (ax + b) + C$

(6) $\int e^{ax} \, \mathrm{d}x = \dfrac{1}{a} e^{ax} + C$

(a and b are constants, C the arbitrary constant).

Examples:

(i) Evaluate $\int_0^2 (2x^2 + x - 1) \, \mathrm{d}x$.

$$\int_0^2 (2x^2 + x - 1) \, \mathrm{d}x = \left[\frac{2}{3} x^3 + \frac{1}{2} x^2 - x \right]_0^2$$

$$= (\tfrac{2}{3} \times 2^3 + \tfrac{1}{2} \times 2^2 - 2) - 0 = \underline{5\tfrac{1}{3} \text{ units}}$$

(ii) Evaluate $\int_2^5 \dfrac{1}{x} \, \mathrm{d}x$.

$$\int_2^5 \frac{1}{x} \, \mathrm{d}x = \left[\log_e x \right]_2^5 = \log_e 5 - \log_e 2$$

$$= \log_e \left(\frac{5}{2} \right) = \log_e 2 \cdot 5 = \underline{0 \cdot 9163}$$

(iii) $\int_0^{0 \cdot 5} e^x \, \mathrm{d}x = \left[e^x \right]_0^{0 \cdot 5} = e^{0 \cdot 5} - e^0 = 1 \cdot 6487 - 1 = \underline{0 \cdot 6487}$

(iv) $\displaystyle\int_{-1}^{+1} (2x + 3)^4 \, \mathrm{d}x = \left[\frac{(2x + 3)^5}{5 \times 2}\right]_{-1}^{+1} = \frac{1}{10}[5^5 - 1^5]$

$\qquad\qquad = \frac{1}{10}(3125 - 1) = \frac{1}{10}(3124) = \underline{312\cdot4}$

(v) $\displaystyle\int_0^3 \frac{1}{4x + 5} \, \mathrm{d}x = \left[\frac{1}{4}\log_e (4x + 5)\right]_0^3 = \frac{1}{4}[\log_e 17 - \log_e 5]$

$\qquad\qquad = \frac{1}{4}\log_e\left(\frac{17}{5}\right) = \frac{1}{4}\log_e 3\cdot4 = \frac{1}{4} \times 1\cdot2238$

$\qquad\qquad = \underline{0\cdot3059}$

(vi) $\displaystyle\int_{-0\cdot5}^{0\cdot5} \mathrm{e}^{2x} \, \mathrm{d}x = \left[\frac{1}{2}\mathrm{e}^{2x}\right]_{-0\cdot5}^{0\cdot5} = \frac{1}{2}[\mathrm{e}^1 - \mathrm{e}^{-1}]$

$\qquad\qquad = \frac{1}{2}(2\cdot7183 - 0\cdot3679) = \frac{1}{2}(2\cdot3504) = \underline{1\cdot1752}$

18.3.2 Integration of trigonometric functions

These are tabulated (they can be deduced from corresponding derivatives).

(7) $\displaystyle\int \cos x \, \mathrm{d}x = \sin x + C$ [x in radians]

(8) $\displaystyle\int \sin x \, \mathrm{d}x = -\cos x + C$

(9) $\displaystyle\int \sec^2 x \, \mathrm{d}x = \tan x + C$

(10) $\displaystyle\int \operatorname{cosec}^2 x \, \mathrm{d}x = -\cot x + C$

This list may be extended as follows:

(11) $\displaystyle\int \cos (ax) \, \mathrm{d}x = \frac{1}{a} \sin (ax) + C$

(12) $\displaystyle\int \sin (ax) \, \mathrm{d}x = -\frac{1}{a} \cos (ax) + C$

(13) $\displaystyle\int \sec^2 (ax) \, \mathrm{d}x = \frac{1}{a} \tan (ax) + C$

(14) $\displaystyle\int \operatorname{cosec}^2 (ax) \, \mathrm{d}x = -\frac{1}{a} \cot (ax) + C$

Examples:

(i) $\displaystyle\int_0^{\pi/3} \cos x \, \mathrm{d}x = \left[\sin x\right]_0^{\pi/3} = \sin \pi/3 - \sin 0 = \sin 60° - \sin 0$

$$= 0 \cdot 8660 - 0 = \underline{0 \cdot 866}$$

(ii) $\displaystyle\int_0^{\pi/4} \sin x \, \mathrm{d}x = \left[-\cos x\right]_0^{\pi/4} = (-\cos \pi/4) - (-\cos 0)$

$$= -\cos 45° + 1 = 1 - 0 \cdot 7071 = 0 \cdot 2929 \simeq \underline{0 \cdot 293}$$

(iii) $\displaystyle\int_{\pi/4}^{\pi/3} \sec^2 x \, \mathrm{d}x = \left[\tan x\right]_{\pi/4}^{\pi/3} = \tan \pi/3 - \tan \pi/4$

$$= \tan 60° - \tan 45° = 1 \cdot 7321 - 1 \cdot 0000 = \underline{0 \cdot 7321}$$

(iv) $\displaystyle\int_{0 \cdot 5}^{1 \cdot 2} \mathrm{cosec}^2 x \, \mathrm{d}x = \left[-\cot x\right]_{0 \cdot 5}^{1 \cdot 2} = -\cot 1 \cdot 2 + \cot 0 \cdot 5$

$$= -\cot 68°45' + \cot 28°39'$$

$$= 1 \cdot 8303 - 0 \cdot 3889 = \underline{1 \cdot 4414}$$

(v) $\displaystyle\int_{\pi/6}^{\pi/3} \cos (3x) \, \mathrm{d}x = \left[\frac{1}{3} \sin 3x\right]_{\pi/6}^{\pi/3} = \frac{1}{3}[\sin \pi - \sin \pi/2]$

$$= \frac{1}{3}[\sin 180° - \sin 90°] = \frac{1}{3}[0 - 1] = \underline{-\tfrac{1}{3}}$$

(vi) $\displaystyle\int_{\pi/4}^{\pi/2} \sin (\tfrac{1}{2}x) \, \mathrm{d}x = \left[-2 \cos \left(\frac{1}{2}x\right)\right]_{\pi/4}^{\pi/2} = -2[\cos \pi/4 - \cos \pi/8]$

$$= -2[\cos 45° - \cos 22°30']$$

$$= -2[0 \cdot 7071 - 0 \cdot 9239] = 2 \times 0 \cdot 2168 = \underline{0 \cdot 4336}$$

WORKED EXAMPLES

(*a*) The gradient of the tangent to the curve $y = f(x)$ at the point (x, y) is given by $\mathrm{d}y/\mathrm{d}x = 3x^2 + 1/x - 1$. If the curve passes through the point $(1, 4)$, find the equation of the curve.

$$\frac{\mathrm{d}y}{\mathrm{d}x} = 3x^2 + \frac{1}{x} - 1$$

$$y = \int (3x^2 + 1/x - 1) \, \mathrm{d}x = x^3 + \log_e x - x + C$$

Since the curve passes through $(1, 4)$,

$$\therefore \qquad\qquad 4 = 1^3 + \log_e 1 - 1 + C$$

but $\log_e 1 = 0$,

$$\therefore \qquad 4 = 1 - 1 + C$$

$$\therefore \qquad C = 4$$

$\therefore$ the equation of the curve is

$$y = x^3 + \log_e x - x + 4$$

(b) The velocity v ft/s of a particle at time t is given by the function $0.3t^2 - 1.5t$. If, at time $t = 2$ sec the particle is at a distance of 3 ft from the base point, find the distance s as a function of t.

Velocity $v = \dfrac{ds}{dt} = 0.3t^2 - 1.5t$

$$\therefore \qquad s = \int (0.3t^2 - 1.5t)\,dt = \frac{0.3}{3}t^3 - \frac{1.5}{2}t^2 + C$$

$$s = 0.1t^3 - 0.75t^2 + C$$

$s = 3$ when $t = 2$;

$$\therefore \qquad 3 = 0.1 \times (2)^3 - 0.75 \times (2)^2 + C$$

$$\therefore \qquad C = 3 - 0.8 + 3.0 = \underline{5.2}$$

Hence

$$s = 0.1t^3 - 0.75t^2 + 5.2$$

(c) During an isothermal expansion of a gas in a cylinder, the work done when the volume increases from v_1 to v_2 is given by $W = \displaystyle\int_{v_1}^{v_2} p\,dv$, where $pv = c$ (constant). Find the work done when a gas of initial volume of 3 ft^3 and pressure 70 lbf/in.^{2} expands isothermally to a volume of 8 ft^3.

$$c = pv = p_1 v_1 = 70 \times 144 \times 3 \text{ ft lbf}; \quad p = c/v$$

Work done $W = \displaystyle\int_{v_1}^{v_2} p\,dv = \int_3^8 \frac{c}{v}\,dv$ ft lbf

$$= c\left[\log_e v\right]_3^8 = 70 \times 144 \times 3[\log_e 8 - \log_e 3] \text{ ft lbf}$$

$$= 70 \times 144 \times 3 \times [2.0794 - 1.0986] \text{ ft lbf}$$

$$= 210 \times 144 \times 0.9808 = \underline{29,660 \text{ ft lbf}}$$

(*d*) Evaluate $\int_1^3 x^3 \left(x - \dfrac{1}{x^2} \right)^2 \mathrm{d}x$.

$$\int_1^3 x^3 \left(x - \frac{1}{x^2} \right)^2 \mathrm{d}x = \int_1^3 x^3 \left(x^2 - \frac{2}{x} + \frac{1}{x^4} \right) \mathrm{d}x$$

$$= \int_1^3 \left(x^5 - 2x^2 + \frac{1}{x} \right) \mathrm{d}x = \left[\tfrac{1}{6}x^6 - \tfrac{2}{3}x^3 + \log_e x \right]_1^3$$

$$= [\tfrac{1}{6}(3^6) - \tfrac{2}{3}(3)^3 + \log_e 3] - [\tfrac{1}{6} - \tfrac{2}{3}(1)^3 + \log_e 1], \; \{\log_e 1 = 0\}$$

$$= \tfrac{1}{6}(729) - 18 + \log_e 3 - \tfrac{1}{6} + \tfrac{2}{3} = \tfrac{1}{6}(732) - 18 + \log_e 3$$

$$= 122 - 18 + \log_e 3 = 104 + 1 \cdot 0986 = \underline{105 \cdot 09(86)}$$

(*e*) Evaluate $\int_0^{\pi/4} \sin^2 \theta \, \mathrm{d}\theta$.

$$\cos 2\theta = 1 - 2\sin^2 \theta \quad \therefore \; \sin^2 \theta = \tfrac{1}{2}(1 - \cos 2\theta)$$

$$\therefore \quad \int_0^{\pi/4} \sin^2 \theta \, \mathrm{d}\theta = \int_0^{\pi/4} \tfrac{1}{2}(1 - \cos 2\theta) \, \mathrm{d}\theta = \tfrac{1}{2} \left[\theta - \tfrac{1}{2}\sin 2\theta \right]_0^{\pi/4}$$

$$= \tfrac{1}{2}\{[\pi/4 - \tfrac{1}{2}\sin \pi/2] - [0 - \tfrac{1}{2}\sin 0]\}$$

$$= \tfrac{1}{2}[\pi/4 - \tfrac{1}{2}] = \tfrac{1}{4}[\pi/2 - 1] = \tfrac{1}{4}[1 \cdot 5708 - 1]$$

$$= \tfrac{1}{4}[0 \cdot 5708] = \underline{0 \cdot 1427}$$

(*f*) Evaluate $\int_0^{\pi/3} \sin 3x \cos x \, \mathrm{d}x$.

$\sin 3x \cos x$ must be expressed as the sum of two sines using the formulae established in Chapter 8.

$$\sin 3x \cos x = \tfrac{1}{2}[\sin (\text{sum}) + \sin (\text{difference})]$$
$$= \tfrac{1}{2}[\sin 4x + \sin 2x]$$

$$\therefore \quad \int_0^{\pi/3} \sin 3x \cos x \, \mathrm{d}x = \tfrac{1}{2} \int_0^{\pi/3} (\sin 4x + \sin 2x) \, \mathrm{d}x$$

$$= \tfrac{1}{2} \left[-\tfrac{1}{4}\cos 4x - \tfrac{1}{2}\cos 2x \right]_0^{\pi/3} = -\tfrac{1}{4} \left[+\tfrac{1}{2}\cos 4x + \cos 2x \right]_0^{\pi/3}$$

$$= -\tfrac{1}{4}[(\tfrac{1}{2}\cos 4\pi/3 + \cos 2\pi/3) - (\tfrac{1}{2}\cos 0 + \cos 0)]$$

$$= -\tfrac{1}{4}[-\tfrac{1}{2}\cos \tfrac{1}{3}\pi - \cos \tfrac{1}{3}\pi - \tfrac{1}{2} - 1] = \tfrac{3}{8}[\cos \tfrac{1}{3}\pi + 1]$$

$$= \tfrac{3}{8}[\cos 60° + 1] = \tfrac{3}{8}[\tfrac{1}{2} + 1] = \tfrac{3}{8} \times 3/2 = \underline{\tfrac{9}{16}}$$

Examples 18

1. The slope of a curve at any point (x, y) is given by $\dfrac{dy}{dx} = 2x + 3$. If the curve passes through the point $(2, 1)$ find its equation.

2. The gradient of the tangent to a curve is given by $\dfrac{dy}{dx} = 6x + 4$. If the curve passes through $(2, 22)$ determine its equation.

3. If $\dfrac{dy}{dx} = x - 5$ and $y = -1 \cdot 5$ when $x = 4$ find y as a function of x.

4. Given $\dfrac{dy}{dx} = 3x^2 - 4x + 1$ find the general function for y in terms of x. If $y = -3$ when $x = 1$ find the particular function.

5. At time t sec the velocity v of a particle is given by $v = 2t^2 - t + 3$. If the particle is at a distance of 5 feet from a datum point O at time $t = 2$ sec, express s, the distance from O at time t sec as a function of t.

6. If the acceleration of a particle from a fixed point O at time t sec is $2t + 3$, find the distance s ft of the particle from the fixed point O at time t in terms of t given that when $t = 0$, $s = 5$ ft and the particle is stationary.

7. Write down the indefinite integrals of the following functions:

   (i) $6x^2 + 4\sqrt{x} - 5$, (ii) $\dfrac{5}{x} + 4\,e^{2x}$, (iii) $\dfrac{2x^2 + 3\sqrt{x} + 6}{x^3}$,

   (iv) $\tfrac{1}{2}\sin 2x + \dfrac{1}{2x}$, (v) $\dfrac{x^2 + 1}{x}$, (vi) $(x + \tfrac{1}{2})^2$,

   (vii) $5/(3x + 1)$, (viii) $(x + 1)^2/x$ (ix) $(\sqrt{x} - 2)^2$.

8. Write down the indefinite integrals of the following functions:

   (i) e^{5x}, (ii) $2/(4x + 1)$, (iii) $1/\sqrt{(x + 2)}$, (iv) $3x^{\frac{1}{2}} - 2x^{-\frac{1}{2}}$,

   (v) $3/(1 - 4x)^2$, (vi) $5\cos 2x$, (vii) $e^{0 \cdot 1 x}$, (viii) $e^{-2x} + e^{2x}$.

9. Evaluate the following definite integrals:

   (i) $\displaystyle\int_1^2 x^4 \, dx$, (ii) $\displaystyle\int_1^2 \dfrac{1}{3x - 1} \, dx$, (iii) $\displaystyle\int_1^2 \dfrac{1}{(2x + 1)^2} \, dx$.

10. Evaluate

    (i) $\displaystyle\int_1^3 \dfrac{2\,dt}{3t + 1}$, (ii) $\displaystyle\int_0^2 e^t \, dt$, (iii) $\displaystyle\int_0^{2\pi/3} \cos 2\theta \, d\theta$.

11. Evaluate

    (i) $\displaystyle\int_0^2 (2u + 1)^4 \, du$, (ii) $\displaystyle\int_1^4 \dfrac{x + 1}{\sqrt{x}} \, dx$, (iii) $\displaystyle\int_2^4 \dfrac{dx}{1 + 3x}$,

    (iv) $\displaystyle\int_1^2 \left(2t - \dfrac{1}{t}\right)^2 dt$, (v) $\displaystyle\int_{-1}^{+1} e^{-3x} \, dx$, (vi) $\displaystyle\int_0^{\pi/4} 4\cos 2t \, dt$.

12. Evaluate

    (i) $\displaystyle\int_1^2 \left(x + \dfrac{1}{x}\right)^2 dx$, (ii) $\displaystyle\int_0^2 e^{x/4} \, dx$, (iii) $\displaystyle\int_0^{\pi/3} (\cos 3t + \sin t) \, dt$.

13. Evaluate

(i) $\int_0^2 5\,e^{u/2}\,du$,   (ii) $\int_0^4 \dfrac{dt}{2t+1}$,   (iii) $\int_1^2 \left(\dfrac{t}{\sqrt{2}} + \dfrac{\sqrt{2}}{t}\right)^2 dt$

(iv) $\int_0^{\pi/8} \cos 4x\,dx$,   (v) $\int_4^9 2x^{3/2}\,dx$,   (vi) $\int_1^2 \left(u - \dfrac{1}{u}\right)^2 du$,

14. Evaluate

(i) $\int_2^4 \dfrac{1}{x^3}\,dx$,   (ii) $\int_1^4 \sqrt{(3x+1)}\,dx$,   (iii) $\int_{-3}^{+3} 4(x^2+1)\,dx$,

(iv) $\int_0^4 \sqrt{t(1+t^2)}\,dt$,   (v) $\int_{-1}^{+1} (3x+1)^3\,dx$,   (vi) $\int_0^1 (e^x - x)\,dx$.

15. Find the numerical values of:

(i) $\int_0^2 5\,e^{-\frac{1}{2}x}\,dx$,   (ii) $\int_{-2}^3 (x^2+3x)\,dx$,   (iii) $\int_0^5 (e^{-0.2t})^2\,dt$,

(iv) $\int_0^3 \dfrac{5du}{4u+3}$,   (v) $\int_2^5 (3v-2)^4\,dv$,   (vi) $\int_1^4 (2\sqrt{x}+1)^2\,dx$.

16. Evaluate the following:

(i) $\dfrac{\omega}{\pi}\int_0^{\pi/\omega} \sin \omega t\,dt$,   (ii) $\pi\int_{-a}^a (a^2-x^2)\,dx$,   (iii) $\pi\int_0^h \left(\dfrac{rx}{h}\right)^2 dx$,

(iv) $\int_{v_1}^{v_2} \dfrac{c}{v^n}\,dv$ (c constant),   (v) $\pi\int_a^{2a} 4ax\,dx$,   (vi) $2\pi\int_0^a x(a^2-x^2)\,dx$.

17. Find the values of the following integrals:

(i) $\int_1^4 y\,dx$ when $y = 2\sqrt{x}+1$,   (ii) $\int_2^5 y\,dx$ when $y = x + \dfrac{1}{x}$,

(iii) $\int_0^2 y\,dx$ when $y = 3x^2 - 2x + 1$,   (iv) $\pi\int_4^8 y^2\,dx$ when $xy = 5$.

18. Evaluate

(i) $\int_{\pi/6}^{\pi/3} \cos^2 x\,dx$,   (ii) $\int_0^{\pi/3} \sin 3x\,dx$,   (iii) $\int_0^{\pi/4} (\sin\theta + \cos\theta)^2\,d\theta$,

(iv) $\int_0^{\pi/8} \cos 3x \cos x\,dx$,   (v) $\int_0^{\pi/6} \cos x \sin 3x\,dx$,   (vi) $\int_0^{\pi/6} \cos^2 3x\,dx$,

(vii) $\int_{-\pi/3}^{\pi/3} \sec^2 2x\,dx$,   (viii) $\int_{\pi/4}^{\pi/3} \operatorname{cosec}^2 2x\,dx$.

19. Given that $\dfrac{dy}{dx} = x^3 + x - 1$ and $y = 2$ when $x = -1$ find the value of y when $x = 3$.

20. A gas expands isothermally ($pv = c$) from a volume of 2 ft³ at a pressure of 100 lbf/in² to a volume of 7 ft³. Find the work done by the gas.

19

Integration—Substitution—Partial Fractions—Integration by Parts

Simplification of Integrals

Using various processes it is possible to reduce difficult integrals to ones which are of *standard form*. Three such processes will be described and illustrated in this chapter. They are as follows:

(i) Integration by substitution.
(ii) Integration using partial fractions.
(iii) Integration by parts (for the integration of products).

19.1 Integration by Substitution

Consider the integral $\int f(x)\,dx$ which is not of '*standard form*'. By a suitable substitution for x in terms of another variable u, e.g. $x = x(u)$, then $dx/du = x'(u)$. The *differential* dx is given by $dx = (dx/du)\,du = x'(u)\,du$. After substitution in $f(x)$ a function $F(u)$ is obtained and the original integral transforms into an integral of the form $\int F(u)x'(u)\,du$, which, if the substitution is suitable may now be of *standard form*.

If the limits of the given integral are a and b then the limits for u (c and d) can be put in terms of a and b, hence the definite integral $\int_a^b f(x)\,dx$ transforms to a form $\int_c^d F(u)x'(u)\,du$ which may be evaluated.

Examples:

(i) Using a substitution evaluate $\int_0^4 x\sqrt{(9 + x^2)}\,dx$.

Let $u = 9 + x^2$ [equivalent to $x = \sqrt{(u - 9)}$], then

$$\frac{du}{dx} = 2x, \quad \therefore\ du = 2x\,dx, \quad x\,dx = \tfrac{1}{2}\,du.$$

When $x = 0$, $u = 9$; when $x = 4$, $u = 25$,

$$\therefore\quad \int_0^4 x\sqrt{(9 + x^2)}\,dx = \int_9^{25} \tfrac{1}{2}\sqrt{u}\,du = \tfrac{1}{2}\int_9^{25} u^{\frac{1}{2}}\,du$$

$$= \tfrac{1}{2}\left[\frac{u^{3/2}}{3/2}\right]_9^{25} = \tfrac{1}{3}[25^{3/2} - 9^{3/2}]$$

$$= \tfrac{1}{3}[5^3 - 3^3] = \tfrac{1}{3}[125 - 27] = \underline{\underline{32\tfrac{2}{3}}}$$

(ii) Evaluate $\int_1^3 \frac{1}{3x + 5}\,dx$.

Let $u = 3x + 5$, $du = 3\,dx$, $dx = \tfrac{1}{3}\,du$.

When $x = 1$, $u = 8$; when $x = 3$, $u = 14$,

$$\therefore\quad \int_1^3 \frac{1}{3x + 5}\,dx = \int_8^{14} \frac{1}{u}\,\tfrac{1}{3}\,du = \tfrac{1}{3}\int_8^{14} \frac{1}{u}\,du$$

$$= \tfrac{1}{3}\left[\log_e u\right]_8^{14} = \tfrac{1}{3}[\log_e 14 - \log_e 8]$$

$$= \tfrac{1}{3}\log_e 14/8 = \tfrac{1}{3}\log_e 1\cdot75$$

$$= \tfrac{1}{3} \times 0\cdot5596 = \underline{\underline{0\cdot1865}}$$

(iii) Evaluate $\int_{\pi/6}^{\pi/3} \frac{\sin x}{\cos^3 x}\,dx$.

[The method of substitution requires a change of variable. *Any letter* may be chosen to represent the new variable.]

Let $z = \cos x$, $dz/dx = -\sin x$,

$$\therefore\quad \sin x\,dx = -dz$$

The indefinite integral becomes

$$\int \frac{\sin x}{\cos^3 x}\,dx = -\int \frac{dz}{z^3} = -\int z^{-3}\,dz = \frac{-z^{-2}}{-2} = \frac{1}{2\cos^2 x}$$

Hence

$$\int_{\pi/6}^{\pi/3} \frac{\sin x}{\cos^3 x}\, dx = \left[\frac{1}{2\cos^2 x} \right]_{\pi/6}^{\pi/3} = \tfrac{1}{2} \left[\sec^2 x \right]_{\pi/6}^{\pi/3}$$
$$= \tfrac{1}{2}[\sec^2 \pi/3 - \sec^2 \pi/6] = \tfrac{1}{2}[2^2 - (2/\sqrt 3)^2] = \tfrac{1}{2}[4 - 4/3]$$
$$= 4/3 = \underline{1\tfrac{1}{3}\ \text{unit}}$$

19.2 Integration Using Partial Fractions

In order to integrate certain fractions use is made of the method of breaking a fraction down into partial fractions as shown in Volume I and in Chapter 4 of this volume. The method is illustrated by the following examples.

Examples:

(i) Evaluate $\displaystyle\int_1^3 \frac{3x + 5}{(x + 1)(2x - 1)}\, dx$.

Let $\dfrac{3x + 5}{(x + 1)(2x - 1)} = \dfrac{A}{x + 1} + \dfrac{B}{2x - 1}$

then

$$3x + 5 = A(2x - 1) + B(x + 1)$$

Let $x = -1$; $\ 2 = A(-3)$,

∴ $\underline{A = -2/3}$

Let $x = \tfrac{1}{2}$; $\ 6\tfrac{1}{2} = B(1\tfrac{1}{2})$,

∴ $\underline{B = 13/3}$

∴ $\displaystyle\int_1^3 \frac{3x + 5}{(x + 1)(2x - 1)}\, dx = \int_1^3 \left(\frac{-2/3}{x + 1} + \frac{13/3}{2x - 1} \right) dx$

$$= \left[-\frac{2}{3} \log_e (x + 1) + \frac{13}{3} \times \frac{1}{2} \log_e (2x - 1) \right]_1^3$$

$$= (-\tfrac{2}{3} \log_e 4 + \tfrac{13}{6} \log_e 5) - (-\tfrac{2}{3} \log_e 2 + \tfrac{13}{6} \log_e 1)$$

$$= -\tfrac{2}{3} \log_e 4 + \tfrac{13}{6} \log_e 5 + \tfrac{2}{3} \log_e 2 - \tfrac{13}{6} \log_e 1$$

$$[\log_e 1 = 0, \ \log_e 4 = 2 \log_e 2]$$

$$= \tfrac{13}{6} \log_e 5 - \tfrac{2}{3} \log_e 2 = \tfrac{13}{6} \times 1\cdot6094 - \tfrac{2}{3} \times 0\cdot6931$$

$$= \tfrac{1}{6}[13 \times 1\cdot6094 - 4 \times 0\cdot6931] = \tfrac{1}{6}[20\cdot9222 - 2\cdot7724]$$

$$= \tfrac{1}{6} \times 18\cdot1498 = 3\cdot02497 \simeq \underline{3\cdot025}$$

(ii) Evaluate $\int_3^5 \dfrac{2x + 5}{(x + 1)^2(x - 2)}\,\mathrm{d}x$.

Let

$$\frac{2x + 5}{(x + 1)^2(x - 2)} = \frac{A}{x + 1} + \frac{B}{(x + 1)^2} + \frac{C}{x - 2}$$

$$2x + 5 = A(x + 1)(x - 2) + B(x - 2) + C(x + 1)^2$$

Let $x = 2$; $9 = C(3)^2$,

$\therefore$ $\underline{C = 1}$

Let $x = -1$; $3 = B(-3)$,

$\therefore$ $\underline{B = -1}$

Equate coefficients of x^2:

$$0 = A + C, \quad \therefore \underline{A = -C = -1}$$

$$\therefore \int_3^5 \frac{2x + 5}{(x + 1)^2(x - 2)}\,\mathrm{d}x$$

$$= \int_3^5 \left(-\frac{1}{x + 1} - \frac{1}{(x + 1)^2} + \frac{1}{x - 2}\right)\,\mathrm{d}x$$

$$= \left[-\log(x + 1) + \frac{1}{x + 1} + \log(x - 2)\right]_3^5$$

$$= \left(-\log 6 + \tfrac{1}{6} + \log 3\right) - \left(-\log 4 + \tfrac{1}{4} + \log 1\right)$$

$$= -\log 6 + \tfrac{1}{6} + \log 3 + \log 4 - \tfrac{1}{4} - 0 = \log \frac{4 \times 3}{6} - \frac{1}{12}$$

$$= \log_e 2 - \frac{1}{12} = 0.6931 - 0.0833 = \underline{0.6098}$$

19.3 Integration by Parts

This is a method for integrating the product of two functions of an independent variable. The method is derived from the formula for the derivative of the product of two functions as follows:

$$\frac{\mathrm{d}}{\mathrm{d}x}(uv) = v\frac{\mathrm{d}u}{\mathrm{d}x} + u\frac{\mathrm{d}v}{\mathrm{d}x}$$

integrate with respect to x

$$uv = \int v \frac{du}{dx} \, dx + \int u \frac{dv}{dx} \, dx$$

hence

$$\int v \frac{du}{dx} \, dx = uv - \int u \frac{dv}{dx} \, dx \tag{1}$$

This may often be used more conveniently in its *differential* form, i.e.

$$\int v \, du = uv - \int u \, dv \tag{2}$$

These results may be expressed in words as follows:

(1) $\int$ (1st function)(derivative of 2nd function) dx

$= $ (1st function)(2nd function)

$\quad\quad - \int$ (2nd function)(derivative of 1st function) dx

(2) $\int$ (1st function) d(2nd function)

$= $ (1st function)(2nd function)

$\quad\quad\quad\quad - \int$ (2nd function) d(1st function)

Generally it will be found that (2) is the most useful, particularly if the process has to be repeated.

Examples:

(i) $\int x \cos x \, dx$.

$\quad$ Now $d(\sin x) = \dfrac{d}{dx} (\sin x) \, dx = \cos x \, dx,$

$\quad \therefore \quad \int x \cos x \, dx = \int x \, d(\sin x) = x \sin x - \int \sin x \, dx$

$\quad\quad\quad\quad\quad\quad (v) \quad\quad (u)$

$\quad\quad\quad\quad = x \sin x + \cos x + C$

(ii) $\displaystyle\int x \log_e x \, dx$.

Now $d(x^2) = 2x \, dx$,

$$\therefore \quad \int x \log_e x \, dx = \tfrac{1}{2} \int \log_e x \, d(x^2) = \tfrac{1}{2}[x^2 \log_e x - \int x^2 \, d(\log_e x)]$$

$$= \tfrac{1}{2}\left[x^2 \log_e x - \int x^2 \frac{1}{x} \, dx \right]$$

$$= \tfrac{1}{2}[x^2 \log_e x - \int x \, dx]$$

$$= \tfrac{1}{2}[x^2 \log_e x - \tfrac{1}{2}x^2] + C$$

$$= \underline{\tfrac{1}{2}x^2(\log_e x - \tfrac{1}{2}) + C}$$

(iii) Evaluate $\displaystyle\int_0^{\pi/4} x \sin 2x \, dx$.

$$\int_0^{\pi/4} x \sin 2x \, dx = -\tfrac{1}{2} \int_0^{\pi/4} x \, d(\cos 2x)$$

$$[d(\cos 2x) = -2 \sin 2x \, dx]$$

$$= -\tfrac{1}{2}\left[x \cos 2x - \int \cos 2x \, dx \right]_0^{\pi/4}$$

$$= -\tfrac{1}{2}\left[x \cos 2x - \tfrac{1}{2} \sin 2x \right]_0^{\pi/4}$$

$$= -\tfrac{1}{2}\left[\left(\frac{\pi}{4} \cos \frac{\pi}{2} - \tfrac{1}{2} \sin \frac{\pi}{2} \right) \right.$$

$$\left. - (0 \cos 0 - \tfrac{1}{2} \sin 0) \right]$$

$$= -\tfrac{1}{2}\left[\frac{\pi}{4} \times 0 - \tfrac{1}{2} \times 1 \right] = \tfrac{1}{2} \times \tfrac{1}{2} = \underline{0\cdot25}$$

WORKED EXAMPLES

(a) Evaluate $\displaystyle\int_1^5 \frac{2t}{\sqrt{(2t+1)}} \, dt$.

Let $u^2 = 2t + 1$, $2u \, du = 2 \, dt$, $t = \tfrac{1}{2}(u^2 - 1)$.

When $t = 1$, $u^2 = 3$, $u = \sqrt{3}$; when $t = 5$, $u^2 = 11$, $u = \sqrt{11}$,

$$\therefore \int_1^5 \frac{2t}{\sqrt{(2t+1)}}\, dt = \int_{\sqrt3}^{\sqrt{11}} \frac{\frac{1}{2}(u^2-1)}{u}\, 2u\, du = \int_{\sqrt3}^{\sqrt{11}} (u^2-1)\, du$$

$$= \left[\tfrac{1}{3}u^3 - u\right]_{\sqrt3}^{\sqrt{11}}$$

$$= [\tfrac{1}{3}(\sqrt{11})^3 - \sqrt{11}] - [\tfrac{1}{3}(\sqrt3)^3 - \sqrt3]$$

$$= \tfrac{1}{3}(11\sqrt{11}) - \sqrt{11} - \tfrac{1}{3}(3\sqrt3) + \sqrt3$$

$$= (8/3)\sqrt{11} - \sqrt3 + \sqrt3$$

$$= (8/3)\sqrt{11} = (8/3) \times 3\cdot317 = \underline{8\cdot845}$$

(b) By a suitable substitution, evaluate $\displaystyle\int_0^{\pi/2} \frac{\sin x}{5 + 3\cos x}\, dx.$

Let $u = 5 + 3\cos x$, $du = -3\sin x\, dx$, $\sin x\, dx = -\tfrac{1}{3}\, du$.

When $x = 0$, $u = 5 + 3\cos 0 = 5 + 3 = \underline{8}.$

When $x = \pi/2$, $u = 5 + 3\cos \pi/2 = 5 + 3 \times 0 = \underline{5}.$

$$\therefore \int_0^{\pi/2} \frac{\sin x}{5 + 3\cos x}\, dx = -\int_8^5 \frac{\frac{1}{3}\, du}{u} = -\tfrac{1}{3}\int_8^5 \frac{du}{u}$$

$$= -\tfrac{1}{3}\left[\log u\right]_8^5 = -\tfrac{1}{3}[\log_e 5 - \log_e 8]$$

$$= \tfrac{1}{3}[\log_e 8 - \log_e 5] = \tfrac{1}{3}\log_e (8/5)$$

$$= \tfrac{1}{3}\log_e 1\cdot6 = \tfrac{1}{3} \times 0\cdot4700 = \underline{0\cdot1567}$$

(c) Evaluate $\displaystyle\int_3^8 \frac{3}{4 - v^2}\, dv.$

Let $\dfrac{3}{4 - v^2} = \dfrac{3}{(2+v)(2-v)} = \dfrac{A}{2+v} + \dfrac{B}{2-v},$

then $\qquad\qquad 3 = A(2 - v) + B(2 + v)$

let $v = -2$; $\ 3 = A(4),$

$$\therefore \qquad\qquad\qquad \underline{A = 3/4}$$

let $v = 2$; $\ 3 = B(4),$

$$\therefore \qquad\qquad\qquad \underline{B = 3/4}$$

hence

$$\int_3^8 \frac{3}{4 - v^2}\, dv = \int_3^8 \left[\frac{3}{4(2 + v)} - \frac{3}{4(v - 2)} \right] dv$$

$$= \left[\frac{3}{4} \log_e (2 + v) - \frac{3}{4} \log_e (v - 2) \right]_3^8$$

$$= \tfrac{3}{4}[(\log_e 10 - \log_e 6) - (\log_e 5 - \log_e 1)]$$

$$= \tfrac{3}{4}[\log_e (10/5) - \log_e 6] = \tfrac{3}{4}[\log_e 2 - \log_e 6]$$

$$= -\tfrac{3}{4} \log_e (6/2) = -\tfrac{3}{4} \log_e 3 = -\tfrac{3}{4} \times 1 \cdot 0986$$

$$= -\tfrac{1}{4} \times 3 \cdot 2958 \simeq \underline{-0 \cdot 824}$$

(d) Evaluate $\displaystyle\int_0^1 x^2\, e^x\, dx$.

Consider first the indefinite integral and note that

$$\frac{d}{dx}(e^x) = e^x \quad \text{or} \quad \underline{d(e^x) = e^x\, dx}$$

then

$$\int x^2\, e^x\, dx = \int x^2\, d(e^x) = x^2\, e^x - \int e^x\, d(x^2)$$

$$= x^2\, e^x - \int e^x\, 2x\, dx = x^2\, e^x - 2 \int x\, d(e^x)$$

$$= x^2\, e^x - 2[x\, e^x - \int e^x\, dx] = x^2\, e^x - 2x\, e^x + 2\, e^x + C$$

$$\therefore \int_0^1 x^2\, e^x\, dx = \left[e^x(x^2 - 2x + 2) \right]_0^1 = e^1(1 - 2 + 2) - e^0(0 + 2)$$

$$= e^1 - 2\, e^0 = 2 \cdot 7183 - 2 \cdot 0 = \underline{0 \cdot 7183}$$

(*Note:* Great care must be exercised in the use of brackets at each stage when it is necessary to repeat the process of *integration by parts*.)

Examples 19

1. Using suitable substitutions evaluate the integrals:

(i) $\displaystyle\int_2^4 \frac{x}{x - 1}\, dx$, (ii) $\displaystyle\int_0^5 \frac{x}{\sqrt{(x + 4)}}\, dx$, (iii) $\displaystyle\int_1^2 \frac{3}{(1 - 4x)^2}\, dx$.

2. Obtain the indefinite integrals of the following functions:

(i) $\dfrac{t+1}{\sqrt{(t-1)}}$, (ii) $(2x+1)\sin(x^2+x+2)$, (iii) $\dfrac{x^2}{\sqrt{(x+1)}}$.

3. Evaluate the following definite integrals:

(i) $\displaystyle\int_1^3 \dfrac{2\,dx}{3x+1}$, (ii) $\displaystyle\int_0^2 \dfrac{3x}{\sqrt{(2+x^2)}}\,dx$, (iii) $\displaystyle\int_1^3 \dfrac{4u}{u^2+1}\,du$.

4. Find the following indefinite integrals:

(i) $\displaystyle\int \tan x\,dx$, (ii) $\displaystyle\int x\,e^{x^2}\,dx$, (iii) $\displaystyle\int \sin^2\theta\cos\theta\,d\theta$.

5. Using the method of partial fractions obtain the indefinite integrals of:

(i) $\dfrac{3x+2}{(x-1)(2x+1)}$ (ii) $\dfrac{2x-3}{x^2-5x+6}$ (iii) $\dfrac{5x+7}{x^2+x-20}$.

6. Evaluate the integrals:

(i) $\displaystyle\int_2^3 \dfrac{1}{(x+1)(x-1)^2}\,dx$, (ii) $\displaystyle\int_5^7 \dfrac{t+2}{(t-1)(t-4)}\,dt$, (iii) $\displaystyle\int_3^4 \dfrac{2v-5}{v^2-4}\,dv$.

7. Evaluate:

(i) $\displaystyle\int_1^3 x\sqrt{(x^2-1)}\,dx$, (ii) $\displaystyle\int_1^5 (u+3)\sqrt{(u-1)}\,du$, (iii) $\displaystyle\int_0^4 \dfrac{3t}{\sqrt{(3+t^2)}}\,dt$.

8. Use the method of integration by parts to obtain the indefinite integrals of the functions:

(i) $x\,e^{3x}$ (ii) $x^3\log_e x$ (iii) $x\sin x$.

9. Evaluate the integrals:

(i) $\displaystyle\int_0^2 x\,e^x\,dx$, (ii) $\displaystyle\int_1^2 x^2\log_e x\,dx$, (iii) $\displaystyle\int_0^{\pi/2} x^2\cos x\,dx$.

10. Evaluate the following integrals using any appropriate method:

(i) $\displaystyle\int_0^3 x^2\sqrt{(x^3+9)}\,dx$, (ii) $\displaystyle\int_1^3 \dfrac{3x}{(2x-1)(x+1)}\,dx$, (iii) $\displaystyle\int_{-1}^{+1} t\,e^{2t}\,dt$.

11. Obtain the indefinite integrals:

(i) $\displaystyle\int \dfrac{\cos\theta}{1+2\sin\theta}\,d\theta$, (ii) $\displaystyle\int \dfrac{4x}{9-x^2}\,dx$, (iii) $\displaystyle\int x\sec^2 x\,dx$.

12. Find the numerical values of:

(i) $\displaystyle\int_0^{3/2} z\sqrt{(1+2z)}\,dz$, (ii) $\displaystyle\int_3^5 \dfrac{3du}{u^2-u-2}$, (iii) $\displaystyle\int_0^1 u^3\,e^u\,du$.

13. Express $\dfrac{3x^2+2x+9}{(x-1)^2(4x+3)}$ in partial fractions and hence find the value of the integral of the function between the limits 2 and 3.

14. By means of a substitution evaluate $\displaystyle\int_0^{\pi/3}\cos^3 x\,dx$.

8

15. What method of integration could be used to find the integral of the function $\dfrac{\sin x}{1 + \cos x}$? Hence evaluate $\displaystyle\int_0^{\pi/4} \dfrac{\sin x}{1 + \cos x}\,dx$.

16. The slope of the tangent to the graph of the function $y = f(x)$ is given by $\dfrac{dy}{dx} = (x - 1)\,e^x$. Find the equation of the graph given that the graph passes through the point $(0, 2)$.

17. The first moment of area of the area under the graph of $y = f(x)$ about the y-axis is given by $\int xy\,dx$, between appropriate limits. Evaluate the first moment of area about OY for the areas under the following curves:

   (i) $y = 5 \sin x$ between $x = 0$ and π.
   (ii) $y = e^x$ between $x = 1$ and 2.

18. Solve the 'differential equation':

   $$\frac{dy}{dx} = x\surd(x^2 - 1)$$ given that $y = 4$ when $x = 1$ (i.e. find y in terms of x).

19. Evaluate $\displaystyle\int_0^a x\surd(a^2 - x^2)\,dx$.

20. Express $\sin^2 x$ in terms of $\cos 2x$ and hence, using a suitable method evaluate $\displaystyle\int_0^{\pi/2} x \sin^2 x\,dx$.

20

Applications of Integration (1)—Areas—Mean Values—Volumes—Work done by Forces and Gases

20.1 Applications of Integration

There are many technical applications of integration in scientific, engineering, building, and associated subjects. Some of these applications include: the determination of areas, volumes, mean (average) values, work done by forces, work done by gases, mean forces, mean pressures, positions of centroids, and centres of gravity, moments of inertia, mean currents, root-mean-square currents, and so on.

Basically, however, most applications depend on the method of determining an area under the graph of a function. This application will be dealt with first and subsequently other applications will be illustrated.

20.1.1 Areas by integration

Consider the sketch of the function $y = f(x)$ as x increases [Figure 20.1]. Let it be required to determine the area between the curve, the axis OX and the ordinates at $x = a$ and $x = b$, i.e. the area of ABCD.

First consider an *element* of area MNQP, of width δx. The area between the two ordinates at x and $x + \delta x$ [if y is treated as increasing steadily] will lie between the areas of the two rectangles MNSP and MNQR. Let δA represent the area under the curve (shaded), then

$$\text{area MNSP} < \text{area MNQP} < \text{area MNQR}$$

$$y\delta x < \delta A < (y + \delta y)\,\delta x$$

Dividing by δx:

$$y < \frac{\delta A}{\delta x} < y + \delta y$$

As δx tends to zero and δy also tends to zero, $y + \delta y$ tends to y, whilst $\delta A/\delta x$ tends to dA/dx (the instantaneous rate of change of the area with respect to x), i.e. in the limit it follows that

$$\lim_{\delta x \to 0} \frac{\delta A}{\delta x} = \frac{dA}{dx} = y$$

hence

$$A = \int y\,dx$$

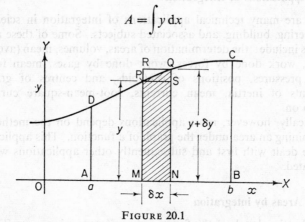

FIGURE 20.1

Clearly the limits involved in this will be a and b, i.e. the area under the graph of $y = f(x)$ above the x-axis between the ordinates $x = a$ and $x = b$ is given by

$$A = \int_a^b y\,dx = \int_a^b f(x)\,dx$$

The area may alternatively be thought of as the limiting value of the sum of rectangles like MNSP as $\delta x \to 0$, i.e.

$$A = \lim_{\delta x \to 0} \sum_{x=a}^{x=b} y\,\delta x = \int_a^b y\,\mathrm{d}x$$

Hence the integral may be thought of as an exact sum of many small elements $y\,\delta x$ to cover the whole area. This concept is used when *Simpson's rule* (or other rules) is used to evaluate integrals approximately. It is also a useful concept in many applications.

(*Note:* When y is *negative*, the integral is also negative, i.e. areas above the x-axis are *positive*, below the x-axis *negative*. If positive and negative occur the integral will give the excess of positive or negative area.)

20.1.2 Mean value of a function

If the area under the graph of $y = f(x)$ between $x = a$ and $x = b$ represents some total effect (e.g. work done by a variable force), then it is possible to estimate a constant value of y which would produce the *same* total effect. Such a value of y is called the mean value of y between $x = a$ and $x = b$ (e.g. a mean or average force).

To calculate the *mean value* of y [$f(x)$] between $x = a$ and $x = b$, consider Figure 20.2.

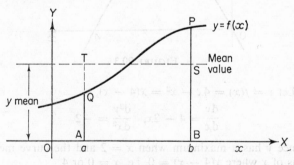

FIGURE 20.2

Let $\bar{y}$ be the mean (average) value of $f(x)$ between $x = a$ and $x = b$, then the area of the rectangle ABST

= area under the curve ABPQ

or

$$\bar{y} \times \text{AB} = \int_a^b f(x)\, dx$$

but $\text{AB} = b - a$,

$\therefore$

$$\bar{y} \times (b - a) = \int_a^b f(x)\, dx$$

$$\text{or mean value of } f(x) = \bar{y} = \frac{1}{(b-a)} \int_a^b f(x)\, dx$$

Examples:

(i) Sketch the graph of the function $4x - x^2$. Determine the area between the curve and the axis of x in the first quadrant and calculate the mean value of the function over this range of values of x.

For the sketch see Figure 20.3.

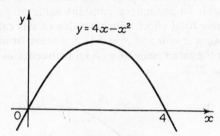

FIGURE 20.3

Let $y = f(x) = 4x - x^2 = x(4 - x)$,

$$\frac{dy}{dx} = 4 - 2x, \quad \frac{d^2y}{dx^2} = -2$$

hence y has a maximum when $x = 2$ and the curve meets the axis of x where $x(4 - x) = 0$, i.e. $\underline{x = 0 \text{ or } 4}$.

Area under the curve in the first quadrant

$$= \int_0^4 y\, dx = \int_0^4 (4x - x^2)\, dx = \left[2x^2 - \frac{1}{3}x^3\right]_0^4$$

$$= 2(4)^2 - \tfrac{1}{3}(4)^3 = 32 - \tfrac{1}{3} \times 64 = \tfrac{1}{3} \times 32 = \underline{10\tfrac{2}{3} \text{ unit}^2}$$

If $\bar{y}$ is the mean value of the function

$$\bar{y} = \frac{1}{(4-0)} \int_0^4 y \, dx = \tfrac{1}{4} \times 10\tfrac{2}{3} = 2\tfrac{2}{3} \text{ unit}$$

(ii) Determine the mean value of $\sin \theta$ between $\theta = 0$ and π (Figure 20.4).

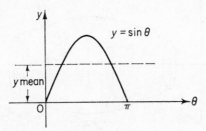

$y = \sin \theta$

y mean

O π θ

FIGURE 20.4

$$\text{Mean value} = \frac{1}{\pi - 0} \int_0^\pi \sin \theta \, d\theta = \frac{1}{\pi} \left[-\cos \theta \right]_0^\pi$$

$$= \frac{1}{\pi} [-\cos \pi - (-\cos 0)]$$

$$= \frac{1}{\pi} [-(-1) + 1] = \frac{2}{\pi}$$

20.2 Volumes by Integration

Let a solid be such that it is bounded by two parallel cross-sections at right angles to an axis OX (Figure 20.5).

Let A be the cross-sectional area at x from O along the axis OX and let the end sections correspond to $x = a$ and $x = b$. Suppose A to be a function of x. The element of volume δV between the cross-section at x and $x + \delta x$ will be approximately $A\delta x$ if δx is small, i.e. $\delta V = A \, \delta x$, hence the total volume between the end sections is deduced as

$$V = \int_a^b A \, dx$$

20.2.1 Volumes of revolution

If a plane area be revolved about an axis in its own plane through 360° a *volume of revolution* is formed and all sections at right angles to the axis are circles (Figure 20.6).

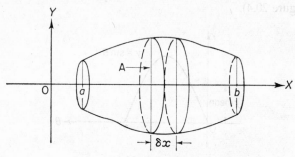

FIGURE 20.5

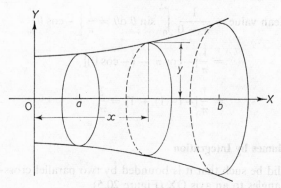

FIGURE 20.6

Let the area under the graph of $y = f(x)$ be rotated through 360° about OX. The solid formed is such that all cross-sections at right angles to the x-axis are circles. The cross-section of the solid at position x has area $A = \pi y^2$. By 20.2, volume

$$V = \int_a^b A \, dx = \int_a^b \pi y^2 \, dx = \pi \int_a^b y^2 \, dx$$

Example:

The area below the curve $y = 2\sqrt{x} + 1$, above the x-axis, between the ordinates $x = 1$ and $x = 4$ is rotated through 360° about the x-axis. Calculate the volume of the solid formed (Figure 20.7).

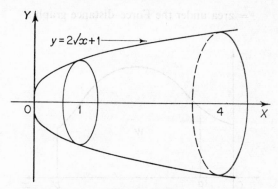

FIGURE 20.7

Volume of revolution

$$V = \pi \int_a^b y^2 \, dx = \pi \int_1^4 (2\sqrt{x} + 1)^2 \, dx$$

$$= \pi \int_1^4 [4x + 4x^{1/2} + 1] \, dx = \left[2x^2 + \frac{8}{3} x^{3/2} + x \right]_1^4$$

$$= [2(4)^2 + (8/3)(4)^{3/2} + 4] - [2 + 8/3 + 1]$$

$$= 32 + (8/3)(8) + 4 - 5\tfrac{2}{3} = 36 + \tfrac{1}{3}(64) - \tfrac{1}{3}(17) = \underline{51\tfrac{2}{3} \text{ unit}^3}$$

20.3 Work Done by Forces and Gases

20.3.1 Work done by a force

This is represented by the area under the graph of force against distance (Figure 20.8).

If $F(x)$ is the force in lbf and x is the distance in ft, in a small displacement δx the work done, δW is given approximately by

$\delta W = F \, \delta x$. Hence, as the point of application moves from $x = a$ to $x = b$ in a straight line parallel to the force F, the work done is

$$W = \lim_{\delta x \to 0} \sum_{x=a}^{x=b} F \, \delta x = \int_a^b F \, \mathrm{d}x \text{ ft lbf}$$

= area under the Force–distance graph

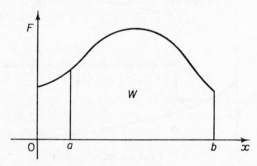

FIGURE 20.8

20.3.2 Mean (average) force

This is the *constant force*, which, if applied over the same distance would do the *same work* and is given by

Mean (average) force = Work done/Distance

$$= \frac{1}{(b-a)} \int_a^b F \, \mathrm{d}x$$

20.3.3 Work done by an expanding gas

Let a gas, at pressure p and volume v be contained in a cylinder by a piston whose face area is A [Figure 20.9]. If the piston moves a distance δx, since the force on the piston exerted by the gas $= pA$, then the work done, δW, in the small expansion $= pA \, \delta x = p \, \delta v$

(approx.). Hence the work done, W, as the gas expands from volume v_1 to v_2 is given by

$$W = \int_{v_1}^{v_2} p \, dv \text{ units}$$

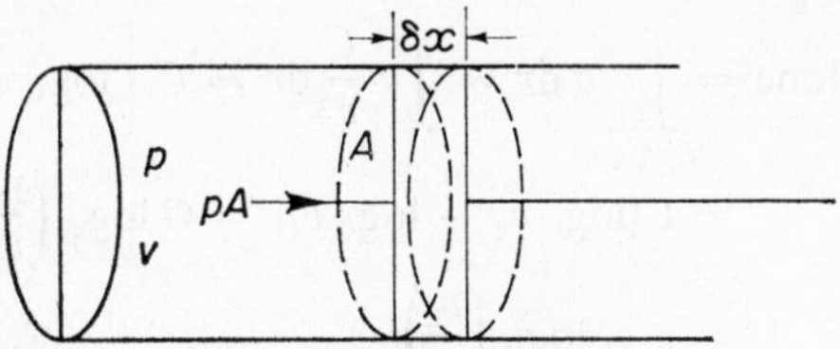

FIGURE 20.9

20.3.4 Mean pressure during an expansion

During the expansion of a gas of volume v_1, pressure p_1 to a volume v_2 at pressure p_2, the *mean pressure* during the expansion (i.e. the constant pressure which would give the same work done) is given by

$$\text{Mean (average) pressure} = \frac{1}{(v_2 - v_1)} \times \text{Work done}$$

$$= \frac{1}{(v_2 - v_1)} \int_{v_1}^{v_2} p \, dv$$

Examples:

(i) The force F lbf acting on a body at a distance x ft from a fixed point is given by $F = x^2 - 4x + 1$ lbf. The body moves in a straight line (along which F acts). Find the work done by the force as x increases from 3 to 8 ft and the mean force during the motion.

Work done by the force

$$= \int_3^8 F \, dx = \int_3^8 (x^2 - 4x + 1) \, dx = \left[\frac{1}{3} x^3 - 2x^2 + x \right]_3^8$$
$$= [\tfrac{1}{3}(8)^3 - 2(8)^2 + 8] - [\tfrac{1}{3}(3)^3 - 2(3)^2 + 3]$$
$$= \tfrac{1}{3}(512) - 128 + 8 - 9 + 18 - 3 = 170\tfrac{2}{3} - 114$$
$$= \underline{56\tfrac{2}{3} \text{ ft lbf}}$$

Mean (average) force = Work done/distance
$$= 56\tfrac{2}{3}/5 = \underline{11\tfrac{1}{3} \text{ lbf}}$$

(ii) A gas obeys the law $pv = C$ and initially has a volume of 1·5 ft³ at a pressure of 120 lbf/in.² Find the work done and the mean pressure during the isothermal expansion of the gas to a volume of 6 ft³.

$$pv = C = p_1v_1 = 120 \times 144 \times 1\cdot5 \text{ ft lbf}$$

$$\text{Work done} = \int_{v_1}^{v_2} p \, dv = \int_{v_1}^{v_2} \frac{C}{v} \, dv = C \left[\log_e v \right]_{v_1}^{v_2}$$

$$= C[\log_e v_2 - \log_e v_1] = C \log_e \left(\frac{v_2}{v_1} \right)$$

$$= p_1v_1 \log_e \left(\frac{v_2}{v_1} \right)$$

$$= 120 \times 144 \times 1\cdot5 \times \log_e (6/1\cdot5) \text{ ft lbf}$$

$$= 180 \times 144 \times \log_e 4 \text{ ft lbf}$$

$$= 180 \times 144 \times 1\cdot3863 = \underline{35{,}940 \text{ ft lbf}}$$

Mean pressure during the expansion

$$= \text{Work done/change of volume}$$

$$= 35{,}940/(6 - 1\cdot5) = 35{,}940/4\cdot5 \text{ lbf/ft}^2$$

$$= \frac{35{,}940}{4\cdot5 \times 144} \text{ lbf/in.}^2 = \underline{55\cdot46 \text{ lbf/in.}^2}$$

WORKED EXAMPLES

(a) Show, by integration, that the area of a semicircle of radius r is $\frac{1}{2}\pi r^2$. If the area be rotated through 360° about the diameter, show that the volume of the sphere formed is $4\pi r^3/3$.

Let the semicircle be as shown in Figure 20.10. If x and y are the

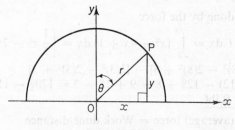

FIGURE 20.10

co-ordinates of P the equation of the circle is $x^2 + y^2 = r^2$, i.e. $y = \sqrt{(r^2 - x^2)}$.

$$\text{Area of semicircle} = \int_{-r}^{+r} y \, dx = \int_{-r}^{+r} \sqrt{(r^2 - x^2)} \, dx$$

Let $x = r \sin \theta$, $dx = r \cos \theta \, d\theta$. By symmetry

$$\text{Area} = 2 \int_{0}^{r} \sqrt{(r^2 - x^2)} \, dx$$

When $x = 0$, $\theta = 0$; when $x = r$, $\theta = \pi/2$.

$$\text{Area} = 2 \int_{0}^{\pi/2} \sqrt{(r^2 - r^2 \sin^2 \theta)} r \cos \theta \, d\theta$$

$$= 2 \int_{0}^{\pi/2} r\sqrt{(1 - \sin^2 \theta)} r \cos \theta \, d\theta = 2r^2 \int_{0}^{\pi/2} \cos^2 \theta \, d\theta$$

$$= 2r^2 \int_{0}^{\pi/2} \tfrac{1}{2}(1 + \cos 2\theta) \, d\theta$$

$$= r^2 [\theta + \tfrac{1}{2} \sin 2\theta]_{0}^{\pi/2} = r^2[(\pi/2 + \tfrac{1}{2} \sin \pi) - (0 + 0)]$$

$$= \tfrac{1}{2}\pi r^2 \quad (\sin \pi = 0)$$

$$\text{Volume of sphere} = \pi \int_{-r}^{+r} y^2 \, dx = \pi \int_{-r}^{+r} (r^2 - x^2) \, dx$$

$$= 2\pi \left[r^2 x - \frac{1}{3}x^3 \right]_{0}^{r} = 2\pi[r^3 - \tfrac{1}{3}r^3]$$

$$= 2\pi \times \tfrac{2}{3}r^3 = \underline{4\pi r^3/3}$$

(b) The current i at time t seconds in an electrical circuit is given by $i = 25 \sin (50\pi t)$ amp. Find the mean value of the current between $t = 0$ and $t = 0.01$ sec.

Mean value of i

$$= \frac{1}{(0.01 - 0)} \int_{0}^{0.01} i \, dt = 100 \int_{0}^{0.01} 25 \sin (50\pi t) \, dt$$

$$= 100 \times 25 \left[-\frac{\cos (50\pi t)}{50\pi} \right]_{0}^{0.01} = \frac{2,500}{50\pi} [-\cos 0.5\pi + \cos 0]$$

$$= \frac{50}{\pi} = \underline{15.92 \text{ amp}} \quad [\cos 0 = 1, \cos 0.5\pi = 0]$$

(c) Derive a formula for the work done during an adiabatic expansion following the law $pv^n = c$ as the volume increases from v_1 to v_2. (p_1 is the initial pressure, p_2 is the final pressure.)

$$\text{Work done} = \int_{v_1}^{v_2} p \, dv = \int_{v_1}^{v_2} \frac{c}{v^n} \, dv = c \int_{v_1}^{v_2} v^{-n} dv$$

$$= c \left[\frac{v^{-n+1}}{-n+1} \right]_{v_1}^{v_2} = \frac{c}{n-1} [-v_2^{-n+1} + v_1^{-n+1}]$$

$$= \frac{c}{n-1} \left[\frac{1}{v_1^{n-1}} - \frac{1}{v_2^{n-1}} \right]$$

Now $c = p_1 v_1^n = p_2 v_2^n$

$$\therefore \quad W = \frac{1}{(n-1)} \left[\frac{p_1 v_1^n}{v_1^{n-1}} - \frac{p_2 v_2^n}{v_2^{n-1}} \right]$$

$$\underline{W = \frac{1}{(n-1)} [p_1 v_1 - p_2 v_2]}$$

Examples 20

1. Sketch the graph of the function $y = 6x - x^2$. Determine, by integration (i) the area enclosed by the graph, the x-axis and the ordinate $x = 5$, (ii) the total area in the first quadrant between the curve and the x-axis, (iii) the volume when the area (ii) is rotated through $360°$ about the x-axis.
2. Sketch the graph of $y = 2 + \sqrt{x}$ between $x = 0$ and $x = 4$. Determine by integration (i) the area between the curve, the x-axis and the ordinates $x = 0$ and $x = 4$, (ii) the volume of revolution when the area is rotated through $360°$ about Ox, (iii) the mean value of y between $x = 0$ and $x = 4$.
3. The rectangular hyperbola $xy = c$ passes through the point $(4, 5)$. Determine (i) the area between the curve the x-axis and the ordinates $x = 4$ and $x = 12$, (ii) the volume when this area is rotated through $360°$ about the x-axis.
4. Determine the area below the curve $y = 2x^{3/2}$ and above the x-axis between the ordinates $x = 1$ and $x = 4$. Find also the volume of revolution when this area is rotated through $360°$ about the x-axis.
5. Find the co-ordinates of the points of intersection of the rectangular hyperbola $xy = 12$ and the line $2x + y = 10$. Hence determine the area between the curve and the line in the first quadrant.
6. Sketch the graphs of $y = \frac{1}{4}x^2$ and $y = 2\sqrt{x}$ in the same figure. Find the area included between the two curves.
7. (i) Determine the area enclosed by the curve $y = x^2 - 12x - 45$ and the x-axis. What is the significance of the result?
(ii) Sketch the curve $y = x^2 + 1$ and the line $y = 2x + 1$. At what points do the two loci intersect. Determine the area between the line and the curve.

8. Sketch the graph of $y = 3 e^x$ from $x = 0$ to 2. Determine (i) the area between the curve, the x-axis and the ordinates $x = 0$ and $x = 2$, (ii) the volume of revolution when this area is rotated through 2π about the x-axis.

9. Find, by integration the mean values of the following functions in the specified ranges:

  (i) $9 - x^2$ between $x = 0$ and $x = 3$,
  (ii) $2\sqrt{x}$ between $x = 1$ and $x = 4$,
  (iii) $x^2 + 3x + 2$ between $x = -1$ and $+2$,
  (iv) $3 e^{-4t}$ between $t = 0$ and $t = \frac{1}{2}$.

10. The diameter of the base of a frustum of a right circular cone is 12 in. and of the upper face is 4 in. If the height of the frustum is 10 in, use integration to calculate the volume of the frustum. Check your result using the formula $V = \frac{1}{3}\pi h(r^2 + Rr + R^2)$.

11. Determine the mean values of the following functions over the stated intervals of the independent variables:

  (i) $100(1 - e^{-0.2t})$ between $t = 0$ and $t = 5$,
  (ii) $6 + 3 \cos x - 4 \sin x$ between $x = 0$ and $x = \pi/2$,
  (iii) $10 \sin \omega t + 20 \cos \omega t$ between $t = 0$ and $t = \pi/\omega$.

12. The force F exerted on a mass by a spring at time t seconds is given by $F = 5 \sin 4t$ lbf. Find the mean value of F between $t = 0$ and $\pi/4$.

13. (i) The current i in a circuit is given by $i = 80 \sin (200\pi t + 0.25)$ amp. Calculate the mean current between $t = 0$ and $t = 1/(400\pi)$ sec.
  (ii) The current i in a circuit is given by $i = 4 \sin 20t$ and the voltage e by $e = 50 \sin (20t - \alpha)$. Find the mean value of the product ei as t varies from 0 to $\pi/40$.

14. An expanding gas obeys the law $pv = c$. Find the work done when 2 ft³ of gas at a pressure of 350 lbf/in² expands to 7 ft³. Find also the mean pressure during the expansion.

15. An amount of steam expands so as to satisfy the law $pv^{1.13} = c$. Find the work done, in ft lbf and the mean pressure during an expansion from a volume of 3 ft³ at a pressure of 150 lbf/in² to a volume of 10 ft³.

21

Applications of Integration (2)—
R.M.S. Values—Centroids—
Approximate Integration

21.1 Root Mean Square Value

In certain cases, where the effect of a quantity is proportional to the *square* of the quantity it may be necessary to use the square root of the *mean* of the *square* of the quantity over a stated interval of the independent variate. Specific *root mean square* values (R.M.S. values) encountered in practice are:

(i) the R.M.S. values of *current* and *voltage* in alternating current circuitry theory,

(ii) the *standard deviation* of a set of discrete values *or* of a continuous distribution of a variate (as used in *statistics* for measuring the *degree of scattering* of data),

(iii) the *radius of gyration* of a body which is the R.M.S. value of distances of all points of a body from a stated axis.

21.1.1 Evaluation of R.M.S. value of a function

Let $y = f(x)$ be the equation of the graph of a function of x and $y = [f(x)]^2$ be the equation of the 'squared function' graph [Figure 21.1].

By previous definition (Chapter 20), the *mean* value of $f(x)$ over the interval $x = a$ to $x = b$ is given by

$$\text{Mean value of } f(x) = \frac{1}{(b-a)} \int_a^b f(x) \, dx$$

Similarly the *mean square* value of $f(x)$ is the area under the 'squared function' graph between $x = a$ and $x = b$ divided by $(b-a)$, i.e.

$$\text{Mean square of } f(x) = \frac{1}{(b-a)} \int_a^b [f(x)]^2 \, dx$$

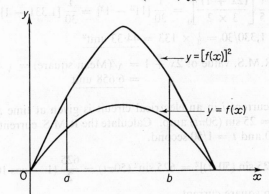

FIGURE 21.1

and the root mean square value of $f(x)$ is given by

$$\underline{\text{R.M.S. value of } f(x) = \sqrt{[\text{Mean square of } f(x)]}}$$

21.1.2 R.M.S. value of a set of discrete values

Let n observations of y ($y_1, y_2, y_3, \ldots y_n$) be given then the mean square of the y values $= (y_1^2 + y_2^2 + \ldots + y_n^2)/n$ which may be abbreviated to

$$\text{Mean square of } y \text{ values} = \frac{1}{n} \sum_{i=1}^{n} y_i^2$$

This evaluation of a mean square is most useful in *statistical analysis* (as is the evaluation of the mean square of y when y is a continuous function of another variate x).

Examples:

(i) Determine the R.M.S. value of $2x + 1$ between $x = 0$ and $x = 5$.

Let $f(x) = 2x + 1$, $[f(x)]^2 = (2x + 1)^2$, then the mean square value of $f(x)$ in the range

$$= \frac{1}{b - a} \int_a^b [f(x)]^2 \, \mathrm{d}x = \frac{1}{5 - 0} \int_0^5 (2x + 1)^2 \, \mathrm{d}x$$

$$= \frac{1}{5} \left[\frac{(2x + 1)^3}{3 \times 2} \right]_0^5 = \frac{1}{30} [11^3 - 1^3] = \frac{1}{30} [1{,}331 - 1]$$

$$= 1{,}330/30 = \tfrac{1}{3} \times 133 = \underline{44 \cdot 33 \text{ unit}^2}$$

$\therefore$ R.M.S. value of $2x + 1 = \sqrt{(\text{Mean square})} = \sqrt{44 \cdot 33}$
$$= \underline{6 \cdot 658 \text{ unit}}$$

(ii) The current i in an electrical circuit is given at time t seconds by $i = 25 \sin (50\pi t)$ amp. Calculate the R.M.S. current between $t = 0$ and $t = 1/50$ second.

$$i^2 = [25 \sin (50\pi t)]^2 = 625 \sin^2 (50\pi t) = \frac{625}{2} (1 - \cos 100\pi t)$$

Mean square current

$$= \frac{1}{1/50 - 0} \int_0^{1/50} 625 \sin^2 (50\pi t) \, \mathrm{d}t$$

$$= 50 \times 625 \times \int_0^{1/50} \tfrac{1}{2}(1 - \cos 100\,\pi t) \, \mathrm{d}t$$

$$= 50 \times 625 \times \tfrac{1}{2} \left[t - \frac{\sin 100\pi t}{100\pi} \right]_0^{1/50}$$

$$= 50 \times 625 \times \tfrac{1}{2} \left[\left(\frac{1}{50} - \frac{\sin 2\pi}{100\pi} \right) - (0 - 0) \right]; \quad [\sin 2\pi = 0]$$

$$= 50 \times 625 \times \tfrac{1}{2} \times 1/50 = \tfrac{1}{2} \times 625$$

$\therefore$ R.M.S. current $i_{\text{R.M.S.}} = \sqrt{(\text{Mean value of } i^2)}$
$$= \sqrt{(\tfrac{1}{2} \times 625)} = 25/\sqrt{2} \text{ amp}$$
$$= \underline{17 \cdot 68 \text{ amp}}$$

(iii) Six observations of a dimension y in. are taken as follows: $y = 5\cdot3,\ 4\cdot8,\ 5\cdot1,\ 4\cdot7,\ 5\cdot2,\ 4\cdot9$ in. Calculate the R.M.S. value of the *deviation* of y from the mean value. [*Note:* Deviation = observation − mean value.]

$$\text{Mean value of } y = \tfrac{1}{6}\Sigma y$$
$$= \tfrac{1}{6}(5\cdot3 + 4\cdot8 + 5\cdot1 + 4\cdot7 + 5\cdot2 + 4\cdot9)$$
$$= \tfrac{1}{6} \times 30\cdot0 = \underline{5 \text{ in.}}$$

Deviations of the observations from the mean are $+0\cdot3$, $-0\cdot2,\ +0\cdot1,\ -0\cdot3,\ +0\cdot2,\ -0\cdot1$ in.

Sum of squares of deviations

$$= (0\cdot3)^2 + (-0\cdot2)^2 + (0\cdot1)^2 + (-0\cdot3)^2 + (0\cdot2)^2 + (-0\cdot1)^2 \text{ in}^2$$
$$= 0\cdot09 + 0\cdot04 + 0\cdot01 + 0\cdot09 + 0\cdot04 + 0\cdot01 = 0\cdot28 \text{ in}^2$$

Mean square deviation from the mean

$$= \tfrac{1}{6} \times (\text{sum of squares}) = \tfrac{1}{6} \times 0\cdot28 = 0\cdot04667 \text{ in}^2$$
$$\therefore \quad \text{R.M.S. value of deviation} = \sqrt{0\cdot04667} = \underline{0\cdot2161 \text{ in.}}$$

[*Note:* (i) The mean square is called the *variance* of the data, (ii) the R.M.S. value is called the *standard deviation*. Both of these are measures of the *degree* to which the values are *scattered* about the *mean value*. They will be dealt with in more detail in the *statistics* in later chapters.]

21.2 Centroids, Centres of Mass, Centres of Gravity

The *centroid* is a point in a body (plane or solid) defined geometrically by the use of moments of area (or volume) about given axes. The *centre of mass* is defined in relation to the *moments* of mass about given axes. The *centre of gravity* of a body is defined in relation to the moments of weights about given axes and is the point through which the *resultant weight* acts.

In *homogeneous* (uniformly dense) bodies of relatively small dimensions the three points coincide.

The points do not coincide if the density varies, e.g. if a rectangular plate is compounded of two parts of different materials then the centroid is the geometrical centre, but the centre of mass and centre of gravity will fall to one side of the centroid.

21.2.1 Determination of the centroid of an area

Consider the area in Figure 21.2, with axes Ox, Oy in the same plane as the area. Let δA be a small element of area at distances y and x from the two axes. Let the co-ordinates of the *centroid* of

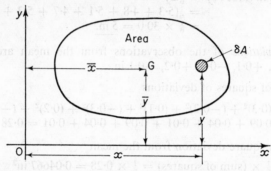

FIGURE 21.2

the area be $\bar{x}$, $\bar{y}$. The moment of the element of area about $Ox = y\,\delta A$ and the moment of the element of area about $Oy = x\,\delta A$, $\bar{x}$ is defined as such that

$$\bar{x} \times \text{total area} = \text{sum of moments of all elements about } Oy$$

i.e.

$$\bar{x} \times \Sigma\,\delta A = \Sigma x\,\delta A$$

or in the limit

$$\bar{x} \times \int dA = \int x\,dA; \qquad \bar{x} = \frac{\int x\,dA}{\int dA}$$

Similarly, $\bar{y}$ is defined as such that

$$\bar{y} \times \text{total area} = \text{sum of moments of all elements about } Ox$$

i.e.

$$\bar{y} \times \Sigma\,\delta A = \Sigma y\,\delta A$$

or in the limit

$$\bar{y} \times \int dA = \int y\,dA; \qquad \bar{y} = \frac{\int y\,dA}{\int dA}$$

Usually a simple geometrical element is chosen as δA.

21.2.2 Determination of centres of mass

Similar formulae may be derived except that moments of mass are considered, e.g. if instead of δA an element of mass δm be taken, then, if $\bar{x}$ and $\bar{y}$ are the co-ordinates of the centre of mass

$$\bar{x} = \frac{\int x \, dm}{\int dm}; \quad \bar{y} = \frac{\int y \, dm}{\int dm}$$

21.2.3 Centroids and centres of mass of solids

By considering suitable volume elements (or elements of mass) similar formulae may be developed for the co-ordinates of centroids and centres of mass of solids.

21.2.4 Symmetrical plane areas and solids

When there is an *axis of symmetry* (usually obvious) the centroid (or centre of mass, if the body is homogeneous) will lie on the axis of symmetry.

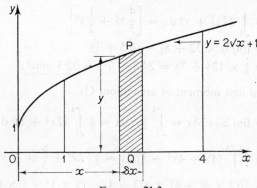

FIGURE 21.3

Examples:

(i) Determine the co-ordinates of the centroid of the area between the curve $y = 2\sqrt{x} + 1$, the x-axis and the ordinates $x = 1$ and $x = 4$.

Consider the element of area under the curve, PQ (Figure 21.3) of width δx and approximate height y.

Area of element $= y\,\delta x$ (approximately).

Moment of area of element about $Oy = xy\,\delta x$.

Moment of area of element about $Ox = \frac{1}{2}y \times y\,\delta x = \frac{1}{2}y^2\,\delta x$ [if δx is very small the shaded area is approximately a rectangle].

Total area under the graph

$$= \lim \Sigma y\,\delta x = \int_1^4 y\,\mathrm{d}x$$

$$= \int_1^4 (2\sqrt{x} + 1)\,\mathrm{d}x = \int_1^4 (2x^{\frac{1}{2}} + 1)\,\mathrm{d}x = \left[\frac{4}{3}x^{\frac{3}{2}} + x\right]_1^4$$

$$= \left(\frac{4}{3} \times 8 + 4\right) - \left(\frac{4}{3} + 1\right) = \frac{28}{3} + 3 = \frac{37}{3}$$

$$= \underline{12\tfrac{1}{3}\ \text{unit}^2}$$

Total first moment of area about Oy

$$= \lim \Sigma xy\,\delta x = \int_1^4 xy\,\mathrm{d}x = \int_1^4 x(2x^{\frac{1}{2}} + 1)\,\mathrm{d}x$$

$$= \int_1^4 (2x^{\frac{3}{2}} + x)\,\mathrm{d}x = \left[\frac{4}{5}x^{\frac{5}{2}} + \frac{1}{2}x^2\right]_1^4$$

$$= [(4/5) \times 32 + 8] - (4/5 + \tfrac{1}{2})$$

$$= \tfrac{1}{5} \times 124 + 7\tfrac{1}{2} = 24{\cdot}8 + 7{\cdot}5 = \underline{32{\cdot}3\ \text{unit}^3}$$

Total first moment of area about Ox

$$= \lim \Sigma \tfrac{1}{2}y^2\,\delta x = \int_1^4 \tfrac{1}{2}y^2\,\mathrm{d}x = \tfrac{1}{2}\int_1^4 (2x^{\frac{1}{2}} + 1)^2\,\mathrm{d}x$$

$$= \tfrac{1}{2}\int_1^4 (4x + 4x^{\frac{1}{2}} + 1)\,\mathrm{d}x = \tfrac{1}{2}\left[2x^2 + \frac{8}{3}x^{\frac{3}{2}} + x\right]_1^4$$

$$= \tfrac{1}{2}[(2 \times 4^2 + 4^{\frac{3}{2}} \times 8/3 + 4) - (2 \times 1^2 + 1 \times 8/3 + 1)]$$

$$= \tfrac{1}{2}[(36 + \tfrac{1}{3} \times 64) - (3 + 8/3)] = \tfrac{1}{2}[33 + \tfrac{1}{3} \times 56]$$

$$= 155/6 = \underline{25\tfrac{5}{6}\ \text{unit}^3}$$

$$\bar{x} \times \text{Area} = 32{\cdot}3, \quad \bar{x} = 32{\cdot}3/12\tfrac{1}{3} = \underline{2{\cdot}619\ \text{unit}}$$

$$\bar{y} \times \text{Area} = 25\tfrac{5}{6}, \quad \bar{y} = 25\tfrac{5}{6}/12\tfrac{1}{3} = \underline{2{\cdot}094\ \text{unit}}$$

(ii) Show, by integration that the centre of mass of a uniform solid cone of height h, base radius r, and density m is $\frac{3}{4}h$ from the vertex. (See Figure 21.4.)

By *symmetry* the centre of mass lies on the altitude ON which is an *axis of symmetry*. Let G be the centre of mass at $\bar{x}$ from

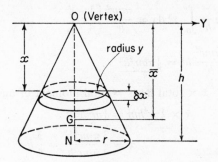

O (Vertex)

radius y

FIGURE 21.4

the vertex O, OY an axis parallel to the base. Consider a thin slice of the cone at x from O, of radius y and thickness δx (parallel to the base), then

$$\frac{y}{x} = \frac{r}{h}, \quad y = \frac{r}{h}x$$

Volume of thin slice $= \pi y^2 \,\delta x$.

Mass $= m\pi y^2 \,\delta x$.

Total mass $= \lim \Sigma m\pi y^2 \,\delta x = \int_0^h \pi m y^2 \,\mathrm{d}x$

$$= \pi m \int_0^h \left(\frac{r}{h}x\right)^2 \mathrm{d}x = \pi m \frac{r^2}{h^2} \int_0^h x^2 \,\mathrm{d}x$$

$$= \pi m \frac{r^2}{h^2} \left[\frac{1}{3}x^3\right]_0^h = \frac{\pi m r^2}{h^2} \times \tfrac{1}{3}h^3 = \underline{\tfrac{1}{3}\pi r^2 h m}$$

Moment of mass of element about axis $OY = x \times \pi y^2 m\ \delta x$.
Total moment of mass about OY

$$= \pi m \int_0^h xy^2\ \mathrm{d}x = \pi m \int_0^h x\frac{r^2}{h^2}x^2\ \mathrm{d}x$$

$$= \pi m \frac{r^2}{h^2} \int_0^h x^3\ \mathrm{d}x = \frac{\pi m r^2}{h^2}\left[\frac{1}{4}x^4\right]_0^h$$

$$= \frac{1}{4}\frac{\pi m r^2}{h^2}h^4 = \tfrac{1}{4}\pi m r^2 h^2$$

$$\bar{x} \times \text{total mass} = \text{total moment of mass}$$

$$\bar{x} \times \tfrac{1}{3}\pi r^2 hm = \tfrac{1}{4}\pi r^2 h^2 m$$

$$\therefore \qquad \bar{x} = \frac{3}{4}\frac{\pi m r^2 h^2}{\pi m r^2 h} = \frac{3}{4}h$$

i.e. the centre of mass lies on the altitude at $\frac{3}{4}h$ from the vertex.

21.3 Approximate Integration Using Simpson's Rule

Since the definite integral of a function between two limits basically represents the area under the graph of the function it is possible to evaluate integrals approximately using an approximate method for finding the area, e.g. mid-ordinate rule, trapezoidal rule, or Simpson's rule.

Simpson's Rule

If n ordinates (n odd), $y_1, y_2, y_3, \ldots, y_n$ are given at equal intervals h of the independent variable x, then the approximate area between the end ordinates is given by

$$A \simeq \tfrac{1}{3}h[\text{sum of first and last} + 4(\text{sum of even})$$
$$+ 2(\text{sum of odds})]$$
$$\simeq \tfrac{1}{3}h[F + L + 4E + 2O]$$

Example:

Tabulate the values of $1/(1 + 2x)$ for x from 0 to 4 in steps of 0·5 and hence find the approximate numerical value of

$\int_0^4 \dfrac{1}{1+2x}\,dx$. Evaluate the integral exactly and hence state the error to 3 decimal places in using the approximation.

x	0	0·5	1·0	1·5	2·0	2·5	3·0	3·5	4·0
$(2x+1)$	1	2	3	4	5	6	7	8	9
$\dfrac{1}{2x+1}$	1	0·5000	0·3333	0·2500	0·2000	0·1667	0·1429	0·1250	0·1111
	y_1	y_2	y_3	y_4	y_5	y_6	y_7	y_8	y_9

Sum of first and last ordinates $= y_1 + y_9 = 1·1111$.
Sum of even ordinates $= y_2 + y_4 + y_6 + y_8 = 1·0417$; $h = 0·5$.
Sum of odd ordinates $= y_3 + y_5 + y_7 = 0·6762$,

$$\therefore \int_0^4 \frac{1}{2x+1}\,dx \simeq \tfrac{1}{3}h[F + L + 4E + 2O]$$

$$\simeq \tfrac{1}{3}(0·5)[1·1111 + 4 \times 1·0417 + 2 \times 0·6762]$$

$$\simeq \tfrac{1}{6}[1·1111 + 4·1668 + 1·3524]$$

$$\simeq \tfrac{1}{6} \times 6·6303 \simeq \underline{1·105(05)}$$

$$\int_0^4 \frac{1}{2x+1}\,dx = \tfrac{1}{2}\left[\log_e(2x+1)\right]_0^4 = \tfrac{1}{2}[\log_e 9 - \log_e 1]$$

$$= \tfrac{1}{2}\log_e 9 = \log_e 3 = 1·0986$$

Error $= +(1·10505 - 1·09860) = +0·00645$
$\simeq \underline{+0·006}$ (3 decimal places)

WORKED EXAMPLES

(a) Determine the R.M.S. value of the function $e^{0·5t}$ between $t = -2$ and $+2$.

Let $f(t) = e^{0·5t}$, $[f(t)]^2 = [e^{0·5t}]^2 = e^t$

Mean square of $f(t)$

$$= \frac{1}{[2-(-2)]}\int_{-2}^2 e^t\,dt = \frac{1}{4}\left[e^t\right]_{-2}^2 = \tfrac{1}{4}[e^2 - e^{-2}]$$

$$= \tfrac{1}{4}[7·3891 - 0·1353] = \tfrac{1}{4} \times 7·2538 = \underline{1·8134(5)}$$

$\therefore$ R.M.S. value of $f(t) = \sqrt{}$(Mean square) $= \sqrt{1·81345} = \underline{1·346}$

(b) The voltage V volt in a circuit at time t seconds is given by $V = 5 \cos 2\pi t + 10 \sin 3\pi t$ volt. Determine the R.M.S. value of V between $t = 0$ and $t = \frac{1}{2}$ sec.

$$
\begin{aligned}
V^2 &= [5 \cos 2\pi t + 10 \sin 3\pi t]^2 \\
&= 25 \cos^2 2\pi t + 100 \cos 2\pi t \sin 3\pi t + 100 \sin^2 3\pi t \\
&= 25[\tfrac{1}{2}(1 + \cos 4\pi t) + 4 \cos 2\pi t \sin 3\pi t + 2(1 - \cos 6\pi t)] \\
&= 25[5/2 + \tfrac{1}{2} \cos 4\pi t + 2(\sin 5\pi t + \sin \pi t) - 2 \cos 6\pi t]
\end{aligned}
$$

Mean square of V

$$
\begin{aligned}
&= \frac{1}{\frac{1}{2} - 0} \int_0^{1/2} V^2 \, dt \\
&= 50 \int_0^{1/2} [5/2 + \tfrac{1}{2} \cos 4\pi t + 2 \sin 5\pi t + 2 \sin \pi t - 2 \cos 6\pi t] \, dt \\
&= 50 \left[\frac{5}{2}t + \frac{1}{8\pi} \sin 4\pi t - \frac{2}{5\pi} \cos 5\pi t - \frac{2}{\pi} \cos \pi t - \frac{2}{6\pi} \sin 6\pi t \right]_0^{\frac{1}{2}} \\
&= 50 \left[\left(\frac{5}{2} \times \tfrac{1}{2} + \frac{1}{8\pi} \sin 2\pi - \frac{2}{5\pi} \cos \frac{5\pi}{2} - \frac{2}{\pi} \cos \frac{\pi}{2} - \frac{1}{3\pi} \sin 3\pi \right) \right. \\
&\qquad \left. - \left(\frac{5}{2} \times 0 + \frac{1}{8\pi} \sin 0 - \frac{2}{5\pi} \cos 0 - \frac{2}{\pi} \cos 0 - \frac{1}{3\pi} \sin 0 \right) \right] \\
&= 50[(5/4) + (2/5\pi) + (2/\pi)] = 50[(5/4) + (12/5\pi)] \\
&= (250/4) + (120/\pi) = 62 \cdot 5 + 38 \cdot 19 = \underline{100 \cdot 69} \\
\therefore \quad & \text{R.M.S. value of } V = \sqrt{100 \cdot 69} = \underline{10 \cdot 03 \text{ volt}}
\end{aligned}
$$

(c) Evaluate the R.M.S. value of the following discrete values of W: 17, 18, 19, 20, 21, 16.

$$
\begin{aligned}
S^2 &= \text{Mean square of } W = \tfrac{1}{6}(\text{sum of squares}) \\
&= \tfrac{1}{6}(17^2 + 18^2 + 19^2 + 20^2 + 21^2 + 16^2) \\
&= \tfrac{1}{6}(289 + 324 + 361 + 400 + 441 + 256) \\
&= \tfrac{1}{6} \times 2{,}071 = 345 \cdot 17 \\
S &= \text{R.M.S. value} = \sqrt{345 \cdot 17} = \underline{18 \cdot 58}
\end{aligned}
$$

(d) Determine the centroid of the area between the curve $y = x(4 - x)$ and the x-axis for that part of the area above the x-axis (Figure 21.5).

The curve meets the axis of x where $x(4 - x) = 0$, i.e. $x = 0$ or 4. The curve is a parabola with its axis parallel to the y-axis and this axis passes through the vertex, hence the centroid of the area lies on this axis and $\bar{x} = 2$. $\bar{x}$ will also be calculated.

Consider the element of area $y\,\delta x = x(4 - x)\,\delta x$.

First moment of area about $Oy = xy\,\delta x = x^2(4 - x)\,\delta x$.

First moment of area about $Ox = \frac{1}{2}y \times y\,\delta x = \frac{1}{2}x^2(4 - x)^2\,\delta x$.

$$\text{Total area} = \int_0^4 y\,dx = \int_0^4 (4x - x^2)\,dx = \left[2x^2 - \frac{1}{3}x^3\right]_0^4$$

$$= 32 - \tfrac{1}{3} \times 64 = \underline{32/3 \text{ unit}^2}$$

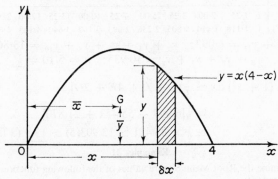

FIGURE 21.5

Total first moment about Oy

$$= \int_0^4 (4x^2 - x^3)\,dx = \left[\frac{4}{3}x^3 - \frac{1}{4}x^4\right]_0^4 = \frac{4}{3} \times 4^3 - \frac{1}{4} \times 4^4$$

$$= \frac{1}{12} \times 4^4 = \frac{64}{3} = \underline{21\tfrac{1}{3} \text{ unit}^3}$$

Total first moment about Ox

$$= \int_0^4 \tfrac{1}{2}x^2(4 - x)^2\,dx = \tfrac{1}{2}\int_0^4 (16x^2 - 8x^3 + x^4)\,dx$$

$$= \frac{1}{2}\left[\frac{16}{3}x^3 - 2x^4 + \frac{1}{5}x^5\right]_0^4$$

$$= \tfrac{1}{2}[(16/3) \times 4^3 - 2 \times 4^4 + (1/5) \times 4^5]$$

$$= \tfrac{1}{2} \times 4^3[(16/3) - 8 + (16/5)] = 32 \times (8/15) = \underline{256/15 \text{ unit}^3}$$

$$\bar{x} \times 32/3 = 64/3 \qquad \therefore \quad \underline{\bar{x} = 2 \text{ unit}}$$

$$\bar{y} \times 32/3 = 256/15 \qquad \therefore \quad \bar{y} = (256/15) \times 3/32 = 8/5$$
$$= 1\cdot6 \text{ unit}$$

$$\therefore \qquad \underline{\text{Centroid is the point } (2, 1\cdot6)}$$

(e) Use Simpson's rule to evaluate $\displaystyle\int_0^5 \sqrt{(1 + x^2)}\,\mathrm{d}x$, approximately taking ten equal intervals for the variable x.

x	0	0·5	1·0	1·5	2·0	2·5	3·0	3·5	4·0	4·5	5·0
$1 + x^2$	1	1·25	2·00	3·25	5·00	7·25	10·00	13·25	17·00	21·25	26·00
$\sqrt{(1 + x^2)}$	1	1·118	1·414	1·803	2·236	2·693	3·162	3·640	4·123	4·609	5·099

$$y_1 + y_{11} = 6\cdot099; \quad y_2 + y_4 + y_6 + y_8 + y_{10} = 13\cdot863$$
$$y_3 + y_5 + y_7 + y_9 = 10\cdot935; \quad h = 5/10 = \tfrac{1}{2}$$

$$\therefore \quad \int_0^5 \sqrt{(1 + x^2)}\,\mathrm{d}x \simeq \frac{h}{3}\,[F + L + 4E + 2O]$$

$$= \tfrac{1}{6}[6\cdot099 + 55\cdot452 + 21\cdot870]$$

$$\simeq \tfrac{1}{6} \times 83\cdot421 \simeq 13\cdot903(5) \simeq \underline{13\cdot9} \text{ (3 figures)}$$

Examples 21

1. Determine the Root Mean Square values of the following functions over the given ranges:

    (i) $2x + 3$ between $x = 0$ and $x = 4$,
    (ii) $t^2 + 2$ between $t = -1$ and $+3$,
    (iii) $\sqrt{x} + 1/\sqrt{x}$ between $x = 1$ and 9.

2. Using integration obtain the R.M.S. values of the functions:

    (i) $5\,e^t$ over the range $t = -0\cdot5$ to $t = 0\cdot5$,
    (ii) $10 \sin 2\theta$ over the range $\theta = 0$ to π,
    (iii) $3 \sin x + 4 \cos x$ over the range 0 to $\pi/2$.

3. The 'form' factor F of a wave form is given by the expression $F = \text{R.M.S.value}/(\text{Mean value})$ where the values are found for the dependent variables. Determine the form factor for the following waveforms:

    (i) $y = I \sin \omega t$ between $t = 0$ and π/ω,
    (ii) $y = x(2 - x)$ between $x = 0$ and $x = 2$,
    (iii) $E = 100 \cos 300t$ between $t = -\pi/600$ and $t = \pi/600$.

4. Determine the R.M.S. value of the current i amp given by $i = 5 \sin 40\pi t + 12 \cos 40\pi t$ amp where t is the time in seconds as t varies from 0 to 1/20th second.

5. The power W developed in a resistor of resistance R ohm, when an electromotive force V volt is applied is given by $W = V^2/R$. If the applied voltage

is sinusoidal and given by $V = 100 \sin 100\pi t$ volt (t in seconds), determine the mean power developed over one cycle of the alternating voltage when $R = 25$ ohm. What is the R.M.S. value of the current in the resistor.

6. Ten independent observations of a variate x are given as 5, 8, 9, 11, 7, 12, 6, 10, 9, 8. Calculate (i) the mean value of x ($\bar{x}$),
(ii) the 'deviations' of the observed values from $\bar{x}$,
(iii) the 'variance' of the observations, i.e. the mean square deviation of the observations from the mean (s^2),
(iv) the 'standard deviation' of the observations from the mean, i.e. the R.M.S. deviation from the mean.

7. Sketch the curve $y = 16 - x^2$ showing that part of the curve in the first quadrant. Obtain the co-ordinates ($\bar{x}, \bar{y}$) of the centroid of the area between the curve and the x-axis in the first quadrant.

8. Determine the coordinates of the centroids of the areas:
(i) bounded by $y = \sqrt{x} + 1$, the x-axis and the ordinates $x = 0$ and $x = 4$,
(ii) bounded by the curve $xy = 10$ and the ordinates $x = 1$ and $x = 4$.

9. Find the x coordinates of the centroids of the following solids of revolution. (*Note:* in each case, by symmetry $\bar{y} = 0$),
(i) the solid formed by rotating the area below $y^2 = 8x$, between $x = 1$ and $x = 6$, through 360° about the x-axis.
(ii) the solid formed by rotating the area below $y = 4/x$ between $x = 2$ and $x = 5$ through 360° about the x-axis.

10. Determine the position of the centre of mass of a frustum of a uniform cone, if the altitude of the frustum is 10 in., the end radii 5 in. and 8 in.

11. Show that the centroid of a semicircular area of radius r is at a distance of $4r/(3\pi)$ from the bounding diameter.

12. Show, by integration, that the centre of mass of a uniform solid hemisphere of radius r is at a distance $3r/8$ from the plane base.

13. Tabulate the numerical values of the function $\sqrt{(4x + 1)}$ for $x = 2$ to 6 in steps of 0·5. Hence, using Simpson's rule evaluate $\int_2^6 \sqrt{(4x + 1)}\, dx$ approximately correct to 3 places of decimals. By integrating exactly, find the error in using the approximation.

14. Using 10 equal intervals, evaluate the integral $\int_0^5 \dfrac{dx}{\sqrt{(4 + x^2)}}$ approximately using Simpson's rule, correct to 3 decimal places.

15. Evaluate $\int_0^{\pi/2} \sqrt{(1 + \cos\theta)}\, d\theta$ to 3 significant figures, using 6 equal intervals in θ. [*Note:* h in the formula must be in radians.]

22

Moments of Inertia—Theorems of Guldin

22.1 Moments of Inertia

Consider the motion of a rigid body about an axis OX, i.e. assume the body to be rotating about OX with angular velocity of ω rad/second (Figure 22.1).

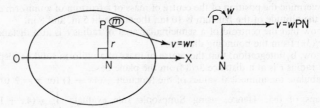

FIGURE 22.1

Let m be the mass of a small element of the body at a point P, where PN is the perpendicular distance of the point P from the axis of rotation (OX). The velocity of the element of mass $= v = r\omega$, perpendicular to PN. The *kinetic energy* of the small mass m is equal to $\frac{1}{2}mv^2 = \frac{1}{2}m(r\omega)^2 = (\omega^2/2)mr^2$, hence the *total kinetic energy* of the body may be represented by $E = \Sigma(\omega^2/2)mr^2$, the sum being made for *all* points P and associated elements of mass.

$$E = \frac{\omega^2}{2} \Sigma mr^2$$

244

The expression Σmr^2 for a body with respect to a given axis is called the *Moment of Inertia* of the body about that axis. It may be denoted by I_{OX} and if the *total mass* of the body is M, I_{OX} may be put in the form $I_{OX} = MK^2$, where K is called the *radius of gyration* of the body about the axis OX. Effectively K is the *R.M.S. distance* (perpendicular) of *all* points of the body from the axis.

By the use of integration it is possible to obtain *moments of inertia* about stated axes (of given bodies). For irregular shapes *approximate* methods may be used to *estimate* moments of inertia.

Two useful theorems which are of great use for finding moments of inertia, when certain moments of inertia are known, will be established at this stage.

22.1.1 Perpendicular axes theorem

This is only applicable to *plane areas* or thin sheets (laminae) (Figure 22.2).

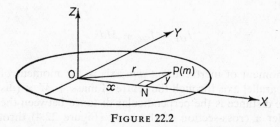

FIGURE 22.2

Let OZ be an axis perpendicular to a lamina, OX, OY two perpendicular axes lying in the plane of the lamina. If P is a point of the lamina and m an element of mass of the area at P, distance OP from the axis OZ, where OP = r, then $I_{OZ} = \Sigma mr^2$. From the triangle OPN, $r^2 = x^2 + y^2$, hence

$$I_{OZ} = \Sigma mr^2 = \Sigma m(x^2 + y^2) = \Sigma mx^2 + \Sigma my^2$$
$$\therefore \quad \underline{I_{OZ} = I_{OY} + I_{OX}}$$

i.e. the *moment of inertia* of a lamina about an axis OZ perpendicular to the lamina = sum of the moments of inertia about two perpendicular axes OX, OY in the plane of the lamina.

22.1.2 Parallel axes theorem

This theorem can be applied to *any* rigid body or *system* of rigid bodies (plane or solid) and is the most useful of the theorems.

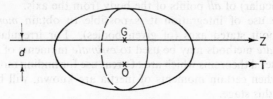

FIGURE 22.3

Let GZ, OT be two axes in the body which are *parallel* (G being the centre of mass) [Figure 22.3]. Let d be the perpendicular distance between the two axes. If the total mass of the body is M the theorem states that

$$I_{OT} = I_{GZ} + Md^2$$

i.e. the moment of inertia about *any* axis = the moment of inertia about a parallel axis through the centre of mass + $M \times$ (distance)^2, where the distance is the perpendicular distance between the axes.

Consider a cross-section of the body (Figure 22.4) through the

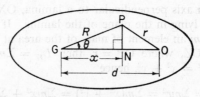

FIGURE 22.4

centre of mass G. Let the axis OT meet the plane section in O and any thin line element of mass m parallel to both axes meet the section in P.

Let the element of mass be m at distance R from GZ and r from OT. Let OG be the distance between the two axes (d) and PN be perpendicular to OG, GN $= x$ and angle PGN $= \theta$.

$$r^2 = R^2 + d^2 - 2dR \cos \theta = R^2 + d^2 - 2dx$$
$$I_{OT} = \Sigma m r^2$$
$$I_{GZ} = \Sigma m R^2$$
$$I_{OT} = \Sigma m r^2 = \Sigma m (R^2 + d^2 - 2dx)$$
$$= \Sigma m R^2 + \Sigma m d^2 - 2\Sigma m dx$$
$$= \Sigma m R^2 + d^2 \Sigma m - 2d\Sigma m x$$
$$= I_{GZ} + d^2 M - 2d\Sigma m x \quad (M = \Sigma m)$$

But $\Sigma m x = M\bar{x}$, where $\bar{x}$ is the distance of the centre of mass from G $= 0$,

$$\therefore \qquad \Sigma m x = M \times 0 = 0$$

Hence

$$\underline{I_{OT} = I_{GZ} + Md^2}$$

(*Notes:* (i) The important *facts* to remember are the results in 22.1.1 and 22.1.2, (ii) when elements of *area* or *volume* are used the results obtained are *second moments of area* or *second moments of volume.*)

Examples:

(i) A thin uniform rod of total mass M has a length of $2l$. Find the moment of inertia of the rod about axes perpendicular to the rod through (a) the centre, (b) one end.

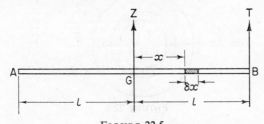

FIGURE 22.5

Let GZ (Figure 22.5) be the axis through the centre of the rod perpendicular to the rod. Let δx be a small element of length of the rod at a distance x from G. If m = mass/unit length ($m = M/2l$), then the element of mass = $m\delta x$.

Moment of inertia of the element about GZ

$$= \text{mass} \times (\text{distance})^2 = m\delta x \times x^2$$

Let this be δI, then

$$\delta I = mx^2\delta x$$

∴   Total moment of inertia about GZ = $\displaystyle\int_{-l}^{+l} mx^2 \, dx$

or by symmetry

$$I_{GZ} = 2 \int_0^l mx^2 \, dx$$

$$I_{GZ} = 2m \left[\frac{1}{3}x^3\right]_0^l = \frac{2m}{3} l^3 = 2ml \times \frac{l^2}{3} = \frac{Ml^2}{3}$$

i.e. $\underline{I_{GZ} = Ml^2/3}$ ($K_G{}^2 = l^2/3$)

Let BT be parallel to GZ, then by the parallel axes theorem

$$I_{BT} = I_{GZ} + Ml^2 = Ml^2/3 + Ml^2 = \underline{4Ml^2/3}$$

(ii) Determine the polar moment of inertia (i.e. the moment of inertia about an axis perpendicular to the plane) of a circular disc of mass M and radius r about the centre. Deduce the moment of inertia about (a) a diameter, (b) a tangential axis, (c) an axis through the edge perpendicular to the plane.

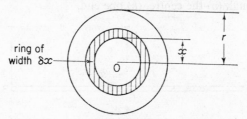

ring of
width δx

FIGURE 22.6

Let the mass/unit area of the disc be m (Figure 22.6). Consider a thin elementary ring of radius x, width δx:

Mass of ring = $2\pi x m \delta x$.

Moment of inertia δI of the ring about O $= (2\pi x m \delta x) \times x^2$, i.e.

$$\delta I = 2\pi m x^3 \delta x$$

$\therefore \quad I_O = \int_0^r 2\pi m x^3 \, dx = 2\pi m \int_0^r x^3 \, dx$

$$= 2\pi m \left[\frac{x^4}{4}\right]_0^r = 2\pi m \frac{1}{4} r^4 = (\pi r^2 m)\frac{r^2}{2}$$

But $M =$ area of disc $\times$ mass/unit area $= \pi r^2 m$

$\therefore \qquad I_O = \dfrac{Mr^2}{2} = \dfrac{Md^2}{8} \quad (d = \text{diameter})$

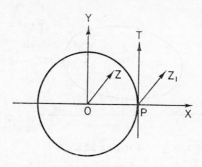

FIGURE 22.7

(a) Let OX, OY be two diameters at right angles. By symmetry $I_{OX} = I_{OY} = I_{DIA}$ (Figure 22.7). By the perpendicular axes theorem:

$$I_{OZ} = I_{OX} + I_{OY} = 2I_{DIA}$$

$\therefore \qquad I_{DIA} = \tfrac{1}{2}I_{OZ} = \tfrac{1}{2}M\dfrac{r^2}{2} = \dfrac{Mr^2}{4}$

(b) Let PT be a tangential axis (parallel to OY). By the parallel axes theorem:

$$I_{PT} = I_{OY} + M(OP)^2 = \frac{Mr^2}{4} + Mr^2 = \frac{5}{4}Mr^2$$

(c) Let PZ_1 be an axis through the perimeter perpendicular to the plane (parallel to OZ). By the parallel axes theorem:

$$I_{PZ_1} = I_{OZ} + M(OP)^2 = \frac{Mr^2}{2} + Mr^2 = \frac{3}{2} Mr^2$$

22.2.1 Theorem of Guldin (or Pappus) (1)

Theorem: If a plane area be rotated about an axis in its own plane, through 2π radians or $360°$, then the volume swept out by the area = area × path of centroid of area.

Proof of Theorem (Figure 22.8)

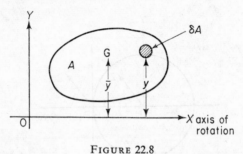

FIGURE 22.8

Consider a small element of area δA at a distance y from OX. When the area is rotated about OX through $360°$, the volume swept out by $\delta A = 2\pi y \delta A$ (approx.),

∴ Total volume swept out by the area

$$= \lim_{\delta A \to 0} \Sigma 2\pi y \delta A = 2\pi \int y \, dA$$

$$= 2\pi \times \text{first moment of area about OX}$$

$$= 2\pi A \times \bar{y} \; (\bar{y} \text{ is the distance of the centroid G from OX})$$

$$= A \times (2\pi \bar{y})$$

$$= \text{Area} \times \text{length of path of centroid of area}$$

22.2.2 Theorem of Guldin (or Pappus) (2)

Theorem: If a plane arc S be rotated about an axis in its own plane through 2π radians or $360°$, then the area swept out by the arc = length of arc × path of centroid of arc.
Proof of Theorem (Figure 22.9)

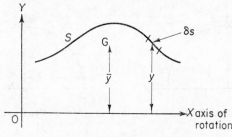

FIGURE 22.9

Consider a small element of arc δs at a distance y from OX. When the arc is rotated about OX through 2π radians or $360°$ the area swept out by $\delta s = 2\pi y \delta s$ (approx.).

∴   Total area swept out by the arc

$$= \lim_{\delta s \to 0} \Sigma 2\pi y \delta s = 2\pi \int y ds$$

$$= 2\pi \times \text{first moment of arc about OX}$$
$$= 2\pi \times S \times \bar{y} \; (\bar{y} \text{ is the distance of the centroid G from OX})$$
$$= S \times (2\pi\bar{y})$$
$$= \text{length of arc} \times \text{length of path of centroid of arc}$$

Examples:

(i) Determine the positions of the centroids of (*a*) a semicircular area of radius r, (*b*) a semicircular arc of radius r.

(*a*) By symmetry the centroid G of the area must lie on the central radius CA (Figure 22.10) at $\bar{y}$ from OX. Let the area be rotated about OX (diameter) through $360°$.
Volume swept out (sphere) $= 4\pi r^3/3$.
Area of semicircle $= \frac{1}{2}\pi r^2$.

By Guldin's theorem (1),

$$\text{Area} \times 2\pi\bar{y} = \text{Volume}$$

$$\tfrac{1}{2}\pi r^2 \times 2\pi\bar{y} = \frac{4}{3}\pi r^3; \quad \therefore \ \bar{y} = \frac{4r}{3\pi}$$

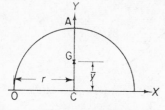

FIGURE 22.10   FIGURE 22.11

(*b*) By symmetry the centroid G_1 of the arc lies on the central radius C_1B (Figure 22.11) at $\bar{y}_1$ from O_1X_1. Let the arc be rotated through 360° about O_1X_1.

Area swept out (surface of a sphere) $= 4\pi r^2$.

Length of arc of semicircle $= \pi r$.

By Guldin's theorem (2),

$$\text{Arc length} \times 2\pi\bar{y}_1 = \text{Area}$$

$$\therefore \qquad \pi r \times 2\pi\bar{y}_1 = 4\pi r^2, \quad \therefore \ \ \bar{y}_1 = \frac{2r}{\pi}$$

(ii) Find, by integration the co-ordinates of the centroid of the area between the curve $y^2 = 4x$, the x-axis and the ordinates $x = 0$ and $x = 4$. Deduce the volume swept out by this area when it is rotated through 360° about (*a*) the axis of x, (*b*) the axis of y.

Let $(\bar{x}, \bar{y})$ be the co-ordinates of the centroid of the area (Figure 22.12), then $\bar{x}$ is given by

$$\bar{x} \times \int y\,\mathrm{d}x = \int xy\,\mathrm{d}x$$

$$\therefore \qquad \bar{x} \times \int_0^4 2\sqrt{x}\,\mathrm{d}x = \int_0^4 2x\sqrt{x}\,\mathrm{d}x$$

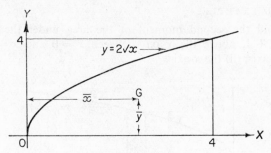

FIGURE 22.12

$$\bar{x} \times \int_0^4 2x^{\frac{1}{2}} \, \mathrm{d}x = \int_0^4 2x^{\frac{3}{2}} \, \mathrm{d}x$$

$$\bar{x} \times \frac{4}{3} \left[x^{\frac{3}{2}} \right]_0^4 = \frac{4}{5} \left[x^{\frac{5}{2}} \right]_0^4$$

$$\bar{x} \times \tfrac{4}{3} \times 8 = \tfrac{4}{5} \times 32$$

$$\bar{x} = \tfrac{3}{5} \times 4 = \underline{2{\cdot}4 \text{ unit}}$$

and $\bar{y}$ is given by

$$\bar{y} \times \int y \, \mathrm{d}x = \int \tfrac{1}{2} y^2 \, \mathrm{d}x$$

$$\bar{y} \times \int_0^4 2x^{\frac{1}{2}} \, \mathrm{d}x = \int_0^4 \tfrac{1}{2}(4x) \, \mathrm{d}x = \int_0^4 2x \, \mathrm{d}x$$

$$\therefore \qquad \bar{y} \times \frac{32}{3} = \left[x^2 \right]_0^4 = 16; \quad \bar{y} = \frac{3}{2} = \underline{1{\cdot}5 \text{ unit}}$$

$$\text{Area} = 32/3 \text{ unit}^2$$

$\therefore$ (a) Volume swept out when the area rotates through 360° about OX

$$= \text{Area} \times 2\pi\bar{y} = \frac{32}{3} \times 2\pi \times 1{\cdot}5 = \underline{32\pi \text{ unit}^3}$$

(b) Volume swept out when the area rotates through 360° about OY

$$= \text{Area} \times 2\pi\bar{x} = \frac{32}{3} \times 2\pi \times 2{\cdot}4 = \underline{51{\cdot}2\pi \text{ unit}^3}$$

WORKED EXAMPLES

(a) Find the second moment of the area under the graph of $y = \sqrt{x} + 1$, above the x-axis between $x = 0$ and $x = 4$ about (i) the y-axis, (ii) the x-axis.

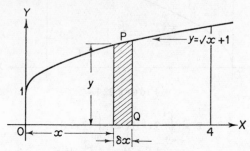

FIGURE 22.13

Consider an element PQ of area $y\delta x$ (Figure 22.13).
Second moment of area about OY $= x^2 \times y\delta x = x^2 y\delta x$.
Second moment of area about OX $= (y^2/3) \times y\delta x = \frac{1}{3}y^3\delta x$.

(i) Total second moment of area about OY

$$= \int_0^4 x^2 y \, dx = \int_0^4 x^2(\sqrt{x} + 1) \, dx = \int_0^4 (x^{\frac{5}{2}} + x^2) \, dx$$

$$= \left[\frac{2}{7}x^{\frac{7}{2}} + \frac{1}{3}x^3\right]_0^4 = \frac{2}{7} \times 4^{\frac{7}{2}} + \frac{1}{3} \times 4^3$$

$$= (2/7) \times 128 + \frac{1}{3} \times 64$$

$$= 64[4/7 + \tfrac{1}{3}] = 64 \times 19/21$$

$$= \underline{57 \cdot 92 \text{ unit}^4}$$

(ii) Total second moment of area about OX

$$= \int_0^4 \tfrac{1}{3}y^3 \, dx = \int_0^4 \tfrac{1}{3}(\sqrt{x} + 1)^3 \, dx$$

$$= \int_0^4 \tfrac{1}{3}(x^{\frac{3}{2}} + 3x + 3x^{\frac{1}{2}} + 1) \, dx$$

$$= \frac{1}{3}\left[\frac{2}{5}x^{\frac{5}{2}} + \frac{3}{2}x^2 + 2x^{\frac{3}{2}} + x\right]_0^4$$

$$= \tfrac{1}{3}[2 \times 4^{5/2}/5 + 3 \times 4^2/2 + 2 \times 4^{3/2} + 4]$$
$$= \tfrac{1}{3}[2 \times 32/5 + 24 + 16 + 4] = \tfrac{1}{3}(12\cdot8 + 44)$$
$$= \underline{18\cdot93 \text{ unit}^4}$$

(b) A cylindrical bar of steel of length 6 in. and diameter 2 in. has two grooves turned symmetrically round the curved surface. The section of one groove is a semicircle of radius $\tfrac{1}{2}$ in. and the section of the other groove is a triangle of base 1 in. and depth $\tfrac{1}{2}$ in. Determine the weight of the bar after the grooves have been cut if the metal weighs 0·28 lbf/in³. Calculate also the total surface area of the solid (Figure 22.14).

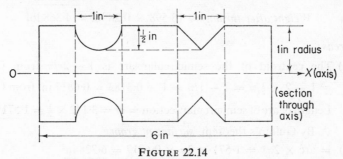

FIGURE 22.14

Let Figure 22.14 be an axial section of the solid after turning. The volumes and surface areas of the grooves may be obtained using the theorems of Guldin.

Volumes:

(i) The volume of the bar before turning

$$= \pi R^2 h = \pi \times 1^2 \times 6 = 6\pi = \underline{18\cdot852 \text{ in}^3}$$

(ii) The centroid of the semicircular section is $1 - 4r/3\pi$ from the axis OX $= 1 - 4 \times \tfrac{1}{2}/(3\pi) = 1 - 2/(3\pi) = 0\cdot788$ in

Area of semicircular section $= \tfrac{1}{2}\pi r^2 = \tfrac{1}{2}\pi(\tfrac{1}{2})^2 = \underline{\tfrac{1}{8}\pi \text{ in}^2}$

∴ By Guldin's theorem, *volume of groove*

$$= \text{Area} \times 2\pi\bar{y} = \tfrac{1}{8}\pi \times 2\pi \times 0\cdot788 \text{ in}^3$$
$$= \tfrac{1}{4} \times \pi^2 \times 0\cdot788 = \tfrac{1}{4}(3\cdot142)^2 \times 0\cdot788 \text{ in}^3$$
$$= \underline{1\cdot945 \text{ in.}^3}$$

(iii) The centroid of the triangular section is $1 - \frac{1}{3} \times h_1 = 1 - \frac{1}{3} \times \frac{1}{2}$ in. from OX, i.e. 5/6 in from OX,

$$\text{Area of triangular section} = \tfrac{1}{2}bh_1 = \tfrac{1}{2} \times 1 \times \tfrac{1}{2} = \underline{\tfrac{1}{4} \text{ in.}^2}$$

∴ By Guldin's theorem, *volume of groove*

$= \text{Area} \times 2\pi\bar{y} = \tfrac{1}{4} \times 2\pi \times 5/6 \text{ in.}^3 = 5\pi/12 \text{ in}^3$

$= 1 \cdot 309 \text{ in}^3$

∴ *Volume of metal left* $= 18 \cdot 852 - (1 \cdot 945 + 1 \cdot 309) \text{ in}^3$

$\qquad\qquad\qquad\qquad = 18 \cdot 852 - 3 \cdot 254 = \underline{15 \cdot 598 \text{ in}^3}$

∴ *Weight after turning* $= 15 \cdot 598 \times 0 \cdot 28 \text{ lbf} = \underline{4 \cdot 368 \text{ lbf}}$

Areas:

(i) The centroid of the semicircular arc is $1 - 2r/\pi$ from OX

$= 1 - 2 \times \tfrac{1}{2}/\pi = 1 - 1/\pi = 1 - 0 \cdot 3183 = 0 \cdot 6817$ in. from OX

Length of arc of semicircular section $= \pi r = 3 \cdot 142 \times \tfrac{1}{2} = \underline{1 \cdot 571 \text{ in.}}$

∴ By Guldin's theorem, *surface of groove*

$= \text{arc} \times 2\pi\bar{y} = 1 \cdot 571 \times 2\pi \times 0 \cdot 6817 = \underline{6 \cdot 728 \text{ in}^2}$

(ii) The centroid of the two sides of the triangle is $1 - \tfrac{1}{2} \times \tfrac{1}{2} = \tfrac{3}{4}$ in. from OX,

length of the two sides $= 2 \times \tfrac{1}{2}\sqrt{2} = \sqrt{2} = \underline{1 \cdot 414 \text{ in.}}$

∴ By Guldin's theorem, *surface of groove*

$= \text{arc} \times 2\pi\bar{y} = 1 \cdot 414 \times 2\pi \times \tfrac{3}{4} = 2 \cdot 121\pi = \underline{6 \cdot 664 \text{ in}^2}$

∴ *Total area of cylinder left*

$= 2\pi R^2 + 2\pi R \times l \qquad [l = 6 - 2 = 4 \text{ in.}]$

$= 2\pi \times 1^2 + 2\pi \times 1 \times 4 = 10\pi = \underline{31 \cdot 42 \text{ in}^2}$

∴ *Total surface of solid after turning*

$= 31 \cdot 420 + 6 \cdot 728 + 6 \cdot 664 = 44 \cdot 812 \simeq \underline{44 \cdot 81 \text{ in}^2}$

Results: (i) Weight after turning $= \underline{4 \cdot 368 \text{ lbf}}$

(ii) Surface area after turning $= \underline{44 \cdot 81 \text{ in}^2}$

(c) Using integration determine the moment of inertia of a uniform solid cone of density 0·26 lb/in³, height 8 in., and base radius 3 in. about (i) the axis of symmetry (the altitude), (ii) an axis through the vertex parallel to the base.

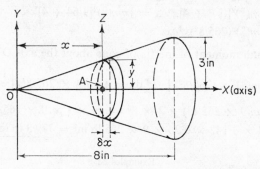

FIGURE 22.15

(i) Consider a circular disc of radius y and thickness δx (Figure 22.15) at x from O (the vertex). Let m = mass/unit volume = 0·26 lb/in³.

Moment of inertia of the disc about OX = mass × (radius)²/2 = $\pi y^2 \delta x \times m \times y^2/2$, i.e. $\delta I = \frac{1}{2}m\pi y^4 \delta x$.

Now $y/x = 3/8$,

$$\therefore \qquad\qquad\qquad y = \tfrac{3}{8}x$$

$$\therefore \qquad\qquad \delta I = \tfrac{1}{2}m\pi(\tfrac{3}{8}x)^4\delta x = \tfrac{1}{2}m\pi(\tfrac{3}{8})^4 x^4 \delta x$$

$\therefore$ Total moment of inertia of cone about OX

$$= \int_0^8 \tfrac{1}{2}m\pi(\tfrac{3}{8})^4 x^4 \, \mathrm{d}x = \tfrac{1}{2}m\pi(\tfrac{3}{8})^4 \int_0^8 x^4 \, \mathrm{d}x = \tfrac{1}{2}m\pi(\tfrac{3}{8})^4 \left[\frac{x^5}{5}\right]_0^8$$

$$= \tfrac{1}{2}m\pi(\tfrac{3}{8})^4 \times 8^5/5 = 4\pi m \times 3^4/5 = 0\cdot8 \times 3\cdot142 \times 0\cdot26 \times 81$$

$$\doteqdot 52\cdot94 \text{ lb in}^2$$

(ii) Moment of inertia of disc about its diameter AZ

$$= \text{mass} \times (\text{radius})^2/4$$

$$\therefore \qquad\qquad \delta I = \pi y^2 \delta x m y^2/4 = (\pi m/4)y^4 \delta x$$

By the parallel axes theorem, the moment of inertia of this disc about OY

$$= I_{AZ} + \text{Mass of disc} \times (x^2) = (\pi m/4)y^4\delta x + \pi m y^2 \delta x \times x^2$$

$$= (\tfrac{1}{4}\pi m)(\tfrac{3}{8}x)^4\delta x + \pi m(\tfrac{3}{8}x)^2 x^2 \delta x$$

$$= \tfrac{1}{4}\pi m(\tfrac{3}{8})^2[(\tfrac{3}{8})^2 + 4]x^4\delta x = \tfrac{1}{4}\pi m(\tfrac{3}{8})^2[9/64 + 4]x^4\delta x$$

$$= \tfrac{1}{4}\pi m(\tfrac{3}{8})^2(265/8^2)x^4\delta x$$

∴ Total moment of inertia of cone about the axis OY

$$= \tfrac{1}{4}\pi m(\tfrac{3}{8})^2 \times (265/8^2)\int_0^8 x^4 \, dx = \tfrac{1}{4}\pi m(9 \times 265/8^4)\left[\tfrac{1}{5}x^5\right]_0^8$$

$$= \tfrac{1}{4}\pi m(9 \times 265/8^4) \times \tfrac{1}{5} \times 8^5 = \pi m \times \tfrac{2}{5} \times 9 \times 265 \text{ lb in}^2$$

$$= 0.4 \times 3.142 \times 0.26 \times 9 \times 265 \text{ lb in}^2 = \underline{779.3 \text{ lb in}^2}$$

Examples 22

1. A rectangular plate has edges of lengths $2a$ and $2b$ and mass M. Find the moment of inertia of the plate about: (i) an edge of length $2a$, (ii) an edge of length $2b$, (iii) an axis through the centre of the plate parallel to an edge of length $2a$, (iv) an axis through the centre of the plate parallel to an edge of length $2b$, (v) the polar axis of the plate through the centre, (vi) an axis through a corner perpendicular to the plate.

2. A lamina consists of a square plate of side 4 in. with a semicircle on one edge as diameter. The material is of uniform thickness and the mass/unit area may be taken as 0.02 lb/in.^{2} Determine the moment of inertia about the diameter of the semicircle.

3. A solid steel shaft consists of 3 cylindrical portions. The central portion has a length of 5 in. and a radius of $1\tfrac{1}{2}$ in., the two end portions are each of length 10 in. and radius 1 in. If the steel has a density of 0.28 lb/in.^{3}, calculate the moment of inertia of the shaft about its principal axis in lb in.^{2}

4. A lamina is in the shape of an annulus of a circle (washer shape) of mass M and internal and external radii r and R. Find the polar moment of inertia of the lamina.

5. A hollow steel shaft of internal diameter 3 in. and external diameter 4 in. has a length of 12 in. and a density of 0.28 lb/in.^{3} Find the moment of inertia of the shaft about its principal axis.

6. Find, by integration, the moment of inertia of a triangular plate of sides 10 in., 10 in. and 12 in. about an axis parallel to the longest side through the opposite vertex if the plate has a mass of 0.03 lb/in.^{2}

7. Determine the second moments of area of the area between the curve $y = 4\sqrt{x}$, the x-axis and the ordinates $x = 0$ and $x = 4$ (the units of both scales are in inches), about the axes of x and y. Find the radius of gyration about each axis.

8. Determine, by integration the second moment of area of the area below the curve $xy = 5$ and above the x-axis between the ordinates $x = 1$ and $x = 4$. Find the radius of gyration of the area about the x and y axes if the units are in centimetres.

9. Find the second moments of area about the x and y axes of the area between the curve $y = 4 - x^2$ and the x-axis (above the axis) and also the radius of gyration of the area about both the x and y axes.

10. Find, by integration the moment of inertia of a lamina in the shape of an equilateral triangle of side 10 in. (the mass being 0·025 lb/in.²) about an axis through a vertex parallel to the opposite side. Deduce (i) the moment of inertia about an axis through the centre of mass parallel to a side, (ii) the moment of inertia about a side.

11. Find the volume and surface area of an anchor ring if its cross-section is a circle of radius 6 in. and the centre of the section is 12 in. from the axis of rotation.

12. Using integration determine the volume of revolution when the area below the curve $y = x + 1/x$, between $x = 1$ and $x = 4$ is rotated through 360° about the x-axis. Hence, using Guldin's theorem find the y co-ordinate of the centroid of the area.

13. A uniform solid cylinder of diameter 3·5 in. and height $\frac{3}{4}$ in. is to be converted into a pulley wheel by cutting a groove in the curved surface. The cross-section of the groove is to be a semicircle of diameter $\frac{1}{2}$ in. Calculate the weight of the pulley if the metal weighs 0·26 lbf/in.³

14. A frustum of a cone has an altitude of 10 in. and end radii of 4 in. and 8 in. The density of the material is 0·23 lb/in.³ Determine, by integration the moment of inertia of the frustum about its principal axis.

15. Determine, by integration, the co-ordinates of the centroid of the area between the curve $y = 3x^2 + 1$, the x-axis and the ordinates $x = 1$ and $x = 3$.

    Using Guldin's theorem deduce the volumes of the solids formed when the area is rotated through 2π radians about (a) the axis of x (b) the axis of y.

23

Elementary Differential Equations—
First and Second Order—
Harmonic Equation

23.1 Differential Equations—Description

In many problems involving the application of mathematics it is often necessary to form and solve a *differential equation*.

Basically a differential equation may be described as an equation which involves the *independent variable* (*x*), the *dependent variable* (*y*), and the derivatives of the dependent variable with respect to the independent variable *x*. In some cases the derivatives may be raised to powers.

23.1.1 Formation of a differential equation

By suitable differentiation of a given function it may be possible to obtain an equation connecting the ordinary derivatives of *y* with respect to *x*, *y*, and *x*. The result is called an *ordinary differential equation*. The coefficients involved may be *constants* or *functions of the independent variable*.

Examples:

(i) If $y = mx + c$, where m and c are constants, $dy/dx = m$ (constant). This is a simple *first order differential equation*.

$\dfrac{d^2y}{dx^2} = 0$. This is a simple *second order differential equation*.

(ii) If $y = A\,e^{2x}$, find, by differentiation and elimination of A, a first order differential equation for y.

$$\frac{dy}{dx} = A \times 2\,e^{2x} = 2(A\,e^{2x}) = 2y$$

i.e. y satisfies the equation

$$\frac{dy}{dx} - 2y = 0$$

(iii) If $s = A\cos 2t + B\sin 2t$, show that s satisfies the differential equation $d^2s/dt^2 + 4s = 0$.

$$\frac{ds}{dt} = -2A\sin 2t + 2B\cos 2t$$

$$\frac{d^2s}{dt^2} = -4A\cos 2t - 4B\sin 2t = -4(A\cos 2t + B\sin 2t)$$

$$= -4s$$

$\therefore$ s satisfies the second order differential equation

$$\frac{d^2s}{dt^2} + 4s = 0$$

23.1.2 The order of a differential equation

The *order* of a differential equation is determined by the highest *order* of derivative present.

Illustrations:

(i) $\dfrac{dy}{dx} + x^2y = 2e^x$ is a *first order* equation.

(ii) $\dfrac{d^2y}{dx^2} + 2\dfrac{dy}{dx} + 3y = \cos 2x$ is a *second order* equation.

(iii) $\dfrac{d^3s}{dt^3} + 3\dfrac{ds}{dt} + 4s = e^{2t}$ is a *third order* equation.

23.1.3 The degree of a differential equation

The *degree* of a differential equation is determined by the highest *power* of the highest order derivative.

Illustrations:

(i) $\left(\dfrac{dy}{dx}\right)^2 + 2\dfrac{dy}{dx} + 3y = x^3$ is a *first order* equation of the *second degree*.

(ii) $\left(\dfrac{d^2s}{dt^2}\right)^3 + \dfrac{ds}{dt} = te^{2t}$ is a *second order* equation of the *third degree*.

23.2 Solution of Differential Equations

If a differential equation be given, the equation is solved if the *dependent variable* is found as a function of the independent variable (either as an *explicit* or *implicit* function). There are many types of differential equations and it would be impossible, at this stage, to deal with many methods of solution. However, some simple methods of solution will be illustrated.

23.2.1 Methods of solution

Basically the methods depend on either (i) integration or (ii) substitution. In more difficult cases certain advanced techniques using *operators* need to be used. These will not be dealt with at this stage. Some methods will be illustrated by examples. Only theory so far dealt with in the text will be used.

Examples:

(i) Solve the equation $dy/dx = 3x^2 + x$, given $y = 5$ when $x = 1$.

This is solved by direct integration as follows:

$$y = \int (3x^2 + x)\, dx = x^3 + \tfrac{1}{2}x^2 + A$$

A being an *arbitrary constant*.

Since $y = 5$ when $x = 1$,

$$\therefore \qquad\qquad 5 = 1 + \tfrac{1}{2} + A; \quad \underline{A = 3\tfrac{1}{2}}$$

∴ the *solution* of the equation is:

$$y = x^3 + \tfrac{1}{2}x^2 + 3\tfrac{1}{2} = \tfrac{1}{2}(2x^3 + x^2 + 7)$$

(ii) Find the solution of the differential equation $d^2s/dt^2 = 2t + 1$, if $s = 4$, $ds/dt = 1$ when $t = 1$.

This is solved by integrating in two stages:

$$\frac{ds}{dt} = \int (2t + 1)\, dt = t^2 + t + A$$

and

$$s = \int (t^2 + t + A)\, dt = \tfrac{1}{3}t^3 + \tfrac{1}{2}t^2 + At + B$$

but $ds/dt = 1$ when $t = 1$,

∴ $\qquad\qquad 1 = 1 + 1 + A, \quad \underline{A = -1}$

and $s = 4$ when $t = 1$,

∴ $\qquad\qquad 4 = \tfrac{1}{3} + \tfrac{1}{2} + A + B$

∴ $\qquad B = 4 - \tfrac{5}{6} - A = 4 - \tfrac{5}{6} + 1 = 4\tfrac{1}{6} = \underline{25/6}$

Hence the *solution* of the equation is

$$s = \tfrac{1}{3}t^3 + \tfrac{1}{2}t^2 - t + 25/6$$

or $\qquad \underline{s = \tfrac{1}{6}[2t^3 + 3t^2 - 6t + 25]}$

(*Notes:* (1) Simple types like (i) and (ii) have, in effect, already been met with as examples of integration. (2) When a *first order* equation is solved *one arbitrary constant* is involved. (3) When a *second order* equation is solved, *two arbitrary constants* are involved.)

It will be seen that the number of arbitrary constants involved in solving any differential equation is equal to the *order* of the equation.

(iii) By a trial *substitution* $y = C\,e^{mx}$, where C and m are constants, find a possible solution of the differential equation $d^2y/dx^2 - 5(dy/dx) + 6y = 0$.

If $y = Ce^{mx}$, $dy/dx = mCe^{mx}$, $d^2y/dx^2 = m^2C\,e^{mx}$. Hence, if this form of y satisfies the differential equation then

$$m^2C\,e^{mx} - 5mC\,e^{mx} + 6C\,e^{mx} = 0$$

or $\qquad\qquad C\,e^{mx}(m^2 - 5m + 6) = 0$

Assuming $C \neq 0$ then m must satisfy

$$m^2 - 5m + 6 = 0 \quad \text{or} \quad (m - 2)(m - 3) = 0$$

i.e. $\underline{m = +2 \text{ or } m = +3}$ are two possible values of m. With each value of m a different constant may be used, i.e. the general solution of the equation may be written as

$$y = A\,e^{2x} + B\,e^{3x}$$

where A and B are arbitrary constants.

23.3 Method of Separation of Variables

By introducing the idea of '*differentials*', if a *first order* equation can be reduced to a form like

$$f(x)\,dx = g(y)\,dy$$

i.e. if the *variables* can be *separated* then the equation may be solved by direct integration of each side.

Examples:

(i) Find the general solution of the equation $x^3(dy/dx) = y + 2$.
    Separating the variables gives

$$\frac{dy}{y + 2} = \frac{dx}{x^3}$$

$$\therefore \qquad \int \frac{dy}{y + 2} = \int \frac{dx}{x^3}$$

$$\log_e (y + 2) = -\frac{1}{2x^2} + C$$

$$y + 2 = e^{-1/(2x^2)+C} = e^C\, e^{-1/(2x^2)}$$

Since C is a constant e^C may be written as A (a constant),

$$\therefore \qquad y + 2 = A\,e^{-1/(2x^2)} \quad \text{or} \quad \underline{y = A\,e^{-1/(2x^2)} - 2}$$

(ii) Solve the equation $v(\mathrm{d}v/\mathrm{d}x) = g - Kv^2$.

Separating the variables gives

$$\frac{v\,\mathrm{d}v}{g - Kv^2} = \mathrm{d}x$$

$$\therefore \quad \int \frac{v\,\mathrm{d}v}{g - Kv^2} = \int \mathrm{d}x; \quad \frac{1}{2}\int \frac{\mathrm{d}(v^2)}{g - Kv^2} = x + C$$

$$-\frac{1}{2K}\log_e(g - Kv^2) = x + C$$

$$\log_e(g - Kv^2) = -2Kx - 2KC$$

$$\therefore \quad g - Kv^2 = \mathrm{e}^{-2Kx-2KC} = \mathrm{e}^{-2KC} \times \mathrm{e}^{-2Kx}$$

Since KC is a constant, e^{-2KC} may be written as A (a constant),

$$\therefore \quad g - Kv^2 = A\,\mathrm{e}^{-2Kx}, \quad v^2 = (g - A\,\mathrm{e}^{-2Kx})/K$$

or

$$v = \sqrt{[(g - A\,\mathrm{e}^{-2Kx})/K]}$$

23.4 The Simple Harmonic Equation

This is a rather special equation and is the final equation to be dealt with. It arises from the motion of a body acted on by a *restoring force proportional to the displacement.*

If s is the distance of a particle from a fixed point O, then the acceleration at time t is given by $f = \mathrm{d}^2s/\mathrm{d}t^2$. If the force acting on the particle is proportional to the displacement and directly towards O then the acceleration may be written as $f = -\omega^2 s$, where ω is a constant.

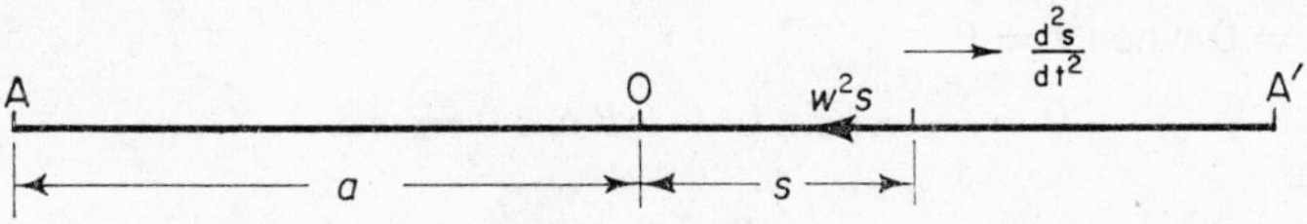

FIGURE 23.1

$\omega^2 s$ represent the force per unit mass, i.e. the equation of motion may be written as

$$\frac{\mathrm{d}^2s}{\mathrm{d}t^2} = -\omega^2 s \quad \text{or} \quad \frac{\mathrm{d}^2s}{\mathrm{d}t^2} + \omega^2 s = 0$$

This equation may be regarded as the *standard harmonic equation*. The motion is a vibration about the point O (the *neutral position* where the force is zero) between two extremes A and A' (Figure 23.1). The motion is described by this *differential equation* and its solution gives the *displacement* s in terms of t (time). The solution will be verified by *substitution*.

Assume

$$s = A \cos \omega t + B \sin \omega t$$

$$\frac{\mathrm{d}s}{\mathrm{d}t} = -\omega A \sin \omega t + \omega B \cos \omega t$$

$$\frac{\mathrm{d}^2 s}{\mathrm{d}t^2} = -\omega^2 A \cos \omega t - \omega^2 B \sin \omega t = -\omega^2 s$$

$$\therefore \qquad\qquad \frac{\mathrm{d}^2 s}{\mathrm{d}t^2} + \omega^2 s = 0$$

Hence a *general* solution of the *standard harmonic equation* may be written as $s = A \cos \omega t + B \sin \omega t$.

In fact both $s = A \cos \omega t$ and $s = B \sin \omega t$ are also separate solutions of the equation.

Solution of the equation, if $s = a$, $\mathrm{d}s/\mathrm{d}t = 0$, when $t = 0$.

$$s = A \cos \omega t + B \sin \omega t; \quad \frac{\mathrm{d}s}{\mathrm{d}t} = -\omega A \sin \omega t + \omega B \cos \omega t$$

$s = a$ when $t = 0$,

$$\therefore \qquad\qquad a = A \cos 0 + B \sin 0 = A$$

$\mathrm{d}s/\mathrm{d}t = 0$ when $t = 0$,

$$0 = -\omega A \sin 0 + \omega B \cos 0 = \omega B$$

Hence

$$B = 0$$

$\therefore$ The solution is

$$s = a \cos \omega t$$

a is called the *amplitude* of the *simple harmonic motion* (or *vibration*).

When $t = 0$, $s = a$; when $t = \pi/\omega$, $s = a \cos \pi = -a$ (at A', $s = a$ and at A, $s = -a$).

The *period* of the vibration is the time taken to perform one complete oscillation between extremes of the path and if T is the period

$$T = 2\pi/\omega$$

The *frequency* N of the vibration $= 1/T$. $N = $ number of vibrations per second, i.e.

$$N = \omega/2\pi$$

23.5 General Note

The object of this chapter has been to introduce students to the topic of *differential equations*. It is not necessary at this stage for students to concern themselves with *general methods*, but it will be useful if they learn to appreciate some of the properties of differential equations for more detailed work which will follow in Higher National Certificate courses. These equations will arise in problems of a practical nature in problems of vibrations, structures, motion under variable forces, chemical rates of reaction, rates of radiation, rates of cooling, and many others.

WORKED EXAMPLES

(*a*) Show that $s = e^t \sin t$ satisfies the equation $(d^2s/dt^2) - 2(ds/dt) + 2s = 0$.

$$\frac{ds}{dt} = e^t \sin t + e^t \cos t$$

$$\frac{d^2s}{dt^2} = e^t \sin t + e^t \cos t + e^t \cos t - e^t \sin t = 2 e^t \cos t$$

$$\therefore \frac{d^2s}{dt^2} - 2\frac{ds}{dt} + 2s = 2 e^t \cos t - 2(e^t \sin t + e^t \cos t) + 2 e^t \sin t$$

$$= 2 e^t \cos t - 2 e^t \sin t - 2e^t \cos t + 2 e^t \sin t$$

$$= 0$$

i.e.

$$\frac{d^2s}{dt^2} - 2\frac{ds}{dt} + 2s = 0$$

(*b*) Find the solution of the equation $(1 + x)(dy/dx) = x$, given $y = 3$ when $x = 0$.

$$\frac{dy}{dx} = \frac{x}{1 + x} = 1 - \frac{1}{1 + x} \quad \text{(by simple division)}$$

$$\therefore \quad y = \int \left(1 - \frac{1}{1 + x}\right) dx = x - \log_e (1 + x) + C$$

$x = 0, y = 3,$

$$\therefore \quad\quad\quad 3 = 0 - \log_e 1 + C \quad [\log_e 1 = 0]$$

$$\therefore \quad\quad\quad\quad\quad C = 3$$

$\therefore$ The solution of the equation is

$$y = x - \log_e (1 + x) + 3$$

(*c*) Given $d^2y/dx^2 = 1/x^2$, express y in terms of x if $y = 2$, $dy/dx = -1$ when $x = 1$.

Integrating once:

$$\frac{dy}{dx} = -\frac{1}{x} + A$$

Integrating again:

$$y = \int \left(-\frac{1}{x} + A\right) dx$$

$$= -\log_e x + Ax + B$$

$dy/dx = -1$ when $x = 1$; $-1 = -1 + A$,

$$\therefore \quad\quad\quad\quad A = 0$$

$y = 2$ when $x = 1$; $2 = -\log_e 1 + A + B$,

$$\therefore \quad\quad\quad\quad B = 2$$

$\therefore$ the solution of the equation is

$$y = 2 - \log_e x$$

(*d*) Using the method of *separation of variables*, solve the equation $e^{-2x}(dy/dx) = 2y + 1$, given $y = 0$ when $x = 0$.

Separating the variables:

$$\frac{dy}{2y + 1} = \frac{dx}{e^{-2x}} = e^{2x}\,dx$$

$$\therefore \quad \int \frac{dy}{2y + 1} = \int e^{2x}\,dx; \quad \tfrac{1}{2}\log_e(2y + 1) + C = \tfrac{1}{2}\,e^{2x}$$

$x = 0$ when $y = 0$,

$$\therefore \qquad \tfrac{1}{2}\log_e 1 + C = \tfrac{1}{2}\,e^0 = \tfrac{1}{2}; \quad \underline{C = \tfrac{1}{2}}$$

$$\tfrac{1}{2}\log_e(2y + 1) + \tfrac{1}{2} = \tfrac{1}{2}\,e^{2x}$$

$$\log_e(2y + 1) + 1 = e^{2x} \qquad (1 = \log_e e)$$

$$\log_e(2y + 1) + \log_e e = e^{2x}$$

$$\log_e[e(2y + 1)] = e^{2x}$$

$$\therefore \qquad e(2y + 1) = \exp(e^{2x})$$

$$\therefore \qquad \underline{y = \tfrac{1}{2}[\exp(e^{2x})/e - 1]}$$

(*e*) If $dx/dt = K(2 - x)(x - 1)$, where K is a constant, express t in terms of x.

Separating the variables:

$$\frac{dx}{(2 - x)(x - 1)} = K\,dt$$

Let

$$\frac{1}{(2 - x)(x - 1)} = \frac{A}{2 - x} + \frac{B}{x - 1}$$

Then

$$1 = A(x - 1) + B(2 - x)$$

let $x = 2$; $1 = A$: let $x = 1$; $1 = B$,

$$\therefore \qquad \int \left(\frac{1}{2 - x} + \frac{1}{x - 1}\right)dx = \int K\,dt = Kt + C$$

$$-\log_e(2 - x) + \log_e(x - 1) = Kt + C$$

$$Kt = \log_e\left(\frac{x - 1}{2 - x}\right) - C$$

$$\therefore \qquad \underline{t = \frac{1}{K}\left[\log_e\left(\frac{x - 1}{2 - x}\right) - C\right]}$$

C would depend on *initial* or *given* conditions.

If $C = -\log_e D$, then

$$t = \frac{1}{K} \log_e \left[D \left(\frac{x-1}{2-x} \right) \right]$$

(*f*) By a substitution solve the equation $(d^2s/dt^2) + 3 (ds/dt) + 2s = 0$.

Let $s = K\,e^{mt}$ (K is a constant),

$$\frac{ds}{dt} = mK\,e^{mt}, \quad \frac{d^2s}{dt^2} = m^2K\,e^{mt}$$

Then

$$m^2K\,e^{mt} + 3mK\,e^{mt} + 2K\,e^{mt} = 0$$
$$K\,e^{mt}(m^2 + 3m + 2) = 0$$

If $K \neq 0$; $m^2 + 3m + 2 = 0$; $(m+1)(m+2) = 0$

$\therefore$ possible values of m are

$$m = -1 \quad \text{or} \quad -2$$

$\therefore$ General solution is

$$s = A\,e^{-t} + B\,e^{-2t}$$

(*g*) The equation of motion of a body projected vertically downwards is given by $dv/dt = g - Kv$ (where Kv represents the resistance to motion). If $v = u$ at $t = 0$ find an expression for the time t required for the velocity to increase to v.

Separating the variables:

$$\frac{dv}{g - Kv} = dt$$

$$\int \frac{dv}{g - Kv} = \int dt; \quad -\frac{1}{K} \log_e (g - Kv) = t + C$$

at $t = 0$, $v = u$,

$$\therefore \qquad\qquad -\frac{1}{K} \log_e (g - Ku) = C$$

$$\therefore \qquad -\frac{1}{K} \log_e (g - Kv) = t - \frac{1}{K} \log_e (g - Ku)$$

$$\therefore \quad t = \frac{1}{K} [\log_e (g - Ku) - \log_e (g - Kv)] = \frac{1}{K} \log_e \left[\frac{g - Ku}{g - Kv} \right]$$

(*h*) A spring is fixed to a point on a smooth horizontal table. At the other end a unit mass is attached. When the spring is extended it exerts a force of 4 × displacement. If the particle is pulled a distance of 5 inches from the neutral position and released, find an expression for the distance *s* at time *t* from the neutral position, i.e. at $t = 0$, $s = 5$, velocity = 0. Find also the period and maximum velocity.

The acceleration of the mass = d^2s/dt^2, restoring force $4s$, hence the equation of motion may be written as

$$\frac{d^2s}{dt^2} = -4s \quad \text{or} \quad \frac{d^2s}{dt^2} + 4s = 0$$

This is a *standard (simple) harmonic equation* ($\omega = 2$). The general solution of the equation is

$$s = A \cos 2t + B \sin 2t$$

$$\frac{ds}{dt} = -2A \sin 2t + 2B \cos 2t$$

$t = 0$, $s = 5$; $5 = A$, i.e.

$$\underline{A = 5}$$

$t = 0$, $ds/dt = 0$; $0 = 2B$, i.e.

$$\underline{B = 0}$$

The solution is

$$s = 5 \cos 2t$$

i.e. the displacement at time *t* is $5 \cos 2t$ inches from the centre of oscillation.

The period of vibration $T = 2\pi/\omega = 2\pi/2 = \underline{\pi \text{ seconds}}$.

Maximum velocity occurs when $\sin 2t = 1$, i.e. $\underline{t = \pi/2 \text{ sec}}$, and Maximum velocity = $2A = 2 \times 5 = \underline{10 \text{ in./s}}$.

Examples 23

1. (i) If $y = 4 e^x + 6 e^{-2x}$ show that *y* satisfies the differential equation $\frac{d^2y}{dx^2} + \frac{dy}{dx} - 2y = 0$.

(ii) Given $y = A e^{-x} + B e^{3x}$ where *A* and *B* are constants, show that *y* satisfies the second order equation $\frac{d^2y}{dx^2} - 2 \frac{dy}{dx} = 3y$.

2. (i) Verify that $s = a \cos t + b \sin t$, where a and b are constants satisfies the equation $\dfrac{d^2 s}{dt^2} + s = 0$.

(ii) Show that the displacement $s = A \cos 3t + B \sin 3t$ satisfies the equation $\dfrac{d^2 s}{dt^2} + 9s = 0$.

3. (i) Verify that the function $y = a\,e^{3x} + b\,e^{-3x}$ where a and b are constants satisfies the equation $\dfrac{d^2 y}{dx^2} = 9y$

(ii) By suitable differentiation and substitution verify that $\theta = 5\,e^{-t} \cos t$ satisfies the second order equation $\dfrac{d^2 \theta}{dt^2} + 2 \dfrac{d\theta}{dt} + 2\theta = 0$.

4. Show that $y = x^{1/2} + x^{-1/2}$ satisfies the homogeneous second order equation:
$4x^2 \dfrac{d^2 y}{dx^2} + 2x \dfrac{dy}{dx} - 2y = -2x^{1/2}$.

5. Show that $y = 1/\sqrt{(x^2 - 1)}$ is a solution of the equation
$(x^2 - 1)\dfrac{d^2 y}{dx^2} + 3x \dfrac{dy}{dx} + y = 0$.

6. Verify, by differentiation and substitution, that $y = (A + Bx)\,e^x$, where A and B are constants satisfies the differential equation $\dfrac{d^2 y}{dx^2} - 2 \dfrac{dy}{dx} + y = 0$.

7. Show that $y = x^3 + C/x$, where C is a constant satisfies the differential equation $\dfrac{dy}{dx} + \dfrac{y}{x} = 4x^2$.

8. By differentiation and substitution, show that $s = \cos t\,(\sin t + A)$, where A is a constant, satisfies the first order equation $\dfrac{ds}{dt} + s \tan t = \cos^2 t$.

9. Solve the differential equation $\dfrac{ds}{dt} = t^2 + t - 1$, given that $s = 5$ when $t = 1$.

10. Given the differential equation $\dfrac{d^2 y}{dx^2} = x(x - 2)$, solve it, given that $y = 3$, $\dfrac{dy}{dx} = 2$ when $x = 1$.

11. Assuming that $y = C\,e^{mx}$ is a solution of the equation $\dfrac{d^2 y}{dx^2} - 3 \dfrac{dy}{dx} - 4y = 0$, obtain the general solution.

12. Using the 'variables separable' method of solution solve the equation: $2y \dfrac{dy}{dx} = x(x - 1)$ given that $y = 4$ when $x = 1$.

13. Solve the first order equation $\dfrac{dy}{dx} = y^2 \cos x$, given that $y = 2$ when $x = 0$.

ELEMENTARY DIFFERENTIAL EQUATIONS 273

14. The equation connecting the tension T lbf in a rope wrapped round a rough cylinder, with the lap angle θ (in radians) when the coefficient of friction is μ is $\dfrac{dT}{d\theta} = \mu T$. Solve the equation to obtain T in terms of θ given that $T = T_0$ lbf when $\theta = 0$.

15. The rate of cooling of a body $-\dfrac{d\theta}{dt}$, where θ is the temperature at time t seconds in degrees Centigrade is given by $\dfrac{d\theta}{dt} = -K(\theta - 20)$ where K is a constant. Find θ at time t if initially (i.e. at $t = 0$) the temperature of the body is 100°C.

16. The charge q coulombs on a condenser of capacitance C farads during discharge through a resistance of R ohms is given at time t seconds by the equation $R\dfrac{dq}{dt} + \dfrac{q}{C} = 0$. If the initial charge on the condenser (at $t = 0$) is q_0, express q in terms of t. What is the time required for the charge to fall to 60 per cent of its initial value when $C = 5 \times 10^{-6}$ farads, $R = 2 \times 10^6$ ohms?

17. The differential equation giving the current i amp at time t seconds in an electrical circuit is given by the equation $L\dfrac{di}{dt} + Ri = E$ where E is the steady applied voltage, L the inductance (in henries) and R the resistance in ohms. Find, using the variables separable method the expression for i at time t, if initially (i.e. at $t = 0$), $i = 0$. If $E = 20$ volts, $R = 5$ ohms, $L = 4 \times 10^{-3}$ henries calculate the time required for the current to grow to 2 amp.

18. In a certain monomolecular reaction the concentration of a compound X is initially n gram molecules per litre, and at time t seconds x gram molecules per litre of X have been decomposed. The speed of the reaction $\dfrac{dx}{dt}$ at any instant is proportional to the concentration of X at that instant. Derive an expression for t in terms of n and x. Show that the time required for the concentration to fall to 40 per cent of its original value is independent of n.

19. The equation of motion of a body fired vertically upwards is given by $\dfrac{dv}{dt} = -g - Kv$, where Kv is the resistance at velocity v ft/s. Find the expression for the velocity v ft/s at time t if the velocity of projection (the initial velocity) is u ft/s. If $g = 32$ ft/s², $K = 0.1$, $u = 150$ ft/s calculate the time required for the velocity to fall to 60 ft/s.

20. A point P of a reciprocating mechanism moves in a straight line so that its distance x inches from its central position at time t seconds is given by the differential equation $\dfrac{d^2x}{dt^2} + 9x = 0$. If $x = 0$ at time $t = 0$ and the velocity is 12 in./s at $t = 0$, find expressions for (i) the displacement x inches at time t seconds, (ii) the velocity at time t seconds. What is the total travel of the point, the period of vibration and the frequency of vibration?

24

Statistics—Probability Laws— Bernoulli's Law

24.1 Statistics

This is a term originally derived from the collection of information (data) for *state* purposes. Statistics, as data, were originally collected for military purposes. In modern times the word is used in two senses:

(i) To describe the general *collection* and *analysis* of *data* of any kind (the science of *statistics*).

(ii) To describe *particular figures* (values), e.g. mean value, standard deviation, relating to a *sample* of values which are used to *estimate* possible similar values for a very large *population*. The word population used generally describes the whole *class* of *possible* values.

24.1.1 Populations

Two main types of *populations* of data are met with in practice, depending on whether the observed values of a *variate* (or observations) are *discrete* or *continuous*.

Some *populations* are *finite* (limited in number) or *infinite* (unlimited in number).

274

Illustrations:

  (i) All the *integers* (whole numbers) from 1 to 100 inclusive would form a *finite* population of a *discrete* variate.

 (ii) All the households in a given town would form a *finite* population of households.

(iii) All the possible values of *x* between 0 and 1 would constitute an *infinite* population of a *continuous* variate.

(iv) Measurements of dimensions which lie in a given *range* would constitute an *infinite* population of a continuous variate (although rounding figures would apparently give a finite population).

24.2.1 Probability

In many analyses involving *statistical data*, the consideration of the *probability* (or chance) that a variate could have been drawn (obtained) from a given *population* is of great importance.

Probability (or *chance*) may be defined in a number of ways.

Illustration:

What is the probability, that in drawing at *random* from a normal pack of cards (*a*) a black will be drawn at the first attempt, (*b*) a King will be drawn at the first attempt?

(*a*) 52 cards (events) may be drawn of which 26 are black (event A). Hence using a simple *ratio* definition, assuming no *bias* the probability of a black at the first draw is

P (black) = Number of black cards/Total possible number of cards
 = 26/52 = $\underline{1/2}$ (or 1 in 2)

(*b*) 4 only of the cards are Kings (event A). Hence using the simple definition above

P (King) = Number of Kings/ Total possible number of cards
 = 4/52 = $\underline{1/13}$ (or 1 in 13)

24.2.2 A definition of probability

One of the most useful concepts of probability is the idea of a *relative frequency* of *occurrence* of an event.

For example, if a large batch of 500 components is known to contain 15 *defective* components only, then for any *random* selection of a component from the batch, the probability of selecting a defective component (*p*) is given by

$$p = \text{frequency of defectives/frequency of components}$$
$$= 15/500 = 3/100 = \underline{0\cdot03}$$

(or the percentage defective $= 3\%$).

Generalized Definition:

If in *n* trials an event A occurs *m* times then the probability (empirical) of A occurring on *any one trial* is denoted by $p(A) = m/n$. If the number of trials could be carried on indefinitely (to infinity), then the probability (absolute) that A could occur on any one occasion (in *random* selection) is defined as

$$p(A) = \lim_{n \to \infty} \left(\frac{m}{n} \right)$$

$p(A) = 0$ corresponds to failure,
$p(A) = 1$ corresponds to certainty.

As can be seen, in general, such probabilities have to be *estimated* using a *finite sized sample*. Sometimes an idea of the probability involved in a problem may be obtained by simple reasoning.

Example:

If a physically unbiased coin be tossed once, what is the probability of obtaining a head?

At each throw there are two possibilities, i.e. a *head* or *tail*. Hence the probability of a head turning up may be taken as $1/2$ (1 in 2).

This does not necessarily mean, that if a coin be tossed 10 times, 5 heads will appear. However there is an implication that, as the number of throws (trials) is increased, *in the long run* the ratio of heads to tails will approach 1 *or* the ratio of heads to the number of trials will tend to $1/2$.

24.2.3 Combinations (or selections)

In *finite* sampling problems the formulae for selections of r objects from n often prove useful (see Chapter 3).

Example:

A drum contains 20 ball bearings, 5 of which are known to be faulty. The balls are thoroughly mixed. What is the probability that, if a sample of 2 is selected at random (i) both will be faulty, (ii) one only will be faulty.

(i) In selecting at random the *total number* of pairs that may be selected

$$= {}^nC_2 = {}^{20}C_2 = \frac{20 \times 19}{2 \times 1} = \underline{190} \text{ pairs}$$

Of these the total number of possible faulty pairs

$$= {}^{n_1}C_2 = {}^5C_2 = \frac{5 \times 4}{2 \times 1} = \underline{10} \text{ pairs}$$

Hence the *probability* that, in a random selection, both of a pair would be faulty

$$= \text{frequency of faulty pairs/frequency of pairs}$$
$$= 10/190 = \underline{1/19} \text{ (1 in 19)}$$

(This may also be obtained as follows: the probability that the first is faulty $= 5/20$; the probability that the second is faulty (given first) $= 4/19$. The probability that both are faulty $= (5/20) \times 4/19 = 1/19$. This is a simple case of the *multiplication law* to be shown in 24.2.4.)

(ii) The total number of pairs which consist of 1 faulty and 1 not $= 5 \times 15 = 75$ (${}^5C_1 \times {}^{15}C_1$), hence the probability that a randomly selected pair will contain only 1 faulty $= 75/190 = \underline{15/38}$.

24.2.4 Probability laws

Two fundamental laws will now be established.

(1) *Addition Law:*

Let the occurrence of an event A be represented by m points in a *sample space* containing N points and the occurrence of an *independent* event B be represented by n points in the same *sample space* where none of the points $m(A)$ and $n(B)$ are common. This situation may be represented as in Figure 24.1.

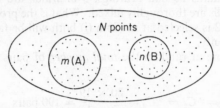

FIGURE 24.1

In a random selection of a point from the *sample space* the probability that a point will represent the occurrence of $A = m/N$ and the probability that a point will represent the occurrence of $B = n/N$. The probability that a point selected will represent the occurrence of *either*

$$A \text{ or } B = (m + n)/N$$

i.e. the probability of either A *or* B occurring on any one occasion is given by

$$P(A \text{ or } B) = (m + n)/N = m/N + n/N = P(A) + P(B)$$

(2) *Multiplication Law:*

Let the occurrence of event A be represented by m points in a *sample space* containing M points and the occurrence of the *independent* event B be represented by n points in a separate space containing N points. (These situations are represented as in Figure 24.2.)

In independent *random* selection of two points, one from M and one from N, the number of possible *joint events* (pairs, one from M and one from N) may be represented by a line joining two such points. The total number of such joint events will be $M \times N$ (since for each point in M there are N joins to points in the second *sample space*). Of these joint events only $m \times n$ would correspond to the joint event of A and B. Hence the probability of A *and* B occurring jointly is given by

$$P(A \text{ and } B) = mn/MN = (m/M) \times (n/N) = p(A) \times p(B)$$

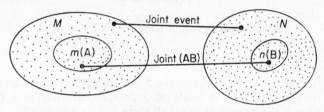

FIGURE 24.2

These two laws are fundamental and may be extended to more than two *independent* events.

Examples:

(i) What is the probability that, in selecting cards at random from a pack (if cards are returned after selection) that (*a*) the first selected will be a Queen *or* Jack, (*b*) the first selected will be a Queen *or* Jack *and* the second selected will be a Jack?

$$P(\text{Queen}) = 4/52 = 1/13 \qquad P(\text{Jack}) = 4/52 = 1/13$$

(*a*) $P(\text{Queen } or \text{ Jack}) = P(\text{Queen}) + P(\text{Jack}) = 1/13 + 1/13$
$$= \underline{2/13}$$

(*b*) $P[(\text{Queen } or \text{ Jack}) \text{ and } (\text{Jack})] = 2/13 \times 1/13 = \underline{2/169.}$

(ii) If a random selection of 3 cards in succession is made from a pack of 52 cards, *without replacement*, what is the probability of a Queen, followed by a Jack, followed by a King?

First choice,
$$P(\text{Queen}) = 4/52 = 1/13$$

If the first is a Queen, there are 51 left containing 4 Jacks.

Second choice,
$$P(\text{Jack}) = 4/51$$

If a Queen and Jack have already appeared, then there are 50 left containing 4 Kings.

Third choice,

$$P(\text{King}) = 4/50$$

Hence

$$\begin{aligned}
P(\text{Queen } and \text{ Jack } and \text{ King}) &= (4/52) \times (4/51) \times 4/50 \\
&= (1/13) \times (4/51) \times 2/25 \\
&= \underline{8/16{,}575}
\end{aligned}$$

(iii) In a batch of 20 components, 5 are known to be defective. If random selection is made of 3 components (one at a time) what is the probability that all three will be defective?

Probability of first being defective = 5/20.
Probability of second also being defective = 4/19.
Probability of third also being defective = 3/18.
∴ Probability of the three being defective

$$= (5/20) \times (4/19) \times (3/18) = \underline{1/114}$$

24.2.5 Bernoulli's law of probability

This gives the probability of exactly r occurrences of an event A in n *independent* trials.

Let $p = constant$ probability that on any one occasion an event A will occur.

(*Note:* The chance of A + chance of failure = 1, i.e. if q represents the chance of failure $p + q = 1$, $q = 1 - p$.)

In n independent trials the probability of r successes followed by $(n - r)$ failures

$$= [p \times p \times p \ldots (r \text{ times})] \times [q \times q \times q \ldots (n - r) \text{ times}]$$
$$= p^r q^{n-r}$$

But the r trials may be selected at random from the n trials in ^nC_r ways and the probability is the same for each selection of r from n.

Hence the *total probability* of obtaining r successes in n trials

$$= p^r q^{n-r} + p^r q^{n-r} \ldots (^nC_r \text{ times})$$
$$= {}^nC_r p^r q^{n-r}$$

i.e. $$\underline{P(r) = {}^nC_r p^r q^{n-r}} \quad (q = 1 - p)$$

24.3 Simple (random) sampling

When objects are selected at *random* from a *population* the *sampling* of the population is said to be *random* if the *chance* of selecting any *one* object is constant. Such sampling is usually referred to as *simple sampling*. General sampling methods may involve varying chance but such problems will only be dealt with in a simple way in this volume.

Examples:

(i) A batch of components is 5% defective. What is the probability that in a random selection of a sample of 5 components (*a*) none are defective, (*b*) 1 is defective, (*c*) 2 are defective, (*d*) up to 2 (inclusive) may be defective?

$$p = 5/100 = 1/20 = 0.05, \quad n = 5, \quad q = 1 - p = 19/20 = 0.95$$

(*a*) Probability that none are defective

$$= q^5 = (0.95)^5 = \underline{0.7736}$$

(*b*) Probability that 1 is defective

$$= {}^5C_1 pq^4 = 5 \times 0.05 \times (0.95)^4$$
$$= 0.25 \times (0.95)^4 = \underline{0.2035}$$

(*c*) Probability that 2 are defective

$$= {}^5C_2 p^2 q^3 = 10 \times (0.05)^2 \times (0.95)^3$$
$$= 0.025 \times (0.95)^3 = \underline{0.02143}$$

(*d*) Probability of 0 or 1 or 2

$$= P(0) + P(1) + P(2) = 0.7736 + 0.2035 + 0.0214(3)$$
$$= \underline{0.9985(3)} \text{ (four figure logarithms used)}$$

(ii) An unbiased coin is tossed 5 times. Calculate the probabilities:

    (*a*) That the first three will be heads, last two tails.
    (*b*) That three heads only will turn up.
    (*c*) That up to two heads will turn up.

p = probability of a head = 1/2.
q = failure to obtain a head = $1 - p = 1/2$ = probability of a tail

(a) The probability of 3 heads followed by 2 tails
$$= p^3 q^2 = (\tfrac{1}{2})^3 \times (\tfrac{1}{2})^2 = (\tfrac{1}{2})^5 = 1/32$$

(b) $n = 5, r = 3, p = \tfrac{1}{2}, q = \tfrac{1}{2},$

$$\therefore \ P(3) = {}^nC_r p^r q^{n-r} = {}^5C_3 (\tfrac{1}{2})^3 (\tfrac{1}{2})^2 = \frac{5 \times 4 \times 3}{1 \times 2 \times 3} \times (\tfrac{1}{2})^5 = \frac{5}{16}$$

(c) Chance of no heads

$$= (\tfrac{1}{2})^5 = 1/32$$

Chance of one head

$$= {}^5C_1 p q^4 = 5(\tfrac{1}{2}) \times (\tfrac{1}{2})^4 = \underline{5/32}$$

Chance of two heads

$$= {}^5C_2 p^2 q^3 = \frac{5 \times 4}{1 \times 2} (\tfrac{1}{2})^5 = \frac{10}{32}$$

$\therefore$ Total chance of up to and including 2 heads in 5 trials

$$= P(0) + P(1) + P(2)$$
$$= 1/32 + 5/32 + 10/32 = 16/32 = \underline{1/2}$$

WORKED EXAMPLES

(a) An unbiased dice is thrown 6 times. What is the probability that:

(i) A six will be thrown on the first occasion?
(ii) That the first three throws only will show a six?
(iii) That only three out of the six throws will show a six?

(i) Chance of a six on any occasion $= 1/6,$

$\therefore$ Chance of a six on the first throw $= 1/6$

(ii) At each throw $p = 1/6, q = 5/6,$
Probability of 3 sixes followed by 3 other numbers

$$= p^3 q^3 = (1/6)^3 \times (5/6)^3 = 125/6^6 = 125/46{,}656 \simeq \underline{0 \cdot 0027}$$

(iii) Using Bernoulli's law the chance of 3 sixes only ($n = 6$, $r = 3$, $p = 1/6$, $q = 5/6$) is given by

$$P(3) = {}^nC_r p^r q^{n-r} = {}^6C_3(1/6)^3 \times (5/6)^3$$
$$= \frac{6 \times 5 \times 4}{1 \times 2 \times 3} \times \frac{125}{6^6} = 20 \times \frac{125}{6^6} \simeq \underline{0.054}$$

(b) 3 unbiased dice are thrown simultaneously. Find the chance that:

(i) 3 of the same number show.
(ii) The total of the three face numbers $= 10$.

(i) The total number of sets of 3 numbers $= 6 \times 6 \times 6 = 216$.

Of these sets of 3, only 6 are possible with all three alike,

∴ the probability that the three dice will show the same number
$$= 6/216 = \underline{1/36} \text{ (1 in 36)}$$

(ii) A total of 10 may be made up as follows:

6, 2, 2, — 3 ways;	6, 1, 3, — 3! = 6 ways
5, 4, 1, — 3! ways = 6;	5, 3, 2, — 3! = 6 ways
4, 4, 2, — 3 ways;	4, 3, 3, — 3 ways

∴ Total number of ways to obtain a total of 10 $= \underline{27}$.

Total number of ways in which the three can fall $= \underline{216}$.

∴ Probability of throwing a total of 10 $= 27/216 = \underline{1/8}$.

(*Note:* In this example the numbers of possibilities are evaluated. This may be necessary in complex examples.)

(c) A bag contains 10 balls of equal size, but 3 are *white*, 5 *red*, and 2 *blue*. If three balls are drawn at random in succession, what are the probabilities that:

(i) The three will all be *red*?
(ii) The first two will be *white* and the third *blue*?
(iii) All three will be of *different* colour?

(i) Using the product law the probability $P(R, R, R)$ is given by
$$P(R, R, R) = (5/10) \times (4/9) \times 3/8 = \underline{1/12}$$

(ii) Probability of a *white* at the first draw = 3/10. If the first is *white*, the probability of second *white* = 2/9. If the first two are *white*, the probability of a *blue* on the third draw = 2/8. By the product law

$$P(W, W, B) = (3/10) \times (2/9) \times 2/8 = \underline{1/60}$$

(iii) If all three are different colour they may occur in 6 ways: (W, R, B), (W, B, R), (B, W, R), (B, R, W), (R, B, W), (R, W, B). Using the multiplication law it follows that the probability that each set occurs is constant.

$$P(W, R, B) = (3/10) \times (5/9) \times 2/8 = 1/24$$
$$P(W, B, R) = (3/10) \times (2/9) \times 5/8 = 1/24$$
$$P(B, W, R) = (2/10) \times (3/9) \times 5/8 = 1/24$$
$$P(B, R, W) = (2/10) \times (5/9) \times 3/8 = 1/24$$
$$P(R, B, W) = (5/10) \times (2/9) \times 3/8 = 1/24$$
$$P(R, W, B) = (5/10) \times (3/9) \times 2/8 = 1/24$$

∴ *total chance* that all three are different is given by the *addition law*

$$= 6 \times 1/24 = \underline{1/4} \ (1 \text{ in } 4)$$

(This may be obtained as follows:

$$\text{Total number of selections of } 3 = {}^{10}C_3 = \frac{10 \times 9 \times 8}{1 \times 2 \times 3} = 120.$$

Of these the number in which all three are different = $3 \times 5 \times 2 = 30$.
∴ Probability that all three are different = 30/120 = 1/4.)

(*d*) Certain components being manufactured are known from past experience to be 10% *defective*. Calculate the probability that up to and including 2 *defectives* may be present in a *random sample* of 5.

The probability of a *defective* on any one occasion = $p = 1/10$ and the probability of *no defective* = $q = 1 - p = 9/10$.

$$P(0) = q^5 = (0 \cdot 9)^5 = \underline{0 \cdot 59049}$$

$$P(1) = {}^5C_1 pq^4 = 5 \times (0 \cdot 1) \times (0 \cdot 9)^4 = 0 \cdot 5 \times 0 \cdot 6561 = \underline{0 \cdot 32805}$$

$$P(2) = {}^5C_2 p^2 q^3 = \frac{5 \times 4}{1 \times 2}(0 \cdot 1)^2 \times (0 \cdot 9)^3 = 0 \cdot 1 \times 0 \cdot 729 = \underline{0 \cdot 07290}$$

Probability that up to 2 may be defective in a random sample of 5

$$= P(0) + P(1) + P(2) = \underline{0.99144}$$

(*Note:* It is seen that, although a relatively small sample is taken the chance of obtaining up to 2 defectives is very large (nearly 1). It is this *characteristic* of a small scale *simple sampling* scheme that makes it so *useful* in practical sampling for purposes of *controlling quality* on a production line.

It will be seen later that when p is *small* useful approximation methods may be used to estimate the chances of obtaining defectives.)

Examples 24

1. An unbiased coin is tossed 5 times. Calculate the probabilities that
   (i) the first three throws produce three heads,
   (ii) the first three throws produce 3 heads *or* 3 tails,
   (iii) only three heads turn up in the five throws,
   (iv) at least *four* heads occur.
2. Three cards are drawn at random in succession from a normal pack of 52. Calculate the probabilities that all three drawn will be aces if
   (i) after each drawing the card is replaced and the pack reshuffled before the next draw,
   (ii) the cards are not replaced after each drawing.
3. An unbiased dice is tossed 5 times. Calculate the probabilities that
   (i) a 5 will appear on the first throw,
   (ii) a 3 *or* 5 will appear on the first throw,
   (iii) four sixes will appear in the five throws.
4. Three equal unbiased dice are tossed simultaneously. Calculate the probabilities that
   (i) two sixes and one five will appear,
   (ii) a total of 14 will be thrown.
5. A drum contains 20 balls, 5 of which are *red*, 7 are *blue* and the rest *white*. If a random sample of 3 balls is made, calculate the probabilities
   (i) that all three will be *white*,
   (ii) that two will be *white* and one *red*,
   (iii) that all three will be of different colour.
6. A batch of manufactured components for a car is known to be 20 per cent *defective*. If a random sample of 4 parts is selected calculate the chances of obtaining 0, 1, 2, 3, 4 defective parts. What is the chance of obtaining up to 2 (inclusive) defectives? What is the sum of all the chances involved (i.e. of 0, 1, 2, 3, 4 defectives). Explain the result.
7. Assuming that the chance that a new car will have a fault of some kind is constant and equal to 0.1, calculate the chance that, in a batch of 20 similar new cars, 4 will have a fault of some kind.

8. A manufacturer making standard resistors finds from a long series of samples that the chance of a faulty resistor is approximately constant at 5 per cent. Calculate the probabilities that, in a random sample of 5 resistors from a batch
   (i) 0, 1 or 2 will be found to be faulty,
   (ii) at least two will be faulty,
   (*Note:* Here it is assumed that $P(0) + P(1) + P(2) + P(3) + P(4) + P(5) = 1$).

9. In manufacturing test tubes of a given size it is found that, on average 1 in 10 are faulty. Assuming that the chance of a faulty tube is constant, calculate the probabilities that
   (i) in a batch of 4, 1 may be faulty,
   (ii) in a batch of 5, up to 2 may be faulty,
   (iii) more than 1 in a batch of 6 may be faulty.

10. In a certain biological process equal sample amounts of a substance are tested for a certain organism. For the size of sample chosen it is known that, on average the chance of the organism occurring in a test sample is 1 in 10. Calculate the probabilities that, in a random batch of 5 such samples (i) all 5 will contain the organism, (ii) none will contain the organism, (iii) up to (and including) 2 samples will contain the organism.

25

Distributions (Discrete and Continuous)—Samples—Data Representation—Parameters of a Sample

25.1 Random Distributions

As already stated, observed variates may be *discrete* or *continuous*. In many cases of testing of *hypotheses* a *mathematical model* in the form of a *random distribution* is assumed. Approximations of *discrete* distributions are also often made by assuming a continuous distribution. Before considering *data collection* in the form of *finite samples* a general *continuous* distribution will be discussed.

25.1.1 A general continuous distribution

Let x be a continuous random variate, capable of assuming all possible values in a given range a to b (these limits may often be *infinite*). Let $y = f(x)$ be the assumed *relative frequency function* for the random variate. Let the graph of $y = f(x)$ be as illustrated in Figure 25.1.

If a value of x is chosen at *random* from this *population*, then the probability that x lies in a given range c to d, say, is measured by the area under the graph between the ordinates at $x = c$ and $x = d$. If the probability that x will lie in a small interval $x - \frac{1}{2}\,dx < x$

$< x + \frac{1}{2}\,dx$ is represented by dp then $dp = f(x)\,dx$, where $f(x)$ is the value of the *relative frequency* function at the centre of the interval. $dp = f(x)\,dx$ is called the *probability differential* of the distribution of the variate x.

If the function is known, the following results hold:

(i) $dp = f(x)\,dx$,

(ii) $P(c < x < d) = \int dp = \int_c^d (x)\,dx$,

(iii) $P(a < x < b) = \int_a^b f(x)\,dx = 1$,

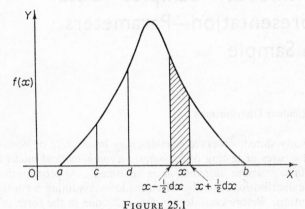

FIGURE 25.1

i.e. it is assumed that a single value of x, when chosen, will *certainly* lie in the range a to b.

[*Note:* In a general manner, the probability that x will assume a value, when chosen at *random* is represented by an area under the graph of the *relative frequency* function. $f(x)$ is sometimes referred to as the *probability density function*, i.e. a comparison is made with mass densities (as in statics).]

25.1.2 Parameters of a distribution

In order to *describe* a population, certain *parameters* may be used. The parameters are *values of the variate x*. Some of these parameters depend on *moments* of the area under the relative frequency curve about specified points.

Mean value (x co-ordinate of the centroid of the whole area): If this is called μ, then

$$Mean = \mu = \int_a^b xf(x)\,\mathrm{d}x \text{ (First moment about the origin)}$$

Mode: This is the value of the variate x for which the relative frequency function is a *maximum*. This value may be found by equating $\mathrm{d}y/\mathrm{d}x$ to zero.

Median: This is the value of x such that the chance of selecting x from a up to the median value or between the median value and b is $1/2$ (or the value of x such that there are equal numbers of the variate above and below it). If M is the *median* value then

$$\int_a^M f(x)\,\mathrm{d}x = \int_M^b f(x)\,\mathrm{d}x = \tfrac{1}{2}$$

Variance (a measure of the degree of *variation* from the *mean*): If this is called σ^2 (sigma squared), then

$$Variance = \sigma^2 = \int_a^b (x - \mu)^2 f(x)\,\mathrm{d}x$$

which is the *second moment* of area about the mean.

Standard deviation (standard error): This is $\sigma = \sqrt{(\text{variance})}$ and is a *linear* measure of the degree of variation (or scattering) of the variate x about the mean.

Other parameters may be defined but these are sufficient for elementary analysis.

Example:

A random variate x is distributed according to the frequency function $f(x) = \frac{3}{4}x(2 - x)$, over the range $x = 0$ to 2:

   (i) Verify that the *total chance* $= 1$,
   (ii) Calculate the mean value of x,
   (iii) Find the *mode* of the distribution,
   (iv) Find the median value,
   (v) Calculate the variance and standard deviation of the distribution (Figure 25.2).
   This is a *symmetrical* distribution.

(i) Total chance (probability)

$$= \int_0^2 f(x)\, dx = \int_0^2 \tfrac{3}{4}x(2-x)\, dx = \tfrac{3}{4} \int_0^2 (2x - x^2)\, dx$$

$$= \frac{3}{4}\left[x^2 - \tfrac{1}{3}x^3 \right]_0^2 = \frac{3}{4}\left[4 - \frac{8}{3} \right] = \frac{3}{4} \times \frac{4}{3} = 1$$

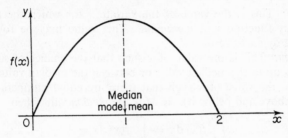

FIGURE 25.2

(ii) The mean μ is given by

$$\mu = \int_0^2 xf(x)\, dx = \int_0^2 \tfrac{3}{4}x^2(2-x)\, dx = \tfrac{3}{4} \int_0^2 (2x^2 - x^3)\, dx$$

$$= \frac{3}{4}\left[\frac{2}{3}x^3 - \frac{1}{4}x^4 \right]_0^2 = \frac{3}{4}\left[\frac{16}{3} - \frac{16}{4} \right] = \frac{3}{4} \times \frac{16}{12} = \frac{3}{4} \times \frac{4}{3} = \underline{1}$$

(iii) The *mode* (or modal value) of x is the value of x when $f(x)$ is a maximum.

$$y = f(x) = \tfrac{3}{4}x(2-x) = \tfrac{3}{4}(2x - x^2)$$

$$\frac{dy}{dx} = f'(x) = \tfrac{3}{4}(2 - 2x) = 3(1-x)/2 = 0$$

for maximum y, when $x = 1$; i.e. *mode* $= \underline{1}$.

(iv) Let the median value $= M$, then

$$\int_0^M f(x)\, dx = \tfrac{1}{2}$$

$$\therefore \int_0^M \tfrac{3}{4}x(2-x)\, dx = \tfrac{3}{4} \int_0^M (2x - x^2)\, dx = \tfrac{3}{4}\left[x^2 - \frac{1}{3}x^3 \right]_0^M = \tfrac{1}{2}$$

or $$\tfrac{3}{4}[M^2 - \tfrac{1}{3}M^3] = \tfrac{1}{2}$$

By trial $M = 1$ satisfies this equation, i.e. *median* = 1.

(*Note:* This distribution is *symmetrical* and for such a distribution *mean = mode = median*.)

(v) *Variance* $\sigma^2 = \displaystyle\int_0^2 (x - \mu)^2 f(x)\, \mathrm{d}x = \int_0^2 (x - 1)^2 \tfrac{3}{4} x(2 - x)\, \mathrm{d}x$

$\therefore \sigma^2 = \tfrac{3}{4} \displaystyle\int_0^2 x(2 - x)(x - 1)^2\, \mathrm{d}x = \tfrac{3}{4} \int_0^2 (-x^4 + 4x^3 - 5x^2 + 2x)\, \mathrm{d}x$

$\quad = \dfrac{3}{4}\left[-\dfrac{1}{5}x^5 + x^4 - \dfrac{5}{3}x^3 + x^2 \right]_0^2$

$\quad = \tfrac{3}{4}[-\tfrac{1}{5} \times 32 + 16 - \tfrac{5}{3} \times 8 + 4]$

$\quad = \tfrac{3}{4}[20 - 296/15] = \tfrac{3}{4} \times 4/15 = \tfrac{1}{5}$

Standard deviation, $\sigma = \sqrt{0{\cdot}2} = 0{\cdot}4472$.

25.2.1 Samples

In order to obtain an idea of a *population*, in practice only *finite sized* samples are usually available. When such *samples* are large the *sample* approximates to the population. Using a given sample of data, by analysing the data, *estimates* of the population parameters can be made.

25.2.2 Frequency distributions

The first step in the analysis of a sample of data is to produce an idea of the *frequency distribution*. This is done by plotting some form of *frequency chart* (or graph). The type of frequency chart will depend on the type of data collected. Some such plots are illustrated in Figures 25.3 to 25.6.

Figure 25.3 shows a frequency *polygon*

Figure 25.4 shows a frequency *curve.*

Figure 25.5 shows a *histogram*. Areas of *cells* of the *histogram* represent the frequencies with which measurements occur in given intervals of *x*. The intervals are often equal and sometimes referred to as *class intervals*.

Figure 25.6 shows a *cumulative* frequency curve (an *ogive*) which shows the frequency of occurrence of data up to a given value.

From a frequency plot it is possible to estimate the *mode* of the distribution of data in the sample.

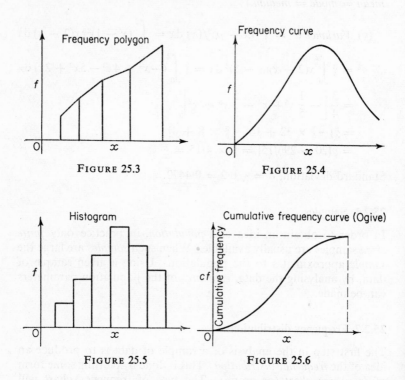

FIGURE 25.3

FIGURE 25.4

FIGURE 25.5

FIGURE 25.6

From an *ogive* estimates can be made of the *median* (50 percentile) and other percentiles, e.g. 10 percentile (the value of x such that there are 10% of the values up to this value of x). The 0 percentile is the lowest value of the variate in the sample and the 100 percentile the greatest value of the variate in the sample.

The *mean*, *variance*, and *standard deviation* of the sample of data may be estimated by calculations.

25.3.1 Calculation of the mean of a sample

Let the values $x_1, x_2, x_3, \ldots, x_n$ in a sample occur with frequencies $f_1 f_2, f_3, \ldots, f_n$, then the mean value $\bar{x}$ is given by

$$\bar{x} = \text{total of all the } x \text{ values/total number of } x \text{ values}$$
$$= (f_1 x_1 + f_2 x_2 + \ldots + f_n x_n)/(f_1 + f_2 + \ldots + f_n)$$
$$= \Sigma f x / \Sigma f$$

Example:

In a sample of data the values of the variate x occur with the frequencies shown in the table. Calculate the sample mean of the variate x.

variate x	1	2	3	4	5	6	7	8
frequency f	2	7	12	16	25	18	13	7

$$\text{Total frequency} = \Sigma f = \underline{100}$$

$$\Sigma f x = 2 \times 1 + 7 \times 2 + 12 \times 3 + 16 \times 4 +$$
$$25 \times 5 + 18 \times 6 + 13 \times 7 + 7 \times 8$$
$$= 2 + 14 + 36 + 64 + 125 + 108 + 91 + 56 = \underline{496}$$

Mean value of the variate $\bar{x} = \Sigma f x / \Sigma f = 496/100 = \underline{4 \cdot 96}$

25.3.2 Calculation of the variance and standard deviation

The variance of the *sample* will be denoted by s^2. For a general sample of data as in 25.3.1, the variance

$$s^2 = \text{Mean square deviation of data from the mean}$$
or $$s^2 = \Sigma f(x - \bar{x})^2 / \Sigma f$$

If $N = \Sigma f = $ total number of sample values then the result may be written as

$$s^2 = \Sigma(x - \bar{x})^2 / N$$

where $\Sigma(x - \bar{x})^2 = $ total sum of squares (of deviations) in the sample.

Example:

Calculate the variance and standard deviation of the numbers $x = 4, 5, 6, 7, 8, 9, 10, 11, 12$.

$$\bar{x} = \Sigma x/N = 72/9 = 8$$

Deviations of the numbers from the *mean* are

$$-4, -3, -2, -1, 0, 1, 2, 3, 4$$

$$\therefore \ \Sigma(x - \bar{x})^2 = 16 + 9 + 4 + 1 + 0 + 1 + 4 + 9 + 16 = 60$$

$$s^2 = \Sigma(x - \bar{x})^2/N = 60/9 = 6 \cdot 667$$
$$s = \sqrt{(6 \cdot 667)} = \underline{2 \cdot 582}$$

i.e. *mean* $= 8$; *variance* $= 6 \cdot 667$; *standard deviation* $= 2 \cdot 582$.

25.3.3 Calculation of the mean using a working mean

When *calculators* are not available it is an advantage to simplify the arithmetic of the calculations of $\bar{x}$ and s. This may be achieved by using a *working mean* (or false zero) and also, if necessary a suitable *class interval* (equivalent to changing the *units*).

Let A be a working mean (or false zero). Using the same general set of data as in 25.3.1, the *deviations* of the data from A are:

$$x_1 - A, x_2 - A, x_3 - A, \ldots, x_n - A$$

Hence the *mean deviation* from A

$$= \Sigma f(x - A)/\Sigma f = \frac{\Sigma fx}{\Sigma f} - A \frac{\Sigma f}{\Sigma f} = \bar{x} - A$$

$$\therefore \qquad \bar{x} = A + \text{mean deviation from } A$$

25.3.4 Calculation of variance and standard deviation using a working mean

Let A be the working mean and $\bar{x}$ the true mean. Considering the same set of data:

Deviations from A are $x_1 - A, x_2 - A, \ldots, x_n - A$.
Deviations from $\bar{x}$ are $x_1 - \bar{x}, x_2 - \bar{x}, \ldots, x_n - \bar{x}$.

Let the mean square deviation from A be S^2, and the mean square deviation from $\bar{x}$ be s^2.

Let the total number of observations $= \Sigma f = N$, then

$$S^2 = \Sigma(x - A)^2/N \quad \text{or} \quad NS^2 = \Sigma(x - A)^2$$
$$s^2 = \Sigma(x - \bar{x})^2/N \quad \text{or} \quad Ns^2 = \Sigma(x - \bar{x})^2$$

$$\begin{aligned}
\therefore \quad N(S^2 - s^2) &= \Sigma(x - A)^2 - \Sigma(x - \bar{x})^2 \\
&= \Sigma\{(x - A)^2 - (x - \bar{x})^2\} \\
&= \Sigma\{(x^2 - 2Ax + A^2) - (x^2 - 2x\bar{x} + \bar{x}^2)\} \\
&= \Sigma 2x(\bar{x} - A) + \Sigma(A^2 - \bar{x}^2) \\
&= 2(\bar{x} - A)\Sigma x + N(A^2 - \bar{x}^2)
\end{aligned}$$

$$\begin{aligned}
\therefore \quad S^2 - s^2 &= 2(\bar{x} - A)\Sigma x/N + (A^2 - \bar{x}^2) \\
&= 2(\bar{x} - A)\bar{x} + A^2 - \bar{x}^2 = \bar{x}^2 - 2A\bar{x} + A^2
\end{aligned}$$

$$\text{or} \quad S^2 - s^2 = (\bar{x} - A)^2$$
$$\underline{S^2 = s^2 + (\bar{x} - A)^2}$$

i.e. $$\underline{s^2 = S^2 - (\bar{x} - A)^2}$$

Note that, from the first of these expressions S^2 is a minimum equal to s^2 when $A = \bar{x}$. This result has a very important application called the *Method of least squares* which is widely used in fitting 'best' straight lines or curves to observed data. This will not, however, be dealt with in this volume.

Example:

Using a suitable *class interval* and a working mean calculate the mean value, variance and standard deviation of the following data:

x	5	10	15	20	25
f	3	5	9	6	2

Using a 'class interval' of 5, the data reduce to $X = x/5$, i.e.

$$X \quad 1 \quad 2 \quad 3 \quad 4 \quad 5$$

The working proceeds using the variable X and the values of the parameters for the variate x are deduced.

Use a working mean A (for X) = 3.

x	f	X	$X - A$	$f(X - A)$	$f(X - A)^2$
5	3	1	-2	-6	12
10	5	2	-1	-5	5
15	9	3	0	0	0
20	6	4	1	6	6
25	2	5	2	4	8
Totals	25			-1	31

$$\Sigma f = 25 = N; \quad \Sigma f(X - A) = -1; \quad \Sigma f(X - A)^2 = 31$$

Mean deviation from $A = \Sigma f(X - A)/\Sigma f = -1/25 = \underline{-0.04}.$

Mean square deviation from $A = \Sigma f(X - A)^2/\Sigma f = 31/25 = \underline{1.24}.$

$\bar{X} = A +$ mean deviation from $A = 3 + (-0.04) = \underline{2.96}$

$s_1^2 =$ mean square deviation from $\bar{X}$
$S_1^2 =$ mean square deviation from A

$$s_1^2 = S_1^2 - (\bar{X} - A)^2 = 1.24 - (-0.04)^2 = 1.24 - 0.0016$$
$$\therefore \quad \underline{s_1^2 = 1.2384}$$

But $X = x/5,$
$$\therefore \qquad\qquad\qquad x = 5X$$

Hence mean value of x

$$= \bar{x} = 5\bar{X} = 5 \times 2.96 = \underline{14.8}$$

Variance of x

$$= s^2 = 5^2 s_1^2 = 1.2384 \times 25 = \underline{30.96}$$

$\therefore$ standard deviation of x

$$= s = \sqrt{30.96} = \underline{5.564}$$

Note: A may often be taken as zero, particularly if calculating machines are available, then

$$\underline{s^2 = S^2 - \bar{x}^2}$$

WORKED EXAMPLES

(a) A variate x in. occurs with frequency f as shown in the table. Assuming that the variate is continuous, plot a frequency curve and an *ogive* for the distribution. From the *ogive* estimate the median value of x and the 25 and 75 percentiles. Calculate the mean and standard deviation of the sample of data:

x (in)	0	2	4	6	8	10	12
f	0	12	18	28	22	15	5
cumulative f	0	12	30	58	80	95	100

The frequency curve is shown in Figure 25.7 and the ogive (cumulative frequency curve) is shown in Figure 25.8.

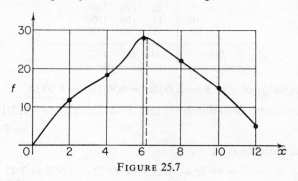

FIGURE 25.7

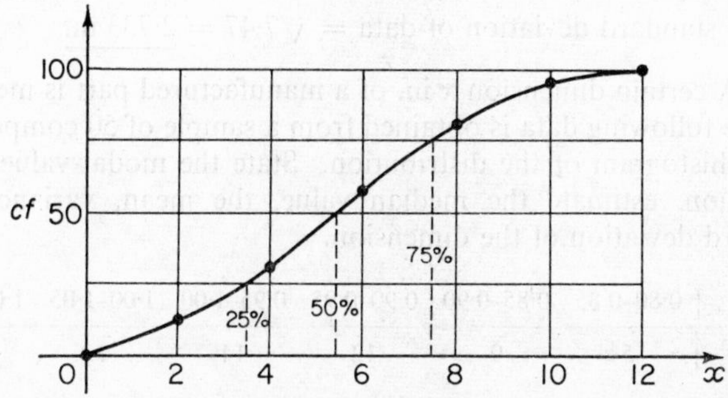

FIGURE 25.8

From the plotted frequency and cumulative frequency curves:

(i) *Mode* (most frequently occurring value of x) = 6·2.
(ii) *Median* (50 percentile) = 5·40.

(iii) 25 *percentile* (lower *quartile*) = 3·46.
(iv) 75 *percentile* (upper quartile) = 7·41.

Taking the working mean $A = 0$, the deviations from A are the x values.

x	f	fx	fx^2
0	0	0	0
2	12	24	48
4	18	72	288
6	28	168	1008
8	22	176	1408
10	15	150	1500
12	5	60	720
Totals	100	650	4972

Mean value of $x = \bar{x} = \Sigma fx/\Sigma f = 650/100 = \underline{6\cdot50 \text{ in.}}$

Mean square deviation from zero $S^2 = \Sigma fx^2/\Sigma f = 4{,}972/100$
$$= \underline{49\cdot72 \text{ in}^2}$$

If the variance is s^2,

$$s^2 = S^2 - \bar{x}^2 = 49\cdot72 - (6\cdot50)^2 = 49\cdot72 - 42\cdot25 = \underline{7\cdot47 \text{ in}^2}$$

$s = $ standard deviation of data $= \sqrt{7\cdot47} = \underline{2\cdot733 \text{ in.}}$

(*b*) A certain dimension x in. of a manufactured part is measured and the following data is obtained from a sample of 50 components. Plot a histogram of the distribution. State the modal value of the dimension, estimate the median value, the mean, variance, and standard deviation of the dimension.

x (in)	0·80–0·85	0·85–0·90	0·90–0·95	0·95–1·00	1·00–1·05	1·05–1·10
f	5	9	12	14	7	3

The *histogram* is shown in Figure 25.9.
The 'modal' range is the range 0·95 to 1·00 in.
The *median* value (such that equal numbers lie above and below it) may be estimated as follows:

$$5 + 9 = 14$$

∴ 11 additional values required to give 25 (50%). Hence using proportion (areas of cells represent frequencies):

Median value $\simeq 0.90 + (11/12) \times 0.05 \simeq 0.90 + 0.046 \simeq \underline{0.946}$,

or Median value $\simeq 0.95 - (1/12) \times 0.05 \simeq 0.95 - 0.004 \simeq \underline{0.946}$

(the interval 0·05 is a class interval).

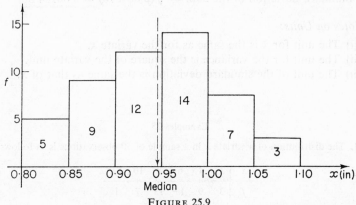

FIGURE 25.9

To compute the mean and variance, the *central values* of the class intervals are used as the means for the cells. Using these mean values take $A = 0.975$.

x (central values)		$x - A$	$f(x - A)$	$f(x - A)^2$
0·825	5	−0·15	−0·75	0·1125
0·875	9	−0·10	−0·90	0·0900
0·925	12	−0·05	−0·60	0·0300
0·975	14	0·00	0·00	0·0000
1·025	7	0·05	0·35	0·0175
1·075	3	0·10	0·30	0·0300
Totals:	50		−2·25	0·2800
			+0·65	
			−1·60	

Mean deviation from $A = -1.60/50 = -0.032$.

Mean value $\bar{x} = A +$ Mean deviation from A
$= 0.975 - 0.032 = \underline{0.943 \text{ in.}}$

Mean square deviation from A,

$$S^2 = 0{\cdot}28/50 = \underline{0{\cdot}0056 \text{ in}^2}$$

Variance $s^2 = S^2 - (\bar{x} - A)^2 = 0{\cdot}0056 - (-0{\cdot}032)^2$
$$= 0{\cdot}005600 - 0{\cdot}001024 = \underline{0{\cdot}004576 \text{ in}^2}$$

Standard deviation of the data $= \sqrt{(0{\cdot}004576)} = \underline{0{\cdot}06764 \text{ in.}}$

Notes on Units:

(i) The unit for $\bar{x}$ is the same as for the variate x.

(ii) The unit for the variance is the square of the variate unit.

(iii) The unit of the standard deviation is the same as that of x.

Examples 25

1. The distribution of a variate x in a sample of 50 observations is as follows:

x	2	4	6	8	10	12
f	3	9	17	13	7	1

Assuming that the variate is continuous, draw a frequency curve to represent the data and a cumulative frequency curve (ogive). From the diagrams estimate (i) the mode of the variate (ii) the median value (iii) the 25 and 75 percentiles.

Calculate the mean and standard deviation of the sample data.

2. Sample observations of a dimension x in. gave the following values in intervals of 0·1 in.

x	0·0 − 0·1	0·1 − 0·2	0·2 − 0·3	0·3 − 0·4	0·4 − 0·5
f	1	6	11	18	22

x	0·5 − 0·6	0·6 − 0·7	0·7 − 0·8	0·8 − 0·9	0·9 − 1·0
f	17	12	8	3	2

Draw a histogram to represent the distribution of the variate. By taking the mid-points of the tops of the 'cells' of the histogram estimate a continuous frequency curve for the distribution of x. From the histogram estimate the median value of the variate and from the curve estimate the mode of the variate.

Estimate the mean and standard deviation of the sample of observations.

3. A variate x is distributed under a binomial form as follows:

x	0	1	2	3	4	5	6
f	1	6	15	20	15	6	1

Calculate the mean and variance of the distribution.

4. A continuous variate takes all values between $x = 0$ and $x = 1$. The probability density function is given by $f(x) = 6x(1 - x)$. Show that this is a 'proper' probability density function, i.e. that the function would give a total probability of 1 that x would be selected between 0 and 1. By integration find the mean and variance of the distribution. Show that the mode of the distribution is the same as the mean.

5. A continuous variate x has a relative frequency function (or probability density function) $f(x) = e^{-x}$ for $0 < x < \infty$. Verify that this is a 'proper' frequency function. Show that the mean value and standard deviation of the distribution are both 1.

6. 150 measurements of a dimension are taken and the following table shows the distribution of deviations of the observations in units of 0·001 in. from 0·510 in.

deviation (x)	−5	−4	−3	−2	−1	0	1	2	3	4	5
frequency (f)	1	8	17	23	35	26	20	14	3	2	1

Calculate the mean and standard deviation of the dimension.

7. In a botanical investigation into the characteristics of a certain plant, samples of mature leaves are selected at random and the mean length measured. The distribution of the lengths of 110 leaves was as follows:

length (in.)	1·6–1·7	1·7–1·8	1·8–1·9	1·9–2·0	2·0–2·1	2·1–2·2	2·2–2·3	2·3–2·4
frequency	6	12	19	25	21	14	9	4

Construct a histogram to represent the distribution of leaf length and calculate the mean length and standard deviation of length.

8. In checking the density of a substance, 20 random samples gave the following results in gram/cm³.

6·03	6·25	5·59	7·03	5·49	6·58	5·89	6·38	5·97	6·21
5·12	6·45	5·44	6·11	5·65	6·78	5·74	6·67	4·87	5·55

Calculate the mean and standard deviation of the density.

9. The lengths of 50 steel rods were measured to the nearest tenth of an inch and the distribution of length was as follows:

length (in.)	5·4	5·5	5·6	5·7	5·8	5·9	6·0	6·1	6·2
frequency	1	2	4	8	14	13	5	2	1

Plot a frequency curve showing the distribution of length, and a cumulative frequency curve. Estimate the mode, median, 20 and 80 percentiles. Calculate the mean and standard deviation of the length.

10. In the manufacture of fuse wire samples are taken of a fuse wire which fuses nominally at 5 amp. Equal lengths are taken at random and fused under similar conditions to test for the fusing current. The distribution of fusing current for the sample of pieces was as follows:

fusing current (amp)	4·6	4·7	4·8	4·9	5·0	5·1	5·2	5·3	5·4
frequency	2	7	19	28	21	17	10	5	1

Calculate the mean and standard deviation of the fusing current in amp.

26

Binomial, Poisson and Normal Distributions

This chapter is devoted to three special distributions of great practical importance in sampling procedures. The three distributions (mathematical *models*) to be considered are

(i) the *binomial* distribution,
(ii) the *Poisson* distribution,
(iii) the *normal* (or Gaussian distribution).

26.1 The Binomial Distribution

This distribution is based on the result obtained in Chapter 24 giving Bernoulli's law of probability. If an event A has a constant probability p of occurring on any one occasion (or trial) and q is the probability of not occurring then the probability of exactly r successes (occurrences of event A) in n *independent* trials is given by $P(r) = {}^nC_r p^r q^{n-r}$.

Consider the function $(q + p\theta)^n$ where θ is an arbitrary parameter then

$$(q + p\theta)^n = q^n + {}^nC_1 pq^{n-1}\theta + {}^nC_2 p^2 q^{n-2}\theta^2 + \ldots + {}^nC_r p^r q^{n-r}\theta^r$$
$$+ \ldots + {}^nC_n p^n \theta^n$$

The coefficient of θ^r in this expansion $= P(r) = {}^nC_r p^r q^{n-r}$, i.e. $(q + p\theta)^n$ *generates* the probabilities of $0, 1, 2, \ldots, n$ successes in n independent trials. The function $(q + p\theta)^n$ is called the *binomial probability generating function*.

26.1.1 Recurrence relation for Binomial probabilities

This is a formula which enables successive binomial probabilities to be evaluated based on probabilities already evaluated.

From 26.1 $\quad P(r) = {}^nC_r p^r q^{n-r} = \dfrac{n!}{r!(n-r)!} p^r q^{n-r}$

$$P(r + 1) = {}^nC_{r+1} p^{r+1} q^{n-r-1} = \dfrac{n!}{(r+1)!(n-r-1)!} p^{r+1} q^{n-r-1}$$

$$\dfrac{P(r+1)}{P(r)} = \dfrac{n!}{(r+1)!(n-r-1)!} \times \dfrac{r!(n-r)!}{n!} \cdot \dfrac{p^{r+1} q^{n-r-1}}{p^r q^{n-r}}$$

$$= \dfrac{n-r}{r+1} \cdot \dfrac{p}{q} \quad \therefore \quad \underline{P(r+1) = \dfrac{n-r}{r+1} \cdot \dfrac{p}{q} \cdot P(r)}$$

i.e. given $n, r, p, q, P(r)$ then $P(r + 1)$ can be calculated. The shape of the distribution depends on p. When $p = q = \frac{1}{2}$ the distribution is *symmetrical*.

Examples:

(i) An unbiased coin is tossed 5 times. Calculate the probabilities that a head will appear on 0, 1, 2, 3, 4, 5 occasions. Sketch a *histogram* of the *relative frequency* distribution.

In this problem $p = q = \frac{1}{2}$.

The probability generating function is $(\frac{1}{2} + \frac{1}{2}\theta)^5 = (\frac{1}{2})^5 (1 + \theta)^5$

$$(\tfrac{1}{2})^5 (1 + \theta)^5 = \tfrac{1}{32}(1 + 5\theta + 10\theta^2 + 10\theta^3 + 5\theta^4 + \theta^5)$$
$$= P(0) + \theta P(1) + \theta^2 P(2) + \theta^3 P(3) + \theta^4 P(4) + \theta^5 P(5)$$

Hence $P(0) = \frac{1}{32}$, $P(1) = \frac{5}{32}$, $P(2) = \frac{10}{32}$, $P(3) = \frac{10}{32}$
$\qquad P(4) = \frac{5}{32}$, $P(5) = \frac{1}{32}$

(*Note:* $P(0) + P(1) + P(2) + P(3) + P(4) + P(5)$

$$= \sum_{r=0}^{r=5} P(r) = \tfrac{32}{32} = 1)$$

Hence, as a *discrete distribution* this is a *proper distribution*. For the *histogram* see Figure 26.1. (The value of the *variate* is placed at the centre of the class-intervals.) In this case the distribution is *symmetrical*.

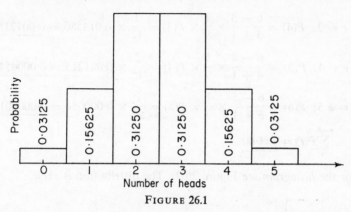

FIGURE 26.1

(ii) In sampling from a large batch of similar components the proportion defective may be assumed constant at 10%. If a sample of six components is drawn at random from the batch calculate the probabilities that the sample will contain 0, 1, 2, 3, 4, 5 or 6 defectives.

The distribution of the number of defectives in the sample is binomial, with $n = 6$, $p = 0.1$, $q = 0.9$, $\dfrac{p}{q} = \dfrac{1}{9}$.

$P(0)$ = probability of 0 defectives in the sample = $(.9)^6 = 0.531441$.

The recurrence formula is $P(r + 1) = \dfrac{n - r}{r + 1} \cdot \dfrac{p}{q} P(r)$

$$= \dfrac{6 - r}{r + 1} \times \dfrac{1}{9} P(r)$$

$r = 0$; $P(1) = \dfrac{6 - 0}{0 + 1} \times \dfrac{1}{9} \times P(0) = \dfrac{6}{9} \times 0.531441$

$$= 2 \times 0.177147 = \underline{0.354294}$$

$r = 1$; $P(2) = \dfrac{6 - 1}{1 + 1} \times \dfrac{1}{9} \times P(1) = \dfrac{5}{18} \times 0.354294$

$$= 5 \times 0.019683 = \underline{0.098415}$$

$r = 2$; $P(3) = \dfrac{6 - 2}{2 + 1} \times \dfrac{1}{9} \times P(2) = \dfrac{4}{27} \times 0.098415$

$$= 4 \times 0.003645 = \underline{0.014580}$$

$$r = 3; \ P(4) = \frac{6 - 3}{3 + 1} \times \frac{1}{9} \times P(3) = \frac{1}{12} \times 0{\cdot}014580 = \underline{0{\cdot}001215}$$

$$r = 4; \ P(5) = \frac{6 - 4}{4 + 1} \times \frac{1}{9} \times P(4) = \frac{2}{45} \times 0{\cdot}001215 = \underline{0{\cdot}000054}$$

$$r = 5; \ P(6) = \frac{6 - 5}{5 + 1} \times \frac{1}{9} \times P(5) = \frac{1}{54} \times 0{\cdot}000054 = \underline{0{\cdot}000001}$$

$$\sum_{r=0}^{r=6} P(r) = \underline{1{\cdot}00}$$

For the *histogram* see Figure 26.2. The distribution is *skew*.

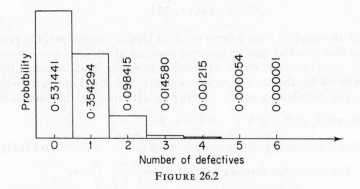

FIGURE 26.2

(*Note:* The probability of up to (and including) 2 defectives $\simeq 1$.)

26.1.2 Mean and variance of the Binomial distribution

These results will be obtained but *it is only necessary to memorize the results*.

For any discrete distribution, if $P(x)$ is the probability of x occurrences of a random event then the mean of the distribution is defined as $\bar{x} = \Sigma x P(x)$. The second moment about zero is defined as $S^2 = \Sigma x^2 P(x)$. The variance σ^2 is given by $\sigma^2 = S^2 - \bar{x}^2$. Consider the generating function $(q + p\theta)^n$,

$$(q + p\theta)^n = P(0) + P(1)\theta + P(2)\theta^2 + \ldots + P(n)\theta^n \qquad (1)$$

Differentiate each side with respect to θ

$$np(q + p\theta)^{n-1} = 1 \cdot P(1) + 2P(2)\theta + \ldots + nP(n)\theta^{n-1} \qquad (2)$$

Let $\theta = 1$, then

$$np(q + p)^{n-1} = 1 \cdot P(1) + 2P(2) + 3P(3) + \ldots + nP(n)$$

$$= \sum_{1}^{n} xP(x)$$

But $q + p = 1$ $\quad \therefore \quad \bar{x} = \Sigma xP(x) = np$

i.e. Mean of Binomial distribution $= np$.

Multiply (2) by θ

$$np\theta(q + p\theta)^{n-1} = 1\theta P(1) + 2P(2)\theta^2 + 3P(3)\theta^3$$
$$+ \ldots + nP(n)\theta^n \qquad (3)$$

Differentiate with respect to θ

$$np[1(q + p\theta)^{n-1} + (n - 1)\theta(q + p\theta)^{n-2}p]$$
$$= 1^2 P(1) + 2^2 P(2)\theta + 3^2 P(3)\theta^2 + \ldots + n^2 P(n)\theta^{n-1} \qquad (4)$$

Let $\theta = 1$, then

$$np[(q + p)^{n-1} + (n - 1)p(q + p)^{n-2}]$$
$$= 1^2 P(1) + 2^2 P(2) + \ldots + n^2 P(n) = \sum_{1}^{n} x^2 P(x)$$

But $q + p = 1$

$$\therefore \quad np[1 + (n - 1)p] = \sum_{1}^{n} x^2 P(x) = S^2$$

$$\therefore \quad \sigma^2 = S^2 - \bar{x}^2 = np + np^2(n - 1) - (np)^2$$
$$= np + n^2 p^2 - np^2 - n^2 p^2 = np - np^2 = np(1 - p)$$

But $1 - p = q$ $\qquad \therefore \qquad \sigma^2 = npq$

i.e. Mean of Binomial distribution $\underline{\bar{x} = np}$

Variance of Binomial distribution $\underline{\sigma^2 = npq}$

Examples:

(i) Referring to the coin tossing problem of 26.1.1

$p = \frac{1}{2}, q = \frac{1}{2}, n = 5$

Mean $= np = 5 \times \frac{1}{2} = \underline{2\cdot5}$ $(\bar{x})$

Variance $= npq = 5 \times \frac{1}{2} \times \frac{1}{2} = \underline{1\cdot25}$ (σ^2)

Standard deviation $\sigma = \sqrt{1\cdot25} = \underline{1\cdot118}$

(*Note:* This mean checks with the *histogram* in Figure 26.1.)

(ii) Referring to the sampling problem in 26.1.1

$p = 0\cdot1, q = 0\cdot9, n = 6$

Mean number of defectives $= np = 6 \times 0\cdot1 = \underline{0\cdot6}$

Variance of number of defectives $= npq = 6 \times 0\cdot1 \times 0\cdot9 = \underline{0\cdot54}$

Standard deviation of number of defectives $= \sqrt{0\cdot54} = \underline{0\cdot7348}$

26.2 The Poisson Distribution

This may be regarded as the distribution which arises from the limiting form of a binomial distribution when p is small (i.e. the event is *rare*). Suppose that p is small and as n tends to infinity np remains finite $= \mu$ $(p = \mu/n,\ q = 1 - p = 1 - \mu/n)$. The probability generating function for the binomial distribution becomes

$$G(\theta) = (q + p\theta)^n = [1 - \mu/n + (\mu/n)\theta]^n = [1 + \mu(\theta - 1)/n]^n$$

The limiting value of this as n tends to infinity (see 4.3.1) may be taken as $e^{\mu(\theta-1)} = e^{-\mu} e^{\mu\theta}$, i.e.

$$G(\theta) = e^{-\mu} \left(1 + \mu\theta + \frac{\mu^2\theta^2}{2!} + \frac{\mu^3\theta^3}{3!} + \ldots + \frac{\mu^r\theta^r}{r!} + \ldots\ ad\ inf.\right)$$

By definition of the probability generating function if $P(r)$ is the probability of exactly r events occurring

$$P(r) = \text{coefficient of } \theta^r = \frac{e^{-\mu}\mu^r}{r!}$$

i.e. the probability of exactly r occurrences of a *statistically rare* event is given by $P(r) = \dfrac{e^{-\mu}\mu^r}{r!}$. This is known as the *Poisson probability function*.

Properties of the Poisson Distribution

(i) It is theoretically possible for any number of occurrences in a given interval.

(ii) The probability of 0 or 1 or 2 or 3 . . . occurrences is

$$P = e^{-\mu}(1 + \mu/1! + \mu^2/2! + \mu^3/3! + \dots \text{ad inf.})$$
$$= e^{-\mu} \times e^{\mu} = e^0 = 1$$

i.e. the distribution is proper.

26.2.1 Recurrence relation for Poisson probabilities

This is a formula which enables successive probabilities to be evaluated based on probabilities already evaluated.

$$P(r) = \frac{e^{-\mu}\mu^r}{r!}, \qquad P(r + 1) = \frac{e^{-\mu}\mu^{r+1}}{(r + 1)!}$$

$$\therefore \qquad P(r + 1) = \frac{e^{-\mu}\mu^r}{r!} \times \frac{\mu}{r + 1} = \frac{\mu}{r + 1} P(r)$$

26.2.2 Mean and variance of the Poisson distribution

$$\text{Poisson mean} \quad = \lim_{n \to \infty} np = \underline{\mu}$$

$$\text{Poisson variance} = \lim_{n \to \infty} npq = \lim_{n \to \infty} np(1 - p) \; [p \text{ small}]$$

$$= \lim_{n \to \infty} np = \underline{\mu}$$

i.e. for a *true* Poisson distribution

$$\underline{\text{Mean} = \text{variance} = \mu}$$

These results may be established by considering the *moments of the distribution*. At this stage it will be sufficient to use the results as established.

For purposes of the evaluation of Poisson probabilities a table of values of $e^{-\mu}$ is useful, but, for any given value of μ these values may be found from standard *Napierian logarithm tables*.

Examples:

(i) Consider the problem discussed in Example (ii) of 26.1.1. This will be treated as an approximate Poisson distribution problem and compared with the exact results.

$n = 6, p = 0\cdot1$, mean $= np = 0\cdot6 = \mu$

The probability of no defectives $= e^{-\mu} = e^{-0\cdot6} = 0\cdot5488 = P(0)$.

Using $P(r + 1) = \dfrac{\mu}{r + 1} P(r)$

$r = 0; \; P(1) = \dfrac{0\cdot6}{1} \times P(0) = 0\cdot6 \times 0\cdot5488 = 0\cdot32928$

$r = 1; \; P(2) = \dfrac{0\cdot6}{2} \times P(1) = 0\cdot3 \times 0\cdot32928 = 0\cdot098784$

$r = 2; \; P(3) = \dfrac{0\cdot6}{3} \times P(2) = 0\cdot2 \times 0\cdot098784 = 0\cdot0197568$

$r = 3; \; P(4) = \dfrac{0\cdot6}{4} \times P(3) = 0\cdot15 \times 0\cdot0197568 = 0\cdot00296352$

These probabilities differ from the Binomial probabilities. The reasons are (a) only a four figure value was used for $P(0)$, (b) p is not small, (c) n is not large. (When the sample is large and p small the probabilities are better approximations and when n is large it is inconvenient to try to calculate Binomial probabilities.)

(ii) During the manufacture of transistors the proportion defective may be taken to be roughly constant at 1%. Estimate the chances that, in a random sample of 100 transistors there will be (a) no defectives, (b) 1 defective, (c) 2 defectives, (d) at least 3 defectives.

$p = 1/100 = 0.01$, $\mu = np = 100 \times 0.01 = \underline{1.0}$

$P(0) = e^{-\mu} = e^{-1} = \underline{0.3679}$

$$P(r + 1) = \frac{\mu}{r + 1} P(r) = \frac{1}{r + 1} P(r)$$

$r = 0$; $P(1) = \frac{1}{1} \times P(0) = \underline{0.3679}$

$r = 1$; $P(2) = \frac{1}{2} \times P(1) = \frac{1}{2} \times 0.3679 = \underline{0.18395}$

$P(0) + P(1) + P(2) = 0.3679 + 0.3679 + 0.18395 = \underline{0.91975}$

But $\sum_{0}^{\infty} P(r) = 1$

$$\therefore \quad P(r > 2) = 1 - P(r \leqslant 2) = 1 - 0.91975$$
$$= 0.08025 \simeq \underline{0.08}$$

i.e. the probability, that in a random sample of 100 at least 3 defectives will occur $= 0.08$ (fairly small).

26.2.3 Poisson probability charts

Charts may be drawn up to read off cumulative probabilities. From these charts it is possible to read off $P(r \geqslant k)$ given μ, i.e. the charts enable the probabilities that the number of *defectives* will exceed or equal any number k to be read off. The charts may also be used to estimate approximately the *probabilities* that the number of defectives will exceed a specified number.

26.3 The Normal Distribution

This is the most important *continuous* distribution. It can be derived in a number of ways, but at this stage it will only be *stated*. It is the basis of many *sampling schemes* and is also used for *hypothesis* testing. In addition it can be used to estimate *approximate* probabilities for *discrete* random distributions.

The general distribution

This may be derived as a limiting form of a binomial distribution on the assumption that the distribution has a mean μ and that there is an *equal chance* of randomly selected values having up to equal *positive and negative* deviations from μ. If the standard deviation (standard error) of a normal (Gaussian) distribution is σ then the *frequency function* (probability density function) is of the form

$$f(x) = \frac{1}{\sigma \sqrt{(2\pi)}}\, e^{-(x-\mu)^2/2\sigma^2}$$

If the frequency function is plotted, a distribution, *symmetrical* about the mean value μ, is obtained. Since the distribution is symmetrical Mean = Mode = Median. The appearance of a general *normal frequency curve* is shown in Figure 26.3. Allowance is made for the

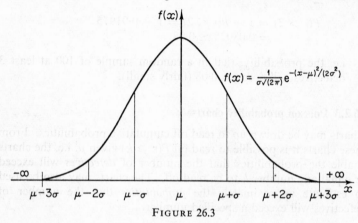

FIGURE 26.3

possibility of *any deviation* arising between $-\infty$ and $+\infty$ (from the mean μ).

From the work done in Chapter 24 on continuous distributions, the probability that, in a random selection, a value of x chosen would lie between a and b is given by

$$P(a \leqslant x \leqslant b) = \int_a^b f(x)\, dx = \frac{1}{\sigma \sqrt{(2\pi)}} \int_a^b e^{-(x-\mu)^2/2\sigma^2}\, dx$$

This integral can only be evaluated by approximate numerical methods (e.g. Simpson's rule), but because of the importance of this distribution values are tabulated in *Standard Normal tables* which enable the approximate values of these integrals to be found for practical purposes.

Note: Since this is a 'proper' distribution

$$\int_{-\infty}^{+\infty} f(x)\,\mathrm{d}x = 1; \quad \int_{-\infty}^{\mu} f(x)\,\mathrm{d}x = \int_{\mu}^{\infty} f(x)\,\mathrm{d}x = \tfrac{1}{2}$$

From tables it can be found that:

$$P(\mu - \sigma \leqslant x \leqslant \mu + \sigma) \quad = 0{\cdot}6826 \ (68{\cdot}26\%)$$
$$P(\mu - 2\sigma \leqslant x \leqslant \mu + 2\sigma) = 0{\cdot}9544 \ (95{\cdot}44\%)$$
$$P(\mu - 3\sigma \leqslant x \leqslant \mu + 3\sigma) = 0{\cdot}9974 \ (99{\cdot}74\%)$$

and also

$$P(\mu - 1{\cdot}96\sigma \leqslant x \leqslant \mu + 1{\cdot}96\sigma) = 0{\cdot}95 \ \ (95\%)$$
$$P(\mu - 3{\cdot}09\sigma \leqslant x \leqslant \mu + 3{\cdot}09\sigma) = 0{\cdot}998 \ (99{\cdot}8\%)$$

Confidence limits

(1) The limits $\pm 1{\cdot}96\sigma$ from the mean are often referred to as the 95% *confidence limits*, i.e. there is a 95% probability that, in a random selection from a normal distribution of mean μ and standard deviation (error) σ the value chosen will lie between $\mu \pm 1{\cdot}96\sigma$.

(2) The limits $\pm 3{\cdot}09\sigma$ from the mean are often referred to as the 99·8% *confidence limits*, i.e. there is a 99·8% probability that, in a random selection from a normal distribution of mean μ and standard deviation (error) σ the value chosen will lie between $\mu \pm 3{\cdot}09\sigma$.

These limits are of *great importance* in problems of production quality control.

26.3.1 The standard normal distribution

For general purposes of estimation of probabilities it is convenient to have a *standard table* to which reference may be made.

The *standard normal distribution* has a mean zero and standard deviation (error) of 1 unit. The probability density function of a standard normal distribution is of the form

$$f(t) = \frac{1}{\sqrt{(2\pi)}}\,\mathrm{e}^{-t^2/2} \ (-\infty < t < +\infty)$$

A *general* normal distribution is reduced to a distribution of this form by substituting $t = (x - \mu)/\sigma$, then if T corresponds to X,

$T = (X - \mu)/\sigma$, then

$$P(-\infty < x < X) = P(-\infty < t < T) = \frac{1}{\sqrt{(2\pi)}} \int_{-\infty}^{T} e^{-t^2/2}\, dt$$

Values of this integral $\left(\text{or } \frac{1}{\sqrt{(2\pi)}} \int_{0}^{T} e^{-t^2/2}\, dt \right)$ are *tabulated* as *standard normal integrals*.

$$\left[\textit{Note: } \frac{1}{\sqrt{(2\pi)}} \int_{\infty}^{0} e^{-t^2/2}\, dt = 0{\cdot}5. \right]$$

The distribution may be used to estimate probabilities and also to estimate numbers likely to fall in any interval. The tables may also be used to *fit* an approximate normal distribution to a given set of data. A normal table is given in the Appendix showing values of

$$P(-\infty \leqslant x \leqslant X) = \frac{1}{\sqrt{(2\pi)}} \int_{-\infty}^{X} e^{-x^2/2}\, dx$$

More detailed tables may be found in standard sets of *Statistical tables*. Interpolation may be used to estimate intermediate probabilities and variate values.

Examples

(i) In the manufacture of a certain component the mean diameter of the component has been established as 3·0 in. The dimension is normally distributed with a standard deviation of 0·01 in. Estimate the probability that the diameter of a particular component will be between 2·980 and 3·015 in.

Assuming a normal distribution, let x = dimension, μ = mean and σ = standard deviation.

Reducing to a *standard normal* variate let

$$T_1 = (x_1 - \mu)/\sigma = (2{\cdot}98 - 3{\cdot}0)/0{\cdot}01 = -0{\cdot}02/0{\cdot}01 = \underline{-2}$$
and $T_2 = (x_2 - \mu)/\sigma = (3{\cdot}015 - 3{\cdot}0)/0{\cdot}01 = +0{\cdot}015/0{\cdot}01 = \underline{+1{\cdot}5}$

$P(2{\cdot}980 < x < 3{\cdot}015) = P(T_1 < t < T_2)$

(*Note:* The symmetry of the distribution is used.)

$P(-2 < t < 0) \;\; = P(0 < t < 2) = 0{\cdot}4772$ (from table)
$P(0 < t < 1{\cdot}5) \;\; = 0{\cdot}4332$

$$\therefore P(2\cdot980 < x < 3\cdot015) = P(-2 < t < 1\cdot5)$$
$$= P(-2 < t < 0) + P(0 < t < 1\cdot5) = 0\cdot4772 + 0\cdot4332$$
$$= \underline{0\cdot9104}$$

or there is approximately a 91 % probability that x is between 2·980 in. and 3·015 in.

(ii) The proportion defective in manufactured bearings may be taken as 2 %. If a sample of 100 is drawn at random estimate the probability that up to 3 defectives may be found assuming that the distribution is approximately normal. (*Note:* Since the distribution of defective is basically a *binomial* distribution and the binomial distribution is represented by a *histogram* with cells 1 unit wide a better approximation is achieved by using $x = 3\cdot5$ in the *normal* computation (see Figure 26.4).

In a sample of 100 the mean number *expected* $= np =$ $100 \times 0\cdot02 = 2$. Variance in the sample $= npq =$ $100 \times 0\cdot02 \times 0\cdot98 = 1\cdot96$. Standard deviation in the sample $= \sqrt{1\cdot96} = \underline{1\cdot4}$.

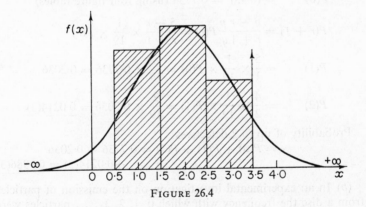

FIGURE 26.4

The probability required is approximately equal to the area under the normal curve from $-\infty$ to $+3\cdot5$. Take $\mu = 2$, $x = 3\cdot5$, $\sigma = 1\cdot4$. For

$$x = 3\cdot5, \quad t = (x - \mu)/\sigma = (3\cdot5 - 2)/1\cdot4 = 1\cdot5/1\cdot4 = \underline{1\cdot07}$$

$P(0 < t < 1\cdot07) = 0\cdot3577$

$P(-\infty < t < 1\cdot07) = 0\cdot5000 + 0\cdot3577 = 0\cdot8577$

or the approximate probability of up to and including 3 defectives = $0\cdot8577$ or approximately $\underline{85\cdot8\%}$.

[*Note:* If the Poisson distribution is used, $\mu = 2$

$P(0) = e^{-\mu} = e^{-2} = 0\cdot1353,\ P(1) = 2P(0) = 0\cdot2706$

$P(2) = \frac{2}{2}P(1) = 0\cdot2706, \qquad P(3) = \frac{2}{3}P(2) = \frac{2}{3} \times 0\cdot2706$
$= 0\cdot1804$

$P(0) + P(1) + P(2) + P(3) = 0\cdot8567$ or approximately $85\cdot7\%$.]

WORKED EXAMPLES

(*a*) In the manufacture of certain resistors it is known from past experience that the proportion defective is approximately constant at 5%. Using the binomial distribution estimate the probability that, in a sample of 5, taken at random up to (and including) 2 may be defective.

$$n = 5,\ p = 5/100 = 0\cdot05,\ q = 0\cdot95$$

$P(0) \qquad = (0\cdot95)^5 = \underline{0\cdot7736}$ (using four figure tables)

$$P(r + 1) = \frac{n - r}{r + 1}\frac{p}{q}\,P(r) = \frac{5 - r}{r + 1} \times \frac{1}{19} \times P(r)$$

$$P(1) \qquad = \frac{5}{1} \times \frac{1}{19} \times P(0) = \frac{5}{19} \times 0\cdot7736 = \underline{0\cdot2036}$$

$$P(2) \qquad = \frac{4}{2} \times \frac{1}{19} \times P(1) = \frac{2}{19} \times 0\cdot2036 = \underline{0\cdot0214(3)}$$

Probability of up to 2 defectives
$$= P(0) + P(1) + P(2) = 0\cdot7736 + 0\cdot2036$$
$$+ 0\cdot0214(3) = \underline{0\cdot9986(3)}$$

(*b*) In an experimental investigation on the emission of particles from a disc the frequency with which 0, 1, 2, 3, . . . particles were emitted were recorded in periods of a given length of time. The numbers of particles emitted during 2000 time periods are recorded and the mean number per period calculated as $3\cdot75$. Assuming that the frequencies follow a Poisson distribution estimate the number of occasions when 0, 1, 2, 3, 4, 5 particles were recorded.

On the assumption of a Poisson distribution

$$P(r) = e^{-\mu}\mu^r/r!, \quad \mu = 3.75, e^{-\mu} = e^{-3.75} = 0.02352$$

$$P(r + 1) = \frac{\mu}{r + 1} P(r)$$

$$P(0) \quad = e^{-\mu} = e^{-3.75} = 0.02352$$

$$P(1) \quad = 3.75 \times P(0) = 3.75 \times 0.02352 = 0.08820$$

$$P(2) \quad = \frac{3.75}{2} \times P(1) = \frac{3.75}{2} \times 0.08820 = 0.16540$$

$$P(3) \quad = \frac{3.75}{3} \times P(2) = 1.25 \times 0.16540 = 0.20660$$

$$P(4) \quad = \frac{3.75}{4} \times P(3) = \frac{3.75}{4} \times 0.20660 = 0.19370$$

$$P(5) \quad = \frac{3.75}{5} \times P(4) = 0.75 \times 0.19370 = 0.14530$$

In the large sample of 2000 the probable frequencies of occurrence of r particles = $2000 \times P(r)$, i.e. the estimated frequencies of 0, 1, 2, 3, 4, 5 emissions are

47·04, 176·40, 330·8, 413·20, 387·4, 290·60

or <u>approximately 47, 176, 331, 413, 387, 291.</u>

(c) Using a normal approximation to the binomial distribution estimate the probabilities that, if a coin is thrown 100 times (i) 54 or more heads will appear, (ii) 47 or less heads will appear.

At each throw the probability of a head $p = \frac{1}{2}$. In 100 throws the *expectation* (mean) number of heads

$$= np = 100 \times \tfrac{1}{2} = \underline{50}$$

Variance of number of heads = $npq = 100 \times \tfrac{1}{2} \times \tfrac{1}{2} = \underline{25.}$

Standard deviation of number of heads = $\sqrt{25} = 5$.

If the histogram is plotted for the binomial distribution the probabilities involved are represented by the areas of the cells (Figure 26.5). The normal curve for a mean of 50 and S.D. = 5 is sketched in.

The probability that the number of heads will be 54 or more = sum of areas of the cells to the right of $x = 53.5$. Let

$$t_1 = (x - \bar{x})/\sigma = (53.5 - 50)/5 = 3.5/5 = \underline{0.7}$$

From normal tables $P(0 < t < 0.7) = 0.2580$

$\therefore$ $\qquad\qquad P(0.7 < t < \infty) = 0.5000 - 0.2580 = \underline{0.242}$

i.e. the probability that the number of heads exceeds $53 \simeq 24.2\%$. The probability that the number of heads will be 47 or less = sum of

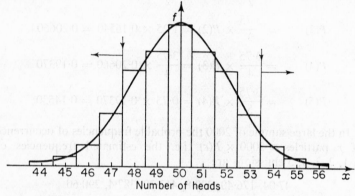

Number of heads

FIGURE 26.5

areas of cells to the *left* of $x = 47.5 \simeq$ area under the normal curve to the *left* of $x = 47.5$. Let

$$t_2 = (x - \bar{x})/\sigma = (47.5 - 50)/5 = -2.5/5 = \underline{-0.5}$$

By symmetry

$$P(-0.5 < t < 0) = P(0 < t < 0.5) = \underline{0.1915}$$

$\therefore$ $\qquad\qquad P(-\infty < t < -0.5) = 0.5000 - 0.1915 = \underline{0.3085}$

i.e. the probability that the number of heads will be 47 or less $\simeq 30.8\%$.

Note: This adjustment of $+\frac{1}{2}$ or $-\frac{1}{2}$ required on integral values to use the normal approximation is called the *correction for continuity* and is always necessary when using the normal *approximation* to a *discrete* distribution.

Examples 26

1. An unbiased coin is tossed 200 times as an experiment. Calculate the mean number of heads expected and the standard deviation of the number of heads. What are the 2σ and 3σ limits of the number of heads expected? (If $\bar{x}$ = mean, σ = standard deviation the limits required are $\bar{x} \pm 2\sigma$ and $\bar{x} \pm 3\sigma$).

2. A large supply of washers is known to be 2 per cent defective. If a sample of 5 is selected at random, using the binomial distribution estimate the probability that not more than 2 will be defective.

3. An unbiased dice is thrown 6 times. (The chance of any number turning up at one throw may be taken as constant $= \frac{1}{6}$.) Using the binomial distribution calculate the probabilities that a 5 will appear on 0, 1, 2, 3, 4, 5 or 6 occasions.

4. Taking the chances of a child being a boy or girl as evens (i.e probability that a child is a boy $= \frac{1}{2}$, probability of being a girl $= \frac{1}{2}$) calculate the chances that (i) in a family of 3 children all are girls,
   (ii) in a family of 5 children 3 are boys, 2 girls,
   (iii) a family of 6 children will have up to and including 2 boys.

5. During the production of a man made fibre equal lengths are subject at random to the same breaking stress. On average it is expected that 1/10th of the test pieces will break.
   If a sample of 5 is tested for breaking, calculate the probabilities that 0, 1, 2, 3, 4 or 5 fibres in the sample will break.

6. If a 3 per cent defective product is sampled in 100's estimate the proportion of samples (using the Poisson distribution) which will contain (i) 0 defectives, (ii) up to and including 2 defectives (iii) more than 3 defectives.

7. A product is 1% defective and is packed in cartons of 50. What percentage of cartons may be (i) free from defectives (ii) contain 2 or more defectives?

8. If in a machine shop, on average the chance of a machine breaking down during any week is 1 in 50, estimate (using the Poisson distribution), the probabilities that in a shop using 100 similar machines that
   (i) 0 will break down during a given week,
   (ii) up to 3 will break down during a given week,
   (iii) more than 3 will break down during a given week.

9. During the manufacture of small electrical components a limit is set to the probability of passing out batches. It is known from past experience that the components are $2\frac{1}{2}$ per cent defective. If the limits of rejection are $\bar{x} \pm 2\sigma$ (where $\bar{x}$ is the mean, σ the standard deviation) using the Poisson distribution, if a batch of 200 contains 8 defectives find whether the batch should be passed or rejected.

10. If x is a normally distributed variate with mean $\mu = 4$ and standard deviation $\sigma = 1$, estimate, using the normal integral tables:
    (i) $P(4 < x < 6)$, (ii) $P(2.5 < x < 5.5)$.

11. A normally distributed dimension has a mean of 15 in. and a standard deviation of 3 in. Out of 500 specimens how many may be expected to have a dimension (i) greater than 18 in., (ii) between 12 and 18 in. For this dimension what are the appropriate 95% and 99.8% confidence limits $(\bar{x} \pm 1.96\sigma, \bar{x} \pm 3.09\sigma)$.

12. A machine is producing components with a mean length of 2·55 in. and a standard deviation of 0·02 in. What proportion of components are likely to be produced with dimensions between 2·52 and 2·60 in. assuming the dimension to be normally distributed?

13. Using a normal approximation to the binomial distribution, if an unbiased coin is tossed a number of times, estimate the probabilities that:

   (i) in 100 tossings 52 or more heads will appear,
   (ii) in a 1000 tossings the number of heads will lie between 475 and 525.

14. In the manufacture of resistors the mean value is 50 ohms and the standard error is 4 ohms. Estimate, using the normal distribution how many, in a batch of 250 are likely to have a resistance (i) greater than 56 ohms, (ii) between 45 and 55 ohms.

15. A box contains 1000 similar items of which 20 are defective. A random sample of 10 is taken from the box. (i) Calculate the probability that 2 of the items in the sample will be defective. (ii) Use the Poisson distribution to estimate this probability.

27

Revision Examples

27.1 Examples on Chapters 1–6

1. (i) Solve the equation $7^{x^2-3} = 5^{2x+1}$.

   (ii) The current i in an electrical circuit at time t is given by $i = 75(1 - e^{-Rt/L})$. Transpose the formula to find t and hence calculate the value of t when $i = 7\cdot5$, $R = 2 \times 10^6$, $L = 5 \times 10^{-3}$.

2. (i) The roots of the equation $3x^2 + 4x - 7 = 0$ are α and β. Without solving the equation, show that the roots are real and calculate the values of (a) $\alpha^2 + \beta^2$, (b) $\dfrac{\alpha}{\beta} + \dfrac{\beta}{\alpha}$, (c) $\alpha^3 + \beta^3$.

   (ii) Without solving the equation, form the equation whose roots are the cubes of the roots of the equation $3x^2 + 4x - 7 = 0$.

3. (i) Using the method of completion of the square (a) find the minimum value of the function $5x^2 - 3x + 4$ and state the value of x at which it occurs. (b) Find the maximum value of the function $15 + 7t - 6t^2$ and state the value of t at which it occurs.

   (ii) Find the range of values of the function $\dfrac{2x + 7}{1 + 3x - x^2}$ for real values of x.

4. (i) Solve the equation $\dfrac{e^x + e^{-x}}{e^x - e^{-x}} = 10$,

   (ii) Determine the co-ordinates of the points of intersection of the straight line $y = 2x + 1$ and the circle $x^2 + y^2 - 6x + 4y - 7 = 0$.

5. (i) Evaluate (a) $^{10}P_3$, $^{7}P_5$, $^{15}P_3$.

      (b) $^{7}C_3$, $^{9}C_4$, $^{10}C_4$.

  (ii) Calculate the number of permutations of the letters of the word RANDOMIZE taking

      (a) 3 letters only (b) 6 letters only (c) all nine letters.

6. A bag contains 12 balls of equal size, 6 white, 4 red and 2 black. Estimate the chances of drawing, at random, a sample of 4 which will contain

  (i) 4 whites, (ii) 3 reds, (iii) 2 whites, 1 red, 1 black.

7. Express $E = \dfrac{4 + 2x - 3x^2}{(1 - x)^2(1 + 2x)}$ in partial fractions and hence expand the function in ascending powers of x as far as the term in x^3. For what range of values of x is the infinite series valid?

8. Using a suitable test for convergence, discuss the absolute convergence of the infinite series

  (i) $\displaystyle\sum_{r=0}^{\infty} (3x)^r$,   (ii) $\displaystyle\sum_{r=0}^{\infty} \frac{x^{2r+1}}{(2r + 1)!}$,   (iii) $\displaystyle\sum_{n=0}^{\infty} \frac{x^n}{(n + 2)^2}$.

9. (i) Using a first order binomial approximation method evaluate

$$F = \frac{(1{\cdot}01)^4 \times \sqrt[3]{(26{\cdot}46)}}{(0{\cdot}985)^5}.$$

  (ii) Using the formula $W = P^2 X^3 / \sqrt{N}$ and a first order binomial approximation, find the percentage change in W when P increases by 1%, X decreases by $0{\cdot}5\%$ and N increases by $1{\cdot}5\%$ stating whether the change is an increase or decrease.

10. Using change of sign of the function $f(x) = 2x^3 - 4x - 1$ show that the equation $2x^3 - 4x - 1 = 0$ has real roots in the ranges -2 to -1, -1 to 0 and 1 to 2. Use a binomial approximation method to estimate the largest root to 2 decimal places.

11. Show that the equation $e^{-2x} - x^2 = 0$ has a root between $0{\cdot}5$ and $0{\cdot}6$. Use the Newton approximation method to estimate the root to two places of decimals.

12. The following values of the variables V and T are suspected to obey a law of the form $V = AT^k$. Verify graphically that this is so and estimate possible values for A and k.

T	30·5	46	61	76	92
V	0·15	0·55	1·19	2·32	4·07

27.2 Examples on Chapters 7–13

1. The vertices of a triangle ABC are A$(-2, -3)$, B$(1, 4)$, and C$(4, -2)$. Find (i) the lengths of the three sides, (ii) the angle ABC, (iii) the equations of the three sides AB, BC and AC.

2. The points P$(3, 5)$ and Q$(-4, 1)$ are the ends of a diameter of a circle. Find the cartesian equation of the circle. What are the co-ordinates of the centre of the circle and the radius of the circle? Find the co-ordinates of the points where the line $y = x + 1$ intersects the circle.

3. (i) Sketch the curve whose polar equation is $r = 5 + 4\cos\theta$.
   (ii) Rewrite the equation of the curve $x^2 + 2y^2 = 2x + 3y + 5$ in polar form.

4. (i) Express $E = 200\sin(50\pi t + \pi/3) + 100\cos(50\pi t - \pi/4)$ in the form $R\cos(50\pi t - \alpha)$. [α in radians and acute.]
   (ii) Prove the identity
   $$\frac{\sin(4x + \alpha) + \sin(4x - \alpha)}{\cos(2x + \alpha) + \cos(2x - \alpha)} = 2\sin 2x.$$

5. Solve the following trigonometric equations giving solutions between $0°$ and $360°$,
   (i) $5\cos 2x = 4\sin x - 1$, (ii) $\cos 3\theta - \cos\theta = 0$,
   (iii) $5\sin\theta - 12\cos\theta = 5\cdot83$.

6. Using appropriate half-angle formulae solve the triangles (i) ABC, given AB $= 8\cdot7$ cm, BC $= 9\cdot6$ cm, AC $= 11\cdot5$ cm, (ii) PQR, given $p = 15\cdot6$ cm, $q = 11\cdot8$ cm and angle PRQ $= 78°16'$.

7. (i) Simplify $(3 + 4j)(5 - j)$. Express the result in the forms $a + jb$, $r(\cos\theta + j\sin\theta)$, $r\,e^{j\theta}$.
   (ii) Solve the quadratic equation $3z^2 + 5z + 4 = 0$ and express the results in the forms $x + jy$, $r\angle\theta$, $r\,e^{j\theta}$.

8. Two forces of magnitudes 15 lbf and 30 lbf act at angles of $40°$ and $65°$ respectively to the positive direction of the x-axis. Using complex numbers find the resultant force in magnitude and direction.

9. (i) The resultant impedance Z of two parallel circuits is given by $\frac{1}{Z} = \frac{1}{Z_1} + \frac{1}{Z_2}$. If $Z_1 = 5 - 3j$ and $Z_2 = 3 + 5j$ express Z in the form $X + jY$.

(ii) If $z = x + \mathrm{j}y$ and $w = u + \mathrm{j}v$ and the z-plane is transformed into the w-plane by the transformation $w = z + 3/z$, express u and v in terms of x and y.

10. The following values of the function $f(x)$, for increasing values of x are given. Show that $f(x)$ is a polynomial in x and find the polynomial.

x	0	1	2	3	4	5	6	7	8
$f(x)$	-5	1	11	25	43	65	91	121	155

11. The following values of a function $f(x)$ occur for increasing values of the argument x. Write down the table of finite differences for the function up to the fourth order. Using the Newton interpolation formula estimate $f(2\cdot3)$ and $f(2\cdot45)$.

x	2·0	2·2	2·4	2·6	2·8	3·0	3·2	3·4
$f(x)$	0·6931	0·7885	0·8755	0·9555	1·0296	1·0986	1·1632	1·2238

12. (i) Using first principles, show that the derivative of $y = 3x^2 + 5/x + 4$ is $6x - 5/x^2$.

(ii) Determine the stationary values of the function

$$2t^3 - 21t^2 + 60t + 19$$

(iii) Determine the equations of the tangent and normal to the curve $y = 2(x - 1)^2$ at the point where $x = 2$.

27.3 Examples on Chapters 14–20

1. If $y = A\,\mathrm{e}^{2x} + B\,\mathrm{e}^{-x}$ where A and B are constants, show that y satisfies the differential equation

$$\frac{\mathrm{d}^2y}{\mathrm{d}x^2} - \frac{\mathrm{d}y}{\mathrm{d}x} - 2y = 0.$$

2. (i) Verify that $s = \mathrm{e}^{-t}\cos 3t$ satisfies the differential equation

$$\frac{\mathrm{d}^2s}{\mathrm{d}t^2} + 2\frac{\mathrm{d}s}{\mathrm{d}t} + 10s = 0.$$

(ii) If the function $y = A \, \mathrm{e}^{mx}$ satisfies the differential equation

$$\frac{\mathrm{d}^2 y}{\mathrm{d}x^2} - 2 \frac{\mathrm{d}y}{\mathrm{d}x} - 15y = 0$$

find the two possible values of m.

3. At time t seconds the distance s feet of a particle moving in a straight line from a fixed point O in the line is given by $s = 2t^3 - 27t^2 + 108t + 35$. Find (i) the times at which the particle comes to rest, (ii) the accelerations at these times, (iii) the distances from O at the times when the particle is at rest, (iv) the distance from O at the instant when the acceleration is zero.

4. Determine the coordinates of the points where the curve $y = 2x^2 - 5x - 3$ cuts the x-axis. Obtain the equations of the tangent and normal to the curve at each of these points.

5. A conical funnel is used to fill a can. The depth of the funnel is 5 inches and the diameter of the top is 6 inches. If fluid is poured slowly into the funnel at a rate of 4·5 in³/s and flows into the can at a rate of 2 in³/s, find the rate at which the level of fluid is rising in the funnel when the depth of fluid in the funnel is 3·5 in.

6. A template is in the form of an equilateral triangle with equal rectangles on each side of the triangle (one edge of each rectangle being a side of the triangle). If the perimeter is to be 48 in., find the dimensions of the template so that the area shall be a maximum.

7. (i) If $U = 2x^3 + 3x^2 y - 6y^2 x$, write down

$$\frac{\partial U}{\partial x}, \quad \frac{\partial^2 U}{\partial x^2}, \quad \frac{\partial^2 U}{\partial y^2}, \quad \frac{\partial^2 U}{\partial x \, \partial y}$$

and verify that

$$x \frac{\partial U}{\partial x} + y \frac{\partial U}{\partial y} = 3U.$$

(ii) If $V = x \sin y + y \cos x$ write down the expressions for

$$\frac{\partial V}{\partial x}, \quad \frac{\partial V}{\partial y}, \quad \frac{\partial^2 V}{\partial x^2}, \quad \frac{\partial^2 V}{\partial y^2}$$

and verify that

$$\frac{\partial^2 V}{\partial y \, \partial x} = \frac{\partial^2 V}{\partial x \, \partial y}.$$

8. (i) The gradient of the tangent to the curve $y = f(x)$ is given by $\dfrac{dy}{dx} = 4x + 3$. Determine the equation of the curve if it passes through the point (3, 15).

   (ii) The acceleration of a particle moving along a fixed straight line from a fixed point O is $(3t + 2)$ ft/s^2 at time t seconds. Find the distance s ft of the particle from the fixed point in terms of t given that, when $t = 0$, $s = 10$ ft and the initial velocity is 15 ft/s.

9. (i) Using suitable substitutions evaluate:

   (a) $\displaystyle\int_2^3 \frac{x}{\sqrt{(x^2 + 5)}}\,dx,$ (b) $\displaystyle\int_0^{\pi/4} \cos^3 \theta \sin \theta \, d\theta,$

   (c) $\displaystyle\int_0^3 \frac{2t + 3}{t^2 + 3t + 5}\,dt.$

   (ii) Using partial fractions evaluate:

   (a) $\displaystyle\int_2^4 \frac{2x + 1}{(x + 2)(x + 1)}\,dx,$ (b) $\displaystyle\int_0^2 \frac{3u + 2}{u^2 + 7u + 12}\,du.$

10. Using integration by parts, find the values of:

   (i) $\displaystyle\int_1^3 x^3 \log_e x \, dx,$ (ii) $\displaystyle\int_0^1 t^2 e^t \, dt,$ (iii) $\displaystyle\int_0^{\pi/4} x \cos 2x \, dx.$

11. Sketch the graph of $y = 2\sqrt{x} + 3$ between $x = 0$ and $x = 4$. Determine, by integration

   (i) the area between the curve, the x-axis and the ordinates $x = 0$ and $x = 4$,

   (ii) the volume of revolution when this area is rotated through 360° about the x-axis,

   (iii) the mean value of y between $x = 0$ and $x = 4$.

12. (i) An expanding gas obeys the law $pv = C$. Find the work done in ft lbf when 1·5 ft^3 of gas at a pressure of 300 lbf/in^2 expands to 6 ft^3. Find also the mean pressure during the expansion.

   (ii) The current at time t seconds in an electrical circuit is given by $i = 100 \sin(100\pi t + 0·5)$ amp. Calculate the mean current in the circuit between $t = 0$ and $t = 1/100$ sec.

27.4 Examples on Chapters 21–24

1. Using integration determine the mean and root mean square values of the functions
   (i) $2\sqrt{x} + 3$ between $x = 0$ and $x = 4$,
   (ii) $50 \sin 3t$ between $t = 0$ and $t = \pi/3$.

2. Determine the co-ordinates $(\bar{x}, \bar{y})$ of the centroid of the area between the curve $y = x(5 - x)$ and the x-axis, lying in the first quadrant.

3. Use suitable tables and Simpson's rule to evaluate

$$\int_0^{\pi/2} \frac{5}{\sqrt{(2 + \sin\theta)}} \, d\theta \text{ (7 ordinates)}$$

giving the result correct to three significant figures.

4. Determine, using integration, the second moment of area of the area below the curve $(y - 1)^2 = 4x$, in the first quadrant between the ordinates $x = 0$ and $x = 9$ about (i) the y-axis, (ii) the x-axis. Hence find the radius of gyration of the area about the y-axis and x-axis.

5. Determine, by integration, the volume formed when the area in the first quadrant, between the curve $y = 2\sqrt{x} + 3$, the ordinates $x = 1$ and $x = 4$, is rotated through $360°$ about the x-axis. Determine the area under the curve and hence, using Guldin's theorem determine $\bar{y}$, the y coordinate of the centroid of the area.

6. Using the 'variables separable' method of solution, solve the differential equation

$$3y^2 \frac{dy}{dx} = x^2 - 1, \text{ given that } y = 3 \text{ when } x = 1.$$

7. Assuming that $y = A e^{mx}$ is a solution of the equation $2\dfrac{d^2y}{dx^2} + 3\dfrac{dy}{dx} - 2y = 0$ find the general solution.

8. The angle θ (radians) turned through by a simple pendulum at time t seconds is given by the differential equation

$$l\frac{d^2\theta}{dt^2} + g\theta = 0$$

where g is the acceleration due to gravity and l is the length of the pendulum. If $g = 32$ ft/s² and $l = 2$ ft find the expression for the angle turned through in time t seconds given that, at $t = 0$, $\theta = \alpha$ and $\dfrac{d\theta}{dt} = 0$. What is the expression for the angular velocity at time t seconds?

9. In a batch of 40 components, 6 are known to be faulty. If a random selection of 5 components is made, what is the probability that up to and including 2 of the selected components will be faulty?

10. Three equal unbiased dice are thrown simultaneously. Calculate the probabilities that (i) 3 sixes will show, (ii) a six, a five and a four will show, (iii) a total of 12 will show.

11. A drum contains 12 balls of equal size, 6 of which are white, 4 red and 2 blue.

    If three balls are drawn at random, in succession, find the probabilities that (i) the three will all be white; (ii) the three will all be red; (iii) the first two will be white and the third blue; (iv) all three will be of different colour.

12. In the manufacture of a certain electrical component it is established that, on average 1 in 10 are faulty. If a random sample of six components is taken from a large batch, calculate the probabilities that (i) 0, 1 or 2 of the sample will be faulty, (ii) at least 2 will be faulty.

    [*Note:* $(0\cdot9)^6 = 0\cdot531441$.]

27.5 Examples on Chapters 25–26

1. A sample of 100 values of a variate x cm gave the values

Variate x(cm)	5	10	15	20	25	30	35	40
Frequency f	1	5	12	23	30	18	9	2

Assuming that the variate is continuous draw a frequency curve to represent the data and a cumulative frequency curve (ogive). From the diagrams estimate (i) the mode of the variate; (ii) the median value; (iii) the 25 and 75 percentiles.

Calculate the mean and standard deviation of the sample data.

2. During the manufacture of a particular component a dimension is measured on 75 components. The ranges of the dimensions and the number falling in these ranges were as follows:

Dimension (cm)	5·0–5·2	5·2–5·4	5·4–5·6	5·6–5·8	5·8–6·0	6·0–6·2
Frequency (f)	4	16	23	15	12	5

   Draw a histogram to represent the distribution of the dimension. From the histogram estimate the median value of the dimension. Estimate the mean and standard deviation of the dimension in the sample.

3. Using a class interval and a working mean, calculate the mean, variance and standard deviation of the following sample of data:

Variable x	6	12	18	24	30	36
Frequency f	5	17	28	32	14	4

4. A random variate x is distributed according to the frequency function $f(x) = Kx(5 - x)$ over the range $x = 0$ to $x = 5$. Determine K so that this is a 'proper' frequency function. Find (i) the mean of the distribution, (ii) the mode of the distribution, (iii) the variance and standard deviation of the distribution.

5. An unbiased coin is tossed 300 times. Estimate the mean number of heads expected and the standard deviation of the number of heads. What are the 2σ and 3σ limits of the number of heads expected?

   (The assumption is made that the distribution is a binomial.)

6. A large supply of screws is known to be 5% defective. If a sample of 5 is selected at random, using the binomial distribution find the probabilities that there will be 0, 1, 2, 3 defectives in the sample. What is the probability that not more than two defectives will be found in the sample?

7. If a 2% defective product is sampled in 100's estimate the proportion of samples (using the Poisson distribution) which will contain
   (i) 0 defectives, (ii) up to 2 defectives, (iii) 3 or more defectives.

8. If x is a normally distributed variate with a mean $\mu = 5$ and standard deviation $\sigma = 2$, estimate, using normal integral tables:
   (i) $P(5 < x < 7 \cdot 5)$, (ii) $P(3 < x < 6 \cdot 5)$.

9. A normally distributed dimension has a mean of 20 cm and a standard deviation of 4 cm. Out of 1000 random specimens, how many may be expected to have a dimension (i) greater than 26 cm, (ii) less than 12 cm, (iii) between 15 and 25 cm?

10. Using a normal approximation to the binomial distribution, if a product is 5% defective, estimate the probabilities that (i) in a random sample of 100 items, more than 7 will be defective, (ii) in a random sample of 500 items the number of defectives will be between 15 and 35.

Answers to Examples

Examples 1

1. 2·586   2. 4·806   3. 19·7   4. 254·3   5. 37·03
6. (i) 1·753 (ii) 0·492 (iii) 2·568
7. (i) 2·303, −1·303 (ii) 2·613, −1·148   8. 80°9′
9. $\dfrac{L}{R}\log[V/(V-IR)]$; 0·267   10. (i) 4·151, 0·241 (ii) 1·6
11. 255·2   12. (i) 26,010 (ii) $5\cdot141 \times 10^7$ 13. $1000\,p^3\sqrt{q}/r$ (ii) 5
14. $M = eN^{1/2}P^{-1/6}$ 15. 6·85
16. (i) 4·994 rad (ii) $T_1 = 764\cdot9$, $T_2 = 1510$
17. $A = 1066\cdot5$, $B = 516\cdot5$ 18. $a = 5\cdot432$, $b = 0\cdot2815$
19. $n = 1\cdot37(3)$, $c = 147\cdot3$ 20. $A = 0\cdot003\,584$, $n = 2\cdot063$

Examples 2

1. (i) $-0\cdot5 \pm 0\cdot866j$ (ii) $0\cdot2 \pm 0\cdot872j$
2. (i) 2·831(5), −1·059(5) (ii) 0·861(4), 0·682(5)
3. (i) ±1, $\pm\frac{1}{2}$ (ii) 4, 36.   4. (i) $x = 5, -3\cdot8$; $y = 1, -3\cdot4$ (ii) 1·3(01)
5. $(-1\cdot132, -4\cdot396)$, $(6\cdot23, 17\cdot69)$ 6. $x = \pm2, y = \pm1$; $x = \pm4/\sqrt{7}$,
   $y = \pm1/\sqrt{7}$ 7. 19/25, 29/25, −72/125   8. $4x^2 - 29x + 25 = 0$
9. $-4\sqrt{3} < K < 4\sqrt{3}$ 10. Outside range $(9 \pm 6\sqrt{2})/4$
15. Min $T = 2\pi\sqrt{(2K/g)}$

Examples 3

1. (i) 210, 3024, 720 (ii) 10, 21, 36   2. Values (i) 56 (ii) 126 (iii) 15
   (iv) $(n + 1)!/[r!(n + 1 - r)!]$ 4. (i) 42 (ii) 210 (iii) 5040   5. 840
6. 120, 20, 24   7. (i) 1/6 (ii) 1/30 (iii) 1/2 (iv) 3/10   8. (i) 60 (ii) 40
9. (i) $(\frac{1}{2})^4$ (ii) $(1/13)^4$
10. (i) $32a^5 + 240a^4b + 720a^3b^2 + 1080a^2b^3 + 810ab^4 + 243b^5$
    (ii) $x^6 - 12x^5y + 60x^4y^2 - 160x^3y^3 + 240x^2y^4 - 192xy^5 + 64y^6$

11. (i) $1 + \frac{3}{2}x + \frac{3}{8}x^2 - \frac{1}{16}x^3 + \frac{3}{128}x^4 - \frac{3}{256}x^5$; $|x| < 1$

(ii) $1 - \frac{3}{2}x - \frac{1}{8}x^2 - \frac{27}{16}x^3 - \frac{405}{128}x^4 - \frac{1701}{256}x^5$; $|x| < \frac{1}{3}$

(iii) $1 + 3x + 6x^2 + 10x^3 + 15x^4 + 21x^5$; $|x| < 1$ 12. 0·9796

13. $1 + 4x + 10x^2 + 16x^3 + 19x^4 + 16x^5 + 10x^6 + 4x^7 + x^8$

14. 1/64, 3/32, 15/64, 5/16, 15/64, 3/32, 1/64

15. 0·05636, 0·1879, 0·2818, 0·2505; 0·77656

Examples 4

4. $8 - 3x + \frac{3}{16}x^2 - \frac{1}{128}x^3$; $|x| < 4$

5. $\dfrac{1}{4(1 + x)} + \dfrac{11}{4(1 - 3x)}$; $3 + 8x + 25x^2 + 74x^3 + 223x^4$; $|x| < 1/3$

6. $\dfrac{4}{3(1 + x)} - \dfrac{4}{3(1 - 2x)} - \dfrac{3}{(1 + x)^2}$; $-3 + \frac{14}{3}x - 13x^2 + \frac{8}{3}x^3 - 35x^4$;

$|x| < \frac{1}{2}$ 7. $1 + x - \dfrac{3}{2}x^2 + \dfrac{5}{6}x^3 - \dfrac{7}{24}x^4$; $\dfrac{1 - 2r}{r!}$

8. $x + \dfrac{x^3}{3!} + \dfrac{x^5}{5!} + \dfrac{x^7}{7!}$; $1/(2r + 1)!$ 9. $1 + \frac{1}{2}x^2 - \frac{1}{3}x^3$

10. 1·1052, 0·8187   11. $(1 - 2x)(1 - x)$; $-3x - \frac{5}{2}x^2 - 3x^3 - \frac{25}{4}x^4$; $-(2^n + 1)/n$; $|x| < \frac{1}{2}$

12. $3x + \frac{3}{2}x^2 + 3x^3 + \frac{15}{4}x^4 + \frac{33}{5}x^5$; $x = 0·1$; 0·3184

13. $1 + x - \frac{1}{3}x^3 - \frac{1}{6}x^4$ 14. $2(1 - 0·02)^{1/3}$; 1·9866

15. (i) Cvgt (ii) Cvgt all x (iii) A.C. $|x| < 1$ (iv) A.C. $|x| < \frac{1}{2}$

Examples 5

1. 0·59   2. (i) $4ml/d^3$ (ii) $\dfrac{4ml}{d^3}\left(1 + \dfrac{2l^2}{d^2}\right)$ 3. 6·5% too high

4. 1·15% increase   5. 0·625% too small

6. $2 + \dfrac{x}{12} - \dfrac{x^2}{576} + \dfrac{x^3}{81 \times 256}$; 2·033   7. 7·25% increase

8. c decreases by 0·5%   9. $x = \pm0·514$ 10. 2·26

11. $(-2, -1), (0, 1), (1, 2)$; 0·43   12. 2·44   13. $(-4, -3), (0, 1), (2, 3)$; 0·44

14. 0·62

Examples 6

1. $-2·5, -0·66, 3·1$; 3·10   2. $-2, 1, 1·5$ 3. 39°

4. 1·6572°, 5·293° or 94°50′, 303°15′   5. 0·51

6. 1·743°, 3·142°, 4·447°; 100°, 180°, 256°   7. 0·675   8. 0, 1·29(3)

9. $a = 15, b = 50$ 10. $a = 4·13, n = 1·73$ 11. $A = 5·55 \times 10^{-7}, m = 4·1$

12. $K = 1·57, n = 0·485$ 13. $a = 9·945, b = 0·277$

14. $T_0 = 14, \mu = 0·093$

Examples 7

1. $AB = 5.831$, $AC = 9.220$, $BC = 6.708$
2. $3x - 4y + 7 = 0$, $4x + 3y + 1 = 0$, $7x - y - 17 = 0$
3. $y + x - 9 = 0$, $y - x = 0$ 4. $x^2 + y^2 - 6x - 8y - 24 = 0$
5. $(-4.7015, -1.7015), (1.7015, 4.7015)$ 6. $m^2x^2 + 2x(mc - 2a) + c^2 = 0$;
   $c = a/m$ 7. circle, centre $(a, 0)$; $x^2 + y^2 - 2ax = 0$
8. $2/(t_1 + t_2)$; $y(t_1 + t_2) = 2(x + at_1t_2)$; $yt = x + at^2$
9. $(2.07, 1.07)$, $(-0.87, -1.87)$ 10. $c = \pm 2\sqrt{5}$ 11. graph
12. (i) $y + tx = at(2 + t^2)$ (ii) $ty - t^3x = c(1 - t^4)$ 13. $(1, 2)$
14. (i) $x + 1 = 0$ (ii) $3(x^2 + y^2) + 56x + 128 = 0$
15. $x^2 + y^2 - 5x - 3y = 0$

Examples 8

1. (i) 0·9239 (ii) 0·96, 0·28, 3·429   2. (i) -0.8616, 0·2158
   (ii) 0·8616, 0·9692, -0.28 3. (i) 0·9917, -0.1288, -7.7 (ii) 0·7076
4. (i) 18°26′, 161°34′, 198°26′, 341°34′
   (ii) 10°27′, 49°33′, 130°27′, 169°33′, 250°27′, 289°33′
   (iii) 40°38′, 160°38′, 280°38′
   (iv) 11°6$\frac{1}{2}$′, 33°53$\frac{1}{2}$′, 101°6$\frac{1}{2}$′, 123°53$\frac{1}{2}$′, 191°6$\frac{1}{2}$′, 213°53$\frac{1}{2}$′, 281°6$\frac{1}{2}$′, 303°53$\frac{1}{2}$′
   (v) 21°49′, 122°11′, 165°49′, 266°11′, 309°49′
5. (i) (a) $2 \sin 40° \cos 10°$ (b) $-2 \sin 45° \sin 25°$ (c) $2 \cos 75° \sin 35°$
   (d) $2 \sin 4x \cos x$ (e) $-2 \sin 3x \sin 2x$ (f) $2 \cos 4\theta \sin \theta$
   (ii) (a) $\sin 60° + \sin 30°$ (b) $\cos 100° + \cos 40°$ (c) $\cos 40° - \cos 60°$
   (d) $\frac{1}{2}[\cos 8x + \cos 2x]$ (e) $\frac{1}{2}[\cos 2t - \cos 4t]$ (f) $\frac{1}{2}[\cos \alpha - \cos 5\alpha]$
8. $\tan \theta$; $\tan \frac{3}{2}\theta$ 9. $17 \cos(\theta + 61°56′)$, $R_{\max} = 17$, $\theta = 298°41′$;
   $R_{\min} = -17$, $\theta = 118°4′$
10. $7.81 \cos(\omega t + 0.6946)$; Max 7·81, $t = 5.589/\omega$; Min -7.81, $t = 2.447/\omega$
11. $50 \sin (50\pi t + 0.6435)$; $50 \sin (50\pi t - 0.6435)$; 50, $\frac{1}{25}$, 0·6435; 50, $\frac{1}{25}$,
    -0.6435
12. $200 \cos (\pi/12) \sin (40\pi t + 5\pi/12)$; $200 \cos \pi/12$, $\frac{1}{20}$, $3\alpha/2$
13. 8, 1/2, 2, $\pi/4$ 15. $11.68 \sin (\theta + 62°27′)$

Examples 9

1. (i) 13°23′, 125°27′, 193°23′, 305°53′ (ii) 66°32′, 127°30′, 246°32′, 307°30′
   (iii) 70°32′, 289°28′ (iv) 33°42′, 116°34′, 213°42′, 296°34′
   (v) 66°6′, 149°26′, 246°6′, 329°26′, (vi) 39°15′, 175°45′, 219°15′, 355°45′
2. (i) 67°, 129°51′, 293°, 309°51′ (ii) 15°, 75°, 195°, 255°
   (iii) 57°31′, 128°21′, 231°39′, 302°29′ (iv) 43°20′, 136°40′
   (v) 76°12′, 146°59′, 213°1′, 283°48′ (vi) 14°29′, 165°31′, 199°28′, 340°32′
3. (i) 120°, 240° (ii) 9°53′, 170°7′ (iii) 6°29′, 122°31′ (iv) 90°, 194°29′,
   345°31′   4. (i) 0°, 90°, 180°, 270°, 360° (ii) 0, 60°, 72°, 144°, 180°
   (iii) 0°, 60°, 90°, 180°
5. (i) 13°17′, 240°27′ (ii) 238°4′, 358°4′ (iii) 24°27′, 261°49′ (iv) 66°4′, 183°56′
   (v) 20°36′, 126°52′ (vi) 117°53′, 351°3′

6. (i) $t = 0.00\ 154$ (ii) $22°48', 284°4'$ (iii) $0.8535, 3.2215$
7. (i) 0.013 (ii) 0.051 (iii) 0.624　　8. $95°34', 341°48'$
9. $33°28', 153°28', 273°28'$　　10. $75°31', 284°29'$　　11. $A = 44°48'$
12. $X = 41°20', Y = 65°28'$　　13. 9.144 in, $69°53', 50.36$ in^2
14. (i) $A = 96°15', B = 49°10', C = 34°35'$ (ii) $P = 75°31', Q = 37°1',$
　　$r = 11.4$ in
15. $BD = 1.396$ ft, $AC = 1.803$ ft; 1.26 ft^2, $ADC = 84°15'$
16. $7.70, 6.53$ miles　　17. $PQ = 520.8, PB = 419.5$ ft
18. 43.83 in, 21.17 in　　19. 115.04 ft; $27°23'$　　20. $21°2'$

Examples 10

1. $9 + 40j$; $41\underline{/77°18'}$　　2. $13(2 + j)$; $13\sqrt{5}, 26°34'$
3. $51 + 16j$; $53.46\underline{/17°25'}$　　4. (i) $35 - 13j$ (ii) $\frac{1}{25}(1 + 18j)$ (iii) $40 + 42j$
　　(iv) $1.2 + 0j$　5. $8.602\underline{/305°32'}, 9.434\underline{/32°}$; $81.16\underline{/337°32'}, 0.912\underline{/273°32'}$
6. $x = 8.66, y = 5$　　7. $-7 + 24j, -117 + 44j$　　8. $u = 3(x^2 - y^2)$,
　　$v = 6xy$　9. $5\underline{/143°8'}$; $-527 - 336j$　　10. $\frac{1}{25}(59 + 87j)$; $4.288\underline{/55°51'}$
11. (i) (a) $\exp(j30°)$ (b) $\exp(2j\alpha)$ (c) $5\exp(j120°)$ (d) $5\exp(j\pi/4)$ (e) $b\exp(ja/2)$
　　(ii) (a) $\cos 95° + j \sin 95°$ (b) $\cos 50° + j \sin 50°$
　　(c) $\cos(8\pi/15) + j \sin(8\pi/15)$
12. (i) (a) $5(\cos \theta + j \sin \theta)$ (b) $5 - 8.66j$ (c) $2.5 + 4.33j$ (d) $-5 - 8.66j$
　　(ii) (a) $0.9434\underline{/58°}$ (b) $3.354\underline{/116°34'}$ (c) $4.472\underline{/243°26'}$ (d) $5.409\underline{/123°41'}$
13. (i) $-0.5 \pm 0.866j$; $1\underline{/240°}, 1\underline{\,\backslash-120°}$ (ii) $1 \pm j\sqrt{2}, 1.732\underline{/54°44'}\ 1.732\underline{/305°16'}$
　　(iii) $-0.1 \pm 0.625j$; $0.6325\underline{/99°6'}, 0.6325\underline{/260°54'}$
14. (i) $u = 2x + 3, v = 2y$ (ii) $u = x/(x^2 + y^2), v = -y/(x^2 + y^2)$
　　(iii) $u = (x + 1)^2 + y^2, v = 2y(x + 1)$
　　(iv) $u = (x - 1)/[(x - 1)^2 + y^2], v = - y/[(x - 1)^2 + y^2]$
　　(v) $u = x(x^2 + y^2 + 1)/[x^2 + y^2], v = y(x^2 + y^2 + 1)/[x^2 + y^2]$
15. $(x + 1)^2 + y^2, (x + 2 - y)^2$　　16. (i) $\cos \theta -j \sin \theta$ (ii) $\cos \theta, \sin \theta$
18. (i) $1.6 - 1.2j$ (ii) $-10 + 95j$ (iii) $1.348 - 1.843j$
19. $|z - 2 - 3j| = 5$; $x^2 + y^2 - 4x - 6y - 12 = 0$
20. (i) circle centre a, radius r (ii) straight line through a, slope $\tan k$

Examples 11

1. $1000(-1 \pm 7j)$; $707\underline{/98°6'}, 707\underline{/261°54'}$
2. $1 + j, 1 - j$; $\sqrt{2}\underline{/45°}, \sqrt{2}\underline{/315°}$　　3. 28.4 lbf, $\theta = 56°55'$
4. 23.39 lbf, $\theta = 112°13'$　　5. $32 + 126j$; $130, I$ lags V by $37°$
6. 6.349 amp; I lags V by $32°9'$
7. $R_2 = R_1/(1 + \omega^2 C_1{}^2 R_1{}^2), C_2 = (1 + \omega^2 C_1{}^2 R_1{}^2)/(\omega^2 C_1{}^2 R_1{}^2)$
8. $\sqrt{(L/C)}$ or $\sqrt{(R/G)}$; $680, 162°40'$ or $-17°20'$
9. 0.665 amp, I lags V by $57°42'$
10. $Z = 3.018 + 0.566j = 3.071\underline{/10°37'}$

Examples 12

1. $y = x^4 + x^2 + 3$ 2. 1·3776, 1·3983   3. 0·5797, 0·5832, 0·5867
4. 1·2900, 1·2783   5. 1·6489, 2·4584, 4·9520   6. $2x^3 + 4x^2 - 3x + 7$
7. 10·39, 8·125 ft/s²   8. 829, 1211, 1695   9. 0·86534, 0·66665
10. 0·8313, 0·7952   11. 704·969, 912·673, 3375, 8000
12. 17·45, 20·144, 20·38   13. $3x^2 - 4x + 5$ 14. $x^3 - 2x^2 + 4x + 7$
15. 2·1943, 2·3276

Examples 13

1. $\Delta y/\Delta x = $ 7·2, 7·6, 8·0, 8·4, 8·8, 9·2
2. $\delta y/\delta x = $ 10·3, 10·24, 10·18, 10·12, 10·06, 10·03, 10·003, 10
3. 1·73, 1·92, 2·12, 2·34, 2·54   4. $6x + 2$; $-1/x^2$; $1/(2\sqrt{x})$
5. 5·1, 5·2, 5·3, 5·4   6. $-1·25$ 7. 2·5, 3·5, 4·5, ft/s
8. (i) $12x^2 + 2$ (ii) $2t + 6$ (iii) $(3\sqrt{x})/2$ (iv) $2z - 6$ (v) $4\theta^3 + 2\theta$
   (vi) $-2/u^3$ (vii) $-15/(2t^{3/2})$ (viii) $-n/v^{n+1}$ (ix) $1 + 1/\sqrt{x}$
   (x) $28y^3 - 18y^2 + 10y$ (xi) $1·4v^{0·4}$ 9. $-c/v^2$, $-nc/v^{n+1}$
10. $(1, 4)$, $(-\frac{3}{2}, 35\frac{1}{4})$ 11. (i) -12 ft/s (ii) 1 or 4 sec (iii) 5; 110 ft
12. $-1·3$, $-0·5$ rad/s; $t = 7·5$ sec.   13. Tangents: $4y + 5x - 20 = 0$,
    $y + 5x + 10 = 0$ Normals: $10y - 8x - 9 = 0$, $5y - x + 24 = 0$
14. Tangent $y - x - 1 = 0$; Normal $y + x - 3 = 0$
15. $\delta i/\delta t = -18·13$, $-14·84$, $-12·15$, $-9·95$, $-8·16$, $-6·67$ amp/s
16. 1·75, 5·0, 9·75

Examples 14

1. 0·1987   2. 0·9801   3. 0·999, 0·989, 0·966, 0·940, 0·901
4. 1·107, 1·352, 1·651, 2·017, 2·464   5. $y = 2·718x$, $y = -0·368x + 3·086$
8. $y_{min} = 3·61$ 11. $3 \sec^2 3x$
14. (i) $-3 \sin 3x$ (ii) $20 \cos 4x$ (iii) $-\frac{1}{2} \sin \frac{1}{2}t$ (iv) $\frac{3}{2} \cos (\frac{3}{2}\theta)$ (v) $30 e^{3x}$
    (vi) $6 e^{2t}$ (vii) $-4 e^{-4x}$ (viii) $0·5 e^{0·5t}$
17. $x = \tan^{-1}(+3/-4)$, $y_{max} = +5$: $x = \tan^{-1}(-3/+4)$, $y_{min} = -5$
18. Tangent, $y + x = \frac{1}{2}\sqrt{3} + \frac{1}{12}\pi$; Normal $y - x = \frac{1}{2}\sqrt{3} - \frac{1}{12}\pi$
19. $\delta y \simeq 0·5436$ 20. $\delta s \simeq 0·922$

Examples 15

1. (i) $15(3x + 1)^4$ (ii) $4(2x + 3)(x^2 + 3x)^3$ (iii) $6x^3/(3x^4 + 6)^{1/2}$
   (iv) $(x + 1)/\sqrt{(x^2 + 2x)}$
2. (i) $12/(1 - 2x)^3$ (ii) $-12/(4x + 3)^4$ (iii) $9x(3x^2 + 1)^{1/2}$ (iv) $2x/\sqrt{(a^2 + 2x^2)}$
3. (i) $2x e^{x^2}$ (ii) $2 e^{2x+3}$ (iii) $(2x + 3) e^{x^2+3x}$ (iv) $\frac{3}{2}(e^{3x} - e^{-3x})$
4. (i) $2/(1 + 2x)$ (ii) $(6t + 1)/(3t^2 + t - 1)$ (iii) $6/(2u + 1)$
   (iv) $(4z + 3)/(z^2 + z)$
5. (i) $\cot x$ (ii) $- \tan x$ (iii) $\sec^2 x \, e^{\tan x}$ (iv) $15x^2 \, e^{x^3}$

6. (i) $12t^2 \cos(t^3 + 1)$ (ii) $6 \sin 3x \cos 3x$ (iii) $\cos t\, e^{\sin t}$
(iv) $3(2x^2 + 3)(2x^2 - 3)/x^4$ (v) $2 \cos 2t\, e^{\sin 2t}$ (vi) $\sec^2 \theta/\tan \theta$
(vii) $\cos 2x/\sqrt{(\sin 2x)}$ (viii) $-3\, e^{1/t}/t^2$ (ix) $-4 \cos^3 u \sin u$
(x) $(e^x - e^{-x})/(e^x + e^{-x})$
7. (i) $(2 + x^2)(2 + 5x^2)$ (ii) $12(6 - x)(2 - x)$ (iii) $(x - 2)(9x + 4)$
(iv) $(3x + 1)/[2\sqrt{(x + 1)}]$
8. (i) $e^{3x}(15x + 14)$ (ii) $4\, e^{2x}(2x + 1)^2(x + 2)$ (iii) $x^2\, e^{2x}(2x + 3)$
(iv) $(x + 1)^2(4x + 7)\, e^{4x}$
9. (i) $t^2(2t \cos 2t + 3 \sin 2t)$ (ii) $(2 + \log_e x)/(2\sqrt{x})$
(iii) $5\, e^x(\sin 3x + 3 \cos 3x)$ (iv) $e^u(\cos u + \sin u)$
10. (i) $2x(\cos 2x - x \sin 2x)$ (ii) $-e^{-x}(3 \sin 3x + \cos 3x)$
(iii) $\cos^2 t - \sin^2 t$ (iv) $t^2 \sin t(2t \cos t + 3 \sin t)$
11. (i) $2 \cos 2t \cos 3t - 3 \sin 2t \sin 3t$ (ii) $\frac{1}{2}x(x \cos \frac{1}{2} x + 4 \sin \frac{1}{2} x)$
(iii) $1/u + \cot u$ (iv) $(2x \sec^2 x + \tan x)/(2\sqrt{x})$ (v) $2\,(e^t \cos t - e^{-t} \sin t)$
(vi) $e^{1/x}[2x - 1]$
12. (i) $2/(x + 1)^2$ (ii) $-17/(3x - 4)^2$ (iii) $4x/(x^2 + 1)^2$
(iv) $(x^2 + 2x - 1)/(x + 1)^2$
13. (i) $e^{2x}(2x - 3)/x^4$ (ii) $2\, e^x/(e^x + 1)^2$ (iii) $(x \cos x - \sin x)/(2x^2)$
(iv) $(1 - \log_e x)/x^2$
14. (i) $x(2 \cos x + x \sin x)/\cos^2 x$ (ii) $(1 - 3 \log_e t)/t^4$
(iii) $(t \cos t - 2 \sin t)/t^2$ (iv) $(\cos^2 x + 3 \sin^2 x)/\cos^4 x$
15. (i) $(x \sec^2 x - \tan x)/x^2$ (ii) $(\log_e x - 1)/(\log_e x)^2$ (iii) $1/(1 + \cos x)$
(iv) $(x \cos x - \sin x)/x^2$
19. $m = 4$ or -5

Examples 16

1. (i) $0.32\pi = 1.006$ in²/min (ii) $\pi = 3.142$ in³/min
2. (i) $2/(5\pi) = 0.1274$ in/s (ii) 8 in²/s   3. 3.15×10^6 lbf ft/min
4. $8/(5\pi) = 0.5096$ in/s (ii) 12.5 lbf/in²/s   5. -12.5 lbf/in²/s
6. (i) $t = 1/3$ or 3 (ii) -8, $+8$ ft/s² (iii) $4\frac{20}{27}$ ft
7. (i) $t = 8$ sec (ii) 2 rad/s²(retard.)   8. $v = 24\pi = 75.41$ in/s;
$f = -1642$ in/s²   9. tangent: $x - y + 1 = 0$, normal $x + y - 3 = 0$;
$(-1, 0)$; $(3, 0)$
10. $5x + 4y - 20 = 0$ (tan), $8x - 10y - 7 = 0$ (norm.)
tan meets $(4, 0)$, $(0, 5)$; normal meets $(7/8, 0)$, $(0, -7/10)$
11. 1; $x - y + 1 = 0$ 12. tan: $3x - 4y + 4 = 0$; norm:
$8x + 6y - 31 = 0$ 13. tan: $x + 4y - 16 = 0$, norm: $4x - y - 30 = 0$
14. tan: $x \cos t + 2y \sin t = 2$; norm: $2x \sin t - y \cos t = 3 \sin t \cos t$
15. tan: $x - 2y + 16 = 0$; norm: $2x + y - 48 = 0$
16. $y_{min} = -74(x = 3)$; $y_{max} = 19\frac{1}{4}(x = -\frac{1}{2})$
17. Min $+4(x = 1)$; Max $-8(x = -1)$ 18. Min $-\sqrt{89}(x = 122°)$;
Max $\sqrt{89}(x = 302°)$
19. i_{max} $50(t = 0.009294)$; i_{min} $-50(t = 0.04069)$ 20. $wl^2/8$
21. For Max W, $I = e/(2r) = 5.555$ amp.   22. $r = h = 9.9$ ft
23. Base width 2.739 ft, depth 3.308 ft   24. $r = 1.15$ in, $h = 7.224$ in
25. 1.924 ft³

Examples 17

2. $\partial^2 u/\partial x^2 = 6$, $\partial^2 u/\partial y^2 = 0$, $\partial^2 u/\partial y\,\partial x = 2$

5. $\partial^2 u/\partial y\,\partial x = \partial^2 u/\partial x\,\partial y = 2(x - y)/(x + y)^3$

6. $\partial u/\partial x = -1/x$, $\partial u/\partial y = 1/y$, $\partial^2 u/\partial x^2 = 1/x^2$, $\partial^2 u/\partial y^2 = -1/y^2$

7. (i) $4d\sqrt{P}$, $d^2/\sqrt{P}$ (ii) $K/(4\pi At)$, $-KA/(4\pi t^2)$ (iii) e^{-Kt}, $-Ku\,\mathrm{e}^{-Kt}$

8. (i) $\pi/\sqrt{(lg)}$, $-\pi\sqrt{(l/g^3)}$ (ii) $-K\sqrt{T}/m^2$, $K/(2m\sqrt{T})$
   (iii) $-1/[4\pi\sqrt{(L^3C)}]$; $-1/[4\pi\sqrt{(LC^3)}]$ (iv) $-\pi\sqrt{[I/(MH^3)]}$

9. $\frac{1}{2}r^2(\theta - \sin\theta)$, $r(\theta - \sin\theta)$, $\frac{1}{2}r^2(1 - \cos\theta)$

10. $\pi(R^2 - r^2)$; $2\pi R$, 2π; $-2\pi r$, -2π 11. $2\cdot8\%$

12. $V = xyz$, $S = 2(yz + xz + xy)$; $\delta V = yz\,\delta x + xz\,\delta y + xy\,\delta z$,
    $\delta S = 2(z + y)\delta x + 2(x + z)\delta y + 2(x + y)\delta z$; $4\cdot5\%$ 13. 5%

14. $1\cdot65$ volt   15. $(\delta l/l) - 3(\delta r/r)$

Examples 18

1. $y = x^2 + 3x - 9$ 2. $y = 3x^2 + 4x + 2$ 3. $y = \frac{1}{2}(x^2 - 10x + 21)$

4. $y = x^3 - 2x^2 + x - 3$ 5. $s = \frac{1}{6}(4t^3 - 3t^2 + 18t - 26)$

6. $s = \frac{1}{6}(2t^3 + 9t^2 + 30)$ 7. (i) $2x^3 + \frac{8}{3}\sqrt{x^3} - 5x + C$
   (ii) $5\log_e x + 2\,\mathrm{e}^{2x} + C$ (iii) $2\log_e x - 2x^{-3/2} - 3x^{-2} + C$
   (iv) $\frac{1}{2}\log_e x - \frac{1}{4}\cos 2x + C$ (v) $\frac{1}{2}x^2 + \log_e x + C$ (vi) $\frac{1}{3}(x + \frac{1}{2})^3 + C$
   (vii) $\frac{5}{3}\log_e(3x + 1) + C$ (viii) $\frac{1}{2}x^2 + 2x + \log_e x + C$
   (ix) $\frac{1}{2}x^2 - \frac{8}{3}x^{3/2} + 4x + C$

8. (i) $\frac{1}{5}\mathrm{e}^{5x}$ (ii) $\frac{1}{2}\log_e(4x + 1) + C$ (iii) $2\sqrt{(x + 2)} + C$
   (iv) $2x^{3/2} - 4x^{1/2} + C$ (v) $3/[4(1 - 4x)] + C$ (vi) $\frac{5}{2}\sin 2x + C$
   (vii) $10\,\mathrm{e}^{0\cdot1x} + C$ (viii) $\frac{1}{2}(\mathrm{e}^{2x} - \mathrm{e}^{-2x}) + C$

9. (i) $6\cdot2$ (ii) $0\cdot3054$ (iii) $1/15$ 10. (i) $0\cdot6109$ (ii) $6\cdot389$ (iii) $-0\cdot433$

11. (i) $312\cdot4$ (ii) $6\frac{2}{3}$ (iii) $0\cdot2064$ (iv) $5\frac{5}{8}$ (v) $6\cdot678$ (vi) 2

12. (i) $4\frac{5}{6}$ (ii) $2\cdot5948$ (iii) $0\cdot5$

13. (i) $17\cdot18$ (ii) $1\cdot0986$ (iii) $4\frac{1}{8}$ (iv) $0\cdot25$ (v) $168\cdot8$ (vi) $5/6$

14. (i) $3/32$ (ii) $8\cdot636$ (iii) 96 (iv) $41\frac{15}{21}$ (v) 20 (vi) $1\cdot218$

15. (i) $6\cdot321$ (ii) $19\frac{1}{6}$ (iii) $2\cdot161(75)$ (iv) $2\cdot011(75)$ (v) $24684\cdot5$ (vi) $51\frac{2}{3}$

16. (i) $2/\pi$ (ii) $4\pi a^3/3$ (iii) $\pi r^2 h/3$ (iv) $\dfrac{c}{n - 1}\left[\dfrac{1}{v_1^{n-1}} - \dfrac{1}{v_2^{n-1}}\right]$
    (v) $6\pi a^3$ (vi) $\frac{1}{2}\pi a^4$ 17. (i) $12\frac{1}{4}$ (ii) $11\cdot416(3)$ (iii) 6 (iv) $9\cdot817$

18. (i) $\pi/12$ (ii) $2/3$ (iii) $\pi/4 + 1/2 = 1\cdot285(4)$ (iv) $0\cdot30175$ (v) $5/16$
    (vi) $\pi/12$ (vii) 1 (viii) $\frac{1}{6}\sqrt{3} = 0\cdot2887$ 19. 22

20. $36\,080$ ft lbf

Examples 19

1. (i) $3\cdot0986$ (ii) $4\frac{2}{3}$ (iii) $1/7$ 2. (i) $\frac{2}{3}(t - 1)^{3/2} + 4(t - 1)^{1/2} + C$
   (ii) $-\cos(x^2 + x + 2) + C$
   (iii) $2[\frac{1}{5}(x + 1)^{5/2} - \frac{2}{3}(x + 1)^{3/2} + (x + 1)^{1/2}] + C$

3. (i) $0\cdot6109$ (ii) $3\cdot105$ (iii) $3\cdot219$ 4. (i) $-\log(\cos x)] + C$
   (ii) $\frac{1}{2}\mathrm{e}^{x^2} + C$ (iii) $\frac{1}{3}\sin^3\theta + C$

5. (i) $\frac{5}{3}\log(x-1) - \frac{1}{6}\log(2x+1) + C$ (ii) $3\log(x-3) - \log(x-2) + C$
   (iii) $2\log(x+5) + 3\log(x-4) + C$ 6. (i) $0\cdot1486$ (ii) $1\cdot7918$
   (iii) $0\cdot2369$ 7. (i) $7\cdot541$ (ii) $34\cdot13$ (iii) $7\cdot881$
8. (i) $\frac{1}{3}e^{3x}(x - \frac{1}{3}) + C$ (ii) $\frac{1}{4}x^4(\log_e x - \frac{1}{4}) + C$ (iii) $-x\cos x + \sin x + C$
9. (i) $8\cdot389$ (ii) $1\cdot0705$ (iii) $4\cdot4675$ 10. (i) 42 (ii) $1\cdot4978$ (iii) $1\cdot9487$
11. (i) $\frac{1}{2}\log_e(1 + 2\sin\theta)$ (ii) $-2\log_e(9 - x^2) + C$
   (iii) $x\tan x + \log_e\cos x + C$ 12. (i) $1\cdot933(3)$ (ii) $0\cdot693(1)$ (iii) $0\cdot5634$
13. $\dfrac{2}{(x-1)^2} + \dfrac{3}{4x+3}$; $1\cdot233$ 14. $0\cdot6495$ 15. $u = 1 + \cos x$; $0\cdot1585$
16. $y = e^x(x - 2) + 4$ 17. (i) 5π (ii) $7\cdot389$
18. $y = \frac{1}{3}(x^2 - 1)^{3/2} + 4$ 19. $\frac{1}{3}a^3$ 20. $0\cdot8669$ (by parts)

Examples 20

1. (i) $33\frac{1}{3}$ unit^2 (ii) 36 unit^2 (iii) $814\cdot3$ unit^3
2. (i) $13\frac{1}{3}$ unit^2 (ii) $142\cdot4$ unit^3 (iii) $3\frac{1}{3}$ unit
3. (i) $21\cdot97$ unit^2 (ii) $209\cdot4$ unit^3 (iii) $2\cdot747$ unit
4. $24\cdot8$ unit^2, $801\cdot1$ unit^3 5. Points $(2, 6)$, $(3, 4)$, Area $= 0\cdot134$ unit^2
6. $5\frac{1}{3}$ unit^2 7. (i) -972 unit^2 (curve below x-axis)
   (ii) Meet at $(0, 1)$, $(2, 5)$, Area $= 1\frac{1}{3}$ unit^2
8. (i) $19\cdot17$ unit^2 (ii) 758 unit^3 9. (i) 6 (ii) $3\frac{1}{6}$ (iii) $4\frac{1}{3}$ (iv) $1\cdot297$
10. $520\pi/3 = 544\cdot5$ in^3 11. (i) $36\cdot79$ (ii) $5\cdot36$ (iii) $20/\pi$
12. $10/\pi$ lbf   13. (i) $37\cdot95$ amp (ii) $100(\pi\cos\alpha - 2\sin\alpha)/\pi$
14. $126\,300$ ft lbf, $175\cdot4$ lbf/in^2 15. $72\,280$ ft lbf, $71\cdot69$ lbf/in^2

Examples 21

1. (i) $7\cdot371$ (ii) $5\cdot053$ (iii) $2\cdot697$ 2. (i) $5\cdot42$ (ii) $7\cdot071$ (iii) $4\cdot487$
3. (i) $\pi/(2\sqrt{2})$ (ii) $1\cdot55$ (iii) $\pi/(2\sqrt{2})$ 4. $9\cdot19$ amp
5. 200 Watt, $2\cdot828$ amp   6. (i) $\bar{x} = 8\cdot5$ (ii) $-$ (iii) $4\cdot25$ (iv) $s = 2\cdot062$
7. $\bar{x} = 3$, $\bar{y} = 3\cdot6$ 8. (i) $\bar{x} = 2\frac{8}{35}$, $\bar{y} = 1\frac{9}{14}$ (ii) $\bar{x} = 2\cdot164$, $\bar{y} = 2\cdot705$
9. (i) $\bar{x} = 4\cdot095$ (ii) $\bar{x} = 3\cdot054$ 10. $1\cdot629$ in from larger end
13. $16\cdot334$, $16\cdot333$, error $+ 0\cdot001$ 14. $1\cdot646$ 15. $1\cdot00$

Examples 22

1. (i) $4Mb^2/3$ (ii) $4Ma^2/3$ (iii) $Mb^2/3$ (iv) $Ma^2/3$ (v) $M(a^2 + b^2)/3$
   (vi) $4M(a^2 + b^2)/3$ 2. $1\cdot832$ lb in^2 3. $19\cdot94$ in^2 5. $M(R^2 + r^2)/2$
5. $57\cdot73$ lb in^2 6. $46\cdot08$ lb in^2
7. $I_{OX} = 273$ in^4, $I_{OY} = 146\cdot3$ in^4; $K_{OX} = 3\cdot578$ in; $K_{OY} = 2\cdot618$ in
8. $I_{OX} = 19\cdot53$ cm^4, $I_{OY} = 37\cdot5$ cm^4, $K_{OX} = 1\cdot678$ cm, $K_{OY} = 2\cdot326$ cm
9. $I_{OX} = 3\cdot901$ unit^4; $I_{OY} = 8\cdot533$ unit^4, $K_{OX} = 1\cdot912$, $K_{OY} = 0\cdot8944$
10. $40\cdot59$ lb in^2 (i) $4\cdot51$ lb in^2 (ii) $13\cdot53$ lb in^2
11. Vol $= 864\pi^2 = 8529$ in^3, Area $= 288\pi^2 = 2843$ in^2
12. Vol $= 87\cdot2$ unit^3, $\bar{y} = 1\cdot563$ 13. $1\cdot613$ lbf   14. 5735 lb in^2

15. $\bar{x} = 16/7$, $\bar{y} = 81/10$;
    vol. of rev about $0x = 1425$ unit^3,
    vol. of rev about $0y = 402\cdot2$ unit^3

Examples 23

9. $s = \frac{1}{6}(2t^3 + 3t^2 - 6t + 31)$ 10. $y = \frac{1}{12}(x^4 - 4x^3 + 32x + 7)$
11. $y = A e^{4x} + B e^{-x}$ 12. $y = \sqrt{[(2x^3 - 3x^2 + 65)/6]}$
13. $y = 2/(1 - 2\sin x)$ 14. $T = T_0 e^{\mu\theta}$ 15. $\theta = 20(1 + 4 e^{-Kt})$
16. $q = q_0 e^{-t/(CR)}$; $5\cdot11$ sec   17. $i = E(1 - e^{-Rt/L})/R$; $5\cdot545 \times 10^{-4}$ sec
18. $t = \dfrac{1}{K} \log [(n/(n - x)]$; $\dfrac{1}{K} \log 2\cdot5$
19. $t = \dfrac{1}{K} \log [(g + Ku)/(g + Kv)]$; $t = 10 \log_e (47/38) = 2\cdot126$ sec
20. (i) $x = 4\sin 3t$ (ii) $v = 12\cos 3t$; 8 in, $2\pi/3$ sec, $\frac{3}{2}\pi$ c/s

Examples 24

1. 1/8, 1/4, 5/16, 3/16   2. 1/2197, 1/5525   3. 1/6, 1/3, 25/7776
4. 1/72, $P(14) = 5/72$ 5. 14/285, 7/72, 14/57
6. $0\cdot32768$, $0\cdot40960$, $0\cdot20480$, $0\cdot05120$, $0\cdot00032$; $0\cdot94208$; 1, certain
7. $^{20}C_4 9^{16}/10^{20}$ 8. $0\cdot7736$, $0\cdot2035$, $0\cdot02143$; $0\cdot0229$
9. (i) $0\cdot2916$ (ii) $0\cdot99144$ (iii) $0\cdot114\,265$
10. (i) $0\cdot00001$ (ii) $0\cdot59049$ (iii) $0\cdot99144$

Examples 25

1. Mode $6\cdot3$, Median $5\cdot6$, $25\% - 4\cdot1$, $75\% - 7\cdot2$, $\bar{x} = 6\cdot6$,
   $s = 2\cdot307$ 2. Median $0\cdot464$, Mode $0\cdot47$, $\bar{x} = 0\cdot475$, $s = 0\cdot1897$
3. $\bar{x} = 3$, $s = 1\cdot5$ 4. $\bar{x} = \frac{1}{2}$, $\sigma^2 = 0\cdot05$, Mode $= \frac{1}{2}$
6. $0\cdot50934$ in, $s = 0\cdot001875$ 7. Mean $= 1\cdot978$ in, $s = 0\cdot1744$
8. Mean $= 5\cdot99$ gm/cm^3, $s = 0\cdot5532$ gm/cm^3
9. Mode $= 5\cdot84$ in, Median $= 5\cdot78$ in, $20\% = 5\cdot64$ in,
   $80\% = 5\cdot88$ in, $\bar{x} = 5\cdot814$ in, $s = 0\cdot1575$ in
10. Mean fusing current $= 4\cdot9644$ amp, $s = 0\cdot1659$ amp

Examples 26

1. $\bar{x} = 100$, $\sigma = 7\cdot071$; $85\cdot86$:–:$114\cdot14$; $78\cdot79$:–:$121\cdot21$
2. $0\cdot9995(6)$ 3. $0\cdot33480$, $0\cdot40176$, $0\cdot20088$, $0\cdot05357$, $0\cdot008035$, $0\cdot000643$,
   $0\cdot000021(4)$ 4. (i) 1/8 (ii) 5/16 (iii) 11/32
5. $0\cdot59049$, $0\cdot32805$, $0\cdot07290$, $0\cdot00810$, $0\cdot00045$, $0\cdot00001$
6. (i) $0\cdot0498$ (ii) $0\cdot4233$ (iii) $0\cdot3526$ 7. (i) $60\cdot65\%$ (ii) $9\cdot025\%$
8. (i) $0\cdot1353$ (ii) $0\cdot8569$ (iii) $0\cdot1431$
9. limits $2\cdot792$, $7\cdot208$, rejected 10. (i) $0\cdot4772$ (ii) $0\cdot8664$
11. (i) $79\cdot35$ (ii) $341\cdot3$; $9\cdot12$:–:$20\cdot88$; $5\cdot73$:–:$24\cdot27$ 12. $92\cdot7\%$

13. (i) 0·3821 (ii) 0·8858   14. (i) 16·7(17) (ii) 197·2(197)
15. (i) 0·01531 (ii) 0·016374

Revision Examples 27

Examples 27.1

1. (i) 2·951, −1·297 (ii) $t = (L/R) \log_e [75/(75 - i)]$; $0·2635 \times 10^{-9}$
2. (i) (a) 58/9 (b) −58/21 (c) −316/27 (ii) $27x^2 + 316x - 343 = 0$
3. (i) (a) Min 3·55, $x = 0·3$ (b) Max 409/24, $t = 7/12$
   (ii) cannot lie between $(20 \pm 2\sqrt{87})/13$
4. (i) $x = \frac{1}{2} \log (11/9)$ (ii) $(−1·4718, −1·9436)$; $(0·2718, 1·5436)$
5. (i) (a) 720, 2520, 2730 (b) 35, 126, 210 (ii) (a) 504 (b) 60 480 (c) 362 880
6. (i) 1/33 (ii) 1/55 (iii) 8/33
7. $\dfrac{1}{(1 - x)^2} + \dfrac{2}{(1 - x)} + \dfrac{1}{1 + 2x} = 4 + 2x + 9x^2 - 2x^3 \ldots$; $-\frac{1}{2} < x < \frac{1}{2}$
8. (i) Cvgt $|x| < \frac{1}{3}$ (ii) A Cvgt all finite x (iii) A Cvgt $|x| \leqslant 1$
9. (i) $E \simeq 3·325$ (ii) W decreases by 0·25%   10. 1·53
11. 0·57   12. $A = 5·62 \times 10^{-6}$, $K = 2·98$

Examples 27.2

1. 7·616 (AB), 6·708 (BC) 6·083 (AC) (ii) 49°46′, (iii) $7x - 3y + 5 = 0$ (AB)
   $2x + y - 6 = 0$ (BC), $x - 6y - 16 = 0$ (AC)
2. $x^2 + y^2 + x - 6y - 7 = 0$; $(-\frac{1}{2}, 3)$; rad $= \frac{1}{2}\sqrt{65}$, $x = \frac{1}{4}(3 \pm \sqrt{105})$,
   $y = \frac{1}{4}(7 \pm \sqrt{105})$
3. (ii) $r^2 (\cos^2 \theta + 2 \sin^2 \theta) = r(2 \cos \theta + 3 \sin \theta) + 5$
4. (i) 297·8 cos $(50\pi t - 0·6106)$ 5. (i) 36°52′, 143°8′, 270°
   (ii) 0°, 90°, 180°, 270°, 360° (iii) 94°2′, 220°44′
6. (i) $A = 54°38′$, $B = 77°42′$, $C = 47°40′$ (ii) $P = 60°32′$, $Q = 41°12′$,
   $PQ = 17·54$ cm   7. (i) $19 + 7j$, $25·5$ (cos 41°49′ + j sin 41°49′),
   $25·5 e^{j(41°49′)}$ (ii) $-0·8333 \pm 0·7993j$, $1·154 /\underline{136°18′}$, $1·154 /\underline{223°42′}$,
   $1·154 e^{j(136°18′)}$, $1·154 e^{j(223°42′)}$ 8. 44·05 lbf at 56°44′ to x-axis
9. (i) $Z = 4 + j$ (ii) $u = x[1 + 3/(x^2 + y^2)]$, $v = y[1 - 3/(x^2 + y^2)]$
10. quadratic, $2x^2 + 4x - 5$ 11. 0·8330, 0·8961
12. (ii) Min 44, $t = 5$; Max 71, $t = 2$ (iii) tan: $4x - y - 6 = 0$,
    norm: $x + 4y - 10 = 0$

Examples 27.3

2. (ii) $m = 5$ or -3 3. (i) $t = 3$ or 6 (ii) $-18, +18$ ft/s²
   (iii) 170, 143 ft (iv) $156\frac{1}{2}$ ft
4. $x = -\frac{1}{2}$ or 3; $(-\frac{1}{2}, 0)$ tan: $2y + 14x + 7 = 0$, Norm: $2x - 14y + 1 = 0$;
   $(3, 0)$ tan: $7x - y - 21 = 0$, norm: $x + 7y - 3 = 0$
5. 0·1804 in/s   6. Max area, side of triangle = 11·25 in, other side of
   rectangle 2·38 in

7. (i) $6x^2 + 6xy - 6y^2$, $3x^2 - 12xy$, $12x + 6y$, $-12x$, $6x - 12y$
   (ii) $\sin y - y \sin x$, $x \cos y + \cos x$, $-y \cos x$, $-x \sin y$
8. (i) $y = 2x^2 + 3x - 12$ (ii) $s = \frac{1}{2}t^3 + t^2 + 15t + 10$
9. (i) (a) 0·742 (b) 3/16 (c) 1·526 (ii) (a) 1·504 (b) 0·4794
10. (i) 17·25 (ii) 0·7183 (iii) 0·143
11. (i) $22\frac{2}{3}$ unit^2 (ii) 132π unit^3 (iii) $5\frac{2}{3}$ unit
12. (i) 89, 840 ft lbf, 138·7 (ii) 55·86 amp

Examples 27.4

1. (i) Mean $5\frac{2}{3}$, R.M.S. 5·745 (ii) Mean 31·83, R.M.S. 35·35
2. $\bar{x} = 2·5$, $\bar{y} = 2·5$ 3. 4·87   4. (i) $I_{\text{OY}} = 1493$ unit^4,
   (ii) $I_{\text{ox}} = 460·2$ unit^4; $K_{\text{OY}} = 5·759$ unit, $K_{\text{ox}} = 3·199$ unit
5. 113π unit^3, 55/3 unit, $\bar{y} = 3·08$ unit
6. $y^3 = \frac{1}{3}(x^3 - 3x + 83)$ 7. $y = A\,e^{\frac{1}{2}x} + B\,e^{-2x}$
8. $\theta = \alpha \cos 4t$, $\dfrac{d\theta}{dt} = -4\alpha \sin 4t$ 9. 0·9731
10. (i) 1/216 (ii) 1/36 (iii) 25/216   11. (i) 1/11 (ii) 1/55 (iii) 1/22
    (iv) 12/55
12. (i) 0·531441, 0·354294, 0·098415 (ii) 0·114265

Examples 27.5

1. (i) 24 (ii) 22·3 (iii) 25%, 17·2; 75%, 26; $\bar{x} = 23·8$ cm,
   $s = 7·036$ cm
2. Median 5·55 cm, Mean 5·58 cm, $s = 0·2733$ cm
3. Mean 20·70, variance 14·67, $s = 3·83$
4. $K = 6/125$ (i) 2·5 (ii) 2·5 (iii) 1·25 unit^2, 1·118 unit
5. 150, 8·66; 132·68 and 167·32; 124·02 and 175·98
6. 0·7736, 0·2035, 0·0214(3), 0·0011(3); 0·9985
7. (i) 0·1353 (ii) 0·6765 (iii) 0·3235   8. 0·3944, 0·6151
9. (i) 66·8 (67) (ii) 22·8(23) (iii) 788·8(789)   10. (i) 12·5%
   (ii) 94·8%

Appendix

Notes on the Distribution of 'Sample Statistics' with Particular Reference to the Distribution of the 'Mean' of Samples Drawn from a Normal Population

Sampling Distributions

In applications of statistics, *sampling techniques* of many kinds are used. Generally the *population* is not known precisely, but assumptions may be made about a population. Samples are taken in order to *estimate* parameters *or* to test whether some change of process has improved a product *or* to compare the *effects* of different *factors* on a particular set of data. Population parameters which may be estimated, may, for example be the *mean, variance* and *standard deviation.*

The distribution of the population *data* may often be assumed to be *Normal.* Assumptions about a distribution are called *Hypotheses* and most statistical tests are used to make *decisions* whether or not hypotheses can, at a reasonable *level of probability*, be accepted or rejected.

In estimating parameters for a population, '*best*' estimates should, as far as possible, be used. *Best* estimates are often *unbiased estimates.* An unbiased estimate of a parameter for a population, using a sample of data is one such that, if an infinite number of samples were taken, in the long run the *mean* of all the estimates (or

343

the *expectation* of the estimate) will be the true value of the population parameter.

By taking a large number of samples, e.g. of the same size, the values of the estimates, when plotted as a *frequency curve* give rise to a *sampling distribution* which will have its own mean, variance and standard deviation. This standard deviation is usually called the *standard error* of the parameter. This is usually an *estimated standard error*.

The full treatment of the distribution of sample statistics is outside the range of this volume, but one distribution will be dealt with as follows:

Distribution of Sample Mean (sampling from a normal population)

If random samples of size n are selected from a normal population of mean μ and standard deviation σ then it can be shown that, if all the mean values ($\bar{x}$) are plotted the distribution of $\bar{x}$ is normal and has a mean $= \mu$ and a *standard error* of $\sigma/\sqrt{n}$ (variance σ^2/n). This means that the statistic $t = (\bar{x} - \mu)/(\sigma/\sqrt{n})$ may be taken as a *standard normal variate* of mean zero and standard error $= 1$ [i.e. an $N(0, 1)$ variate].

Examples:

(i) A random normal variate has a mean $\mu = 4$ and S.D. $\sigma = 1\cdot5$. If a random sample of 9 items has a mean of $4\cdot5$, what is the probability that the sample could have been drawn from the given normal population?
Mean of the distribution of means of samples $= \mu = 4$
S. Error of the distribution of $\bar{x} = \sigma/\sqrt{n} = 1\cdot5/\sqrt{9} = 0\cdot5$
$t = (\bar{x} - \mu)/(\sigma/\sqrt{n})$ is a standard normal variate
$t = (4\cdot5 - 4)/0\cdot5 = 0\cdot5/0\cdot5 = 1$
From the standard normal integral table
$P(-\infty < t < 1) = 0\cdot5000 + 0\cdot3413 = 0\cdot8413$
i.e. there is an $84\cdot13\%$ probability that the sample could have been drawn from the given population.

(ii) A sample of 100 items is drawn at random from a normal population. The population has a mean of $5\cdot0$ cm and S.D. of $2\cdot8$ cm. If the sample mean is $5\cdot7$ cm and it is decided to reject samples if the sample mean differs from the population mean by more than 1·96 standard errors, would this sample be accepted?

Mean of the distribution of sample mean $(\bar{x}) = 5\cdot0$ cm

S.D. of the population $\sigma = 2\cdot8$ cm

S.E. of the mean of samples of size 100

$= \sigma/\sqrt{n} = 2\cdot8/\sqrt{100} = 2\cdot8/10 = 0\cdot28$ cm

Mean of given sample $\bar{x} = 5\cdot7$ cm

$\bar{x} - \mu = 5\cdot7 - 5\cdot0 = 0\cdot7$ cm

$1\cdot96 \times$ S.E. $= 1\cdot96 \, \sigma/\sqrt{n} = 1\cdot96 \times 0\cdot28 = 0\cdot5488$ cm

Since $\bar{x} - \mu > 1\cdot96$ S.E. the sample would be rejected at the stated level.

Statistical Quality Control

When the number of articles being produced is very large, individual inspection may be physically impossible. Instead of individual inspection, random samples (usually of equal numbers) are taken and using a large number of samples limits are set on the assumption of normality. Charts are then used to record data and limit lines are laid on as follows:

Control limits for sample mean:

Assuming that the mean $\bar{X}$ has been set and σ for the population estimated, then $\bar{X} \pm 1\cdot96\sigma/\sqrt{n}$ are the 1/40th limits for the mean $\bar{X} \pm 3\cdot09\sigma/\sqrt{n}$ are the 1/1000th limits for the mean.

A *mean chart* can be set up with these limits and used to maintain control of quality. Often σ may be estimated from the *average sample range*.

Many other control charts may be devised but full discussion of such charts is beyond the scope of this volume.

A few unworked examples are given involving the distribution of the sample mean from a normal population.

Examples (Distribution of Sample Mean)

1. A normal population has a mean $\mu = 12$ cm and S.D. $\sigma = 4$ cm.

    (i) If a random sample of 25 items is selected what is the standard error of the sample mean?

   (ii) What size of sample should be taken so as to reduce the S.E. of the sample mean to $\frac{1}{2}$ that of samples of size 25?

   (iii) Would a mean of 13·7 cm for a sample of 25 be acceptable?

2. The following sample of 5 values of a dimension x cm are obtained by measuring the internal diameters of similar tubes: $x = 2\cdot007$, 2·010, 2·016, 2·018, 2·024. Calculate the mean and standard deviation of the sample.

   If the diameter is supposed normally distributed about a mean $\mu = 2\cdot022$ cm with S.D. $\sigma = 0\cdot007$ cm, would this sample be rejected?

3. A random normal variate has a mean $\mu = 15$ and S.D. $\sigma = 3\cdot8$. Samples are not accepted if their means are more than 2 standard errors from μ.

   (i) If a random sample of 16 items has a mean $= 17\cdot4$, would the sample be accepted or rejected?

   (ii) If a random sample of 25 items has a mean $= 13\cdot65$ would the sample be accepted or rejected?

4. The external radius of a manufactured part is assumed to be normally distributed with a mean of 5·350 cm and S.D. of 0·005 cm. If a random sample of 5 items has a mean of 5·356 cm and the acceptance limits for a sample are $\mu \pm 1\cdot96$S.E. would this sample be acceptable?

5. The percentage impurity of a chemical compound is assumed to be normally distributed with a mean of 1·510% and S.D. of 0·006%. A sample of 5 equal amounts is selected at random from a supply. The percentage impurities of the 5 amounts were as follows:
% impurity: 1·522, 1·517, 1·524, 1·523, 1·496.
Find the mean % impurity and hence decide whether the supply should be rejected assuming a 95% level of acceptance (i.e. accepted if the mean of the sample lies within the limits $\mu \pm 1\cdot96$S.E.).

6. A normal population has a mean of 5·76 cm and standard deviation of 7·5 mm.

   (i) If random samples of 36 items are taken for test purposes what is the standard error of the sample mean $\bar{x}$? What are the 2σ limits for $\bar{x}$?

(ii) How many less items should be taken so that the S.E. of the sample mean would be increased by 20%? What would be the 3σ limits for $\bar{x}$ for such samples?

7. Based on the hypothesis that the variables referred to are normally distributed dimensions, calculate the 1/40th and 1/1000th limits for the sample mean;

(i) if $\mu = 30$, $\sigma = 2\cdot0$, $n = 9$ (sample size)
(ii) if $\mu = 70$, $\sigma = 18\cdot0$, $n = 16$.

8. From a large number of samples taken during the initial running of a production process the 'population' mean of a dimension was estimated to be $11\cdot380$ cm with a standard deviation of 4 cm. Calculate the 1/40th and 1/1000th control limits for the means of samples of size 5.

Answers

1. (i) $0\cdot8$ cm (ii) $n = 100$ (iii) $t = 2\cdot125 > 1\cdot96$, rejected
2. $\bar{x} = 2\cdot015$ cm, $s = 0\cdot006$ cm, $|\bar{x} - \mu| > 2\sigma/\sqrt{n}$, rejected
3. (i) $\bar{x} - \mu > 2$ S.E., rejected (ii) $\bar{x} - \mu < 2$ S.E., accepted
4. $\bar{x} - \mu = 0\cdot006$, S.E. $= 0\cdot002236$, rejected
5. $\mu = 1\cdot510$ per cent, S.E. $= 0\cdot002683$, rejected
6. (i) S.E. $= 1\cdot25$ mm, limits $3\cdot26$, $8\cdot26$ (ii) 11 less items, limits $1\cdot26$, $10\cdot26$
7. (i) 1/40th: $28\cdot693$, $31\cdot307$; 1/1000th: $27\cdot94$, $32\cdot06$
8. 1/40th: $7\cdot875$, $14\cdot885$; 1/1000th: $5\cdot853$, $16\cdot907$

Table of Normal Probability

The integral P of the normal probability function

X	·00	·01	·02	·03	·04	·05	·06
·0	·5000	·5040	·5080	·5120	·5160	·5199	·5239
·1	·5398	·5438	·5478	·5517	·5557	·5596	·5636
·2	·5793	·5832	·5871	·5910	·5948	·5987	·6026
·3	·6179	·6217	·6255	·6293	·6331	·6368	·6406
·4	·6554	·6591	·6628	·6664	·6700	·6736	·6772
·5	·6915	·6950	·6985	·7019	·7054	·7088	·7123
·6	·7257	·7291	·7324	·7357	·7389	·7422	·7454
·7	·7580	·7611	·7642	·7673	·7704	·7734	·7764
·8	·7881	·7910	·7939	·7967	·7995	·8023	·8051
·9	·8159	·8186	·8212	·8238	·8264	·8289	·8315
1·0	·8413	·8438	·8461	·8485	·8508	·8531	·8554
1·1	·8643	·8665	·8686	·8708	·8729	·8749	·8770
1·2	·8849	·8869	·8888	·8907	·8925	·8944	·8962
1·3	·9032	·9049	·9066	·9082	·9099	·9115	·9131
1·4	·9192	·9207	·9222	·9236	·9251	·9265	·9279
1·5	·9332	·9345	·9357	·9370	·9382	·9394	·9406
1·6	·9452	·9463	·9474	·9484	·9495	·9505	·9515
1·7	·9554	·9564	·9573	·9582	·9591	·9599	·9608
1·8	·9641	·9649	·9656	·9664	·9671	·9678	·9686
1·9	·9713	·9719	·9726	·9732	·9738	·9744	·9750
2·0	·9772	·9778	·9783	·9788	·9793	·9798	·9803
2·1	·9821	·9826	·9830	·9834	·9838	·9842	·9846
2·2	·9861	·9864	·9868	·9871	·9875	·9878	·9881
2·3	·9893	·9896	·9898	·99010	·99036	·99061	·99086
2·4	·99180	·99202	·99224	·99245	·99266	·99286	·99305
2·5	·99379	·99396	·99413	·99430	·99446	·99461	·99477
2·6	·99534	·99547	·99560	·99573	·99585	·99598	·99609
2·7	·99653	·99664	·99674	·99683	·99693	·99702	·99711
2·8	·99744	·99752	·99760	·99767	·99774	·99781	·99788
2·9	·99813	·99819	·99825	·99831	·99836	·99841	·99846

Linear interpolation sufficient

X	3·0	3·1	3·2	3·3	3·4	3·5	3·6
P	·99865	·99903	·99931	·99952	·99966	·99977	·99984
Z	·00443	·00327	·00238	·00172	·00123	·00087	·00061

$$P = \frac{1}{\sqrt{2\pi}} \int_{-\infty}^{X} e^{-\frac{1}{2}x^2}\, dx, \quad Z = \frac{1}{\sqrt{2\pi}} e^{-\frac{1}{2}x^2}$$

X is the standardized variable with zero mean and unit standard deviation

For negative values of X note: $P(X) = 1 - P(-X)$

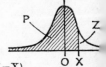

·07	·08	·09
·5279	·5319	·5359
·5675	·5714	·5753
·6064	·6103	·6141
·6443	·6480	·6517
·6808	·6844	·6879
·7157	·7190	·7224
·7486	·7517	·7549
·7794	·7823	·7852
·8078	·8106	·8133
·8340	·8365	·8389
·8577	·8599	·8621
·8790	·8810	·8830
·8980	·8997	·9015
·9147	·9162	·9177
·9292	·9306	·9319
·9418	·9429	·9441
·9525	·9535	·9545
·9616	·9625	·9633
·9693	·9699	·9706
·9756	·9761	·9767
·9808	·9812	·9817
·9850	·9854	·9857
·9884	·9887	·9890
·99111	·99134	·99158
·99324	·99343	·99361
·99492	·99506	·99520
·99621	·99632	·99643
·99720	·99728	·99736
·99795	·99801	·99807
·99851	·99856	·99861

X	·00	·02	·04	·06	·08
·0	·3989	·3989	·3986	·3982	·3977
·1	·3970	·3961	·3951	·3939	·3925
·2	·3910	·3894	·3876	·3857	·3836
·3	·3814	·3790	·3765	·3739	·3712
·4	·3683	·3653	·3621	·3589	·3555
·5	·3521	·3485	·3448	·3410	·3372
·6	·3332	·3292	·3251	·3209	·3166
·7	·3123	·3079	·3034	·2989	·2943
·8	·2897	·2850	·2803	·2756	·2709
·9	·2661	·2613	·2565	·2516	·2468
1·0	·2420	·2371	·2323	·2275	·2227
1·1	·2179	·2131	·2083	·2036	·1989
1·2	·1942	·1895	·1849	·1804	·1758
1·3	·1714	·1669	·1626	·1582	·1539
1·4	·1497	·1456	·1415	·1374	·1334
1·5	·1295	·1257	·1219	·1182	·1145
1·6	·1109	·1074	·1040	·1006	·0973
1·7	·0940	·0909	·0878	·0848	·0818
1·8	·0790	·0761	·0734	·0707	·0681
1·9	·0656	·0632	·0608	·0584	·0562
2·0	·0540	·0519	·0498	·0478	·0459
2·1	·0440	·0422	·0404	·0387	·0371
2·2	·0355	·0339	·0325	·0310	·0297
2·3	·0283	·0270	·0258	·0246	·0235
2·4	·0224	·0213	·0203	·0194	·0184
2·5	·0175	·0167	·0158	·0151	·0143
2·6	·0136	·0129	·0122	·0116	·0110
2·7	·0104	·0099	·0093	·0088	·0084
2·8	·0079	·0075	·0071	·0067	·0063
2·9	·0060	·0056	·0053	·0050	·0047

Linear interpolation sufficient

3·7	3·8	3·9
·99989	·99993	·99995
00042	·00029	·00020

Percentage points

P	50	75	90	95	97·5
X	·0000	·6745	1·2816	1·6449	1·9600

P	99·0	99·5	99·9	99·95
X	2·3263	2·5758	3·0902	3·2905

LOGARITHMS

	0	1	2	3	4	5	6	7	8	9	1	2	3	4	5	6	7	8	9
10	0000	0043	0086	0128	0170	0212	0253	0294	0334	0374	4	8	12	17	21	25	29	33	37
11	0414	0453	0492	0531	0569	0607	0645	0682	0719	0755	4	8	11	15	19	23	26	30	34
12	0792	0828	0864	0899	0934	0969	1004	1038	1072	1106	3	7	10	14	17	21	24	28	31
13	1139	1173	1206	1239	1271	1303	1335	1367	1399	1430	3	6	10	13	16	19	23	26	29
14	1461	1492	1523	1553	1584	1614	1644	1673	1703	1732	3	6	9	12	15	18	21	24	27
15	1761	1790	1818	1847	1875	1903	1931	1959	1987	2014	3	6	8	11	14	17	20	22	25
16	2041	2068	2095	2122	2148	2175	2201	2227	2253	2279	3	5	8	11	13	16	18	21	24
17	2304	2330	2355	2380	2405	2430	2455	2480	2504	2529	2	5	7	10	12	15	17	20	22
18	2553	2577	2601	2625	2648	2672	2695	2718	2742	2765	2	5	7	9	12	14	16	19	21
19	2788	2810	2833	2856	2878	2900	2923	2945	2967	2989	2	4	7	9	11	13	16	18	20
20	3010	3032	3054	3075	3096	3118	3139	3160	3181	3201	2	4	6	8	11	13	15	17	19
21	3222	3243	3263	3284	3304	3324	3345	3365	3385	3404	2	4	6	8	10	12	14	16	18
22	3424	3444	3464	3483	3502	3522	3541	3560	3579	3598	2	4	6	8	10	12	14	15	17
23	3617	3636	3655	3674	3692	3711	3729	3747	3766	3784	2	4	6	7	9	11	13	15	17
24	3802	3820	3838	3856	3874	3892	3909	3927	3945	3962	2	4	5	7	9	11	12	14	16
25	3979	3997	4014	4031	4048	4065	4082	4099	4116	4133	2	3	5	7	9	10	12	14	15
26	4150	4166	4183	4200	4216	4232	4249	4265	4281	4298	2	3	5	7	8	10	11	13	15
27	4314	4330	4346	4362	4378	4393	4409	4425	4440	4456	2	3	5	6	8	9	11	13	14
28	4472	4487	4502	4518	4533	4548	4564	4579	4594	4609	2	3	5	6	8	9	11	12	14
29	4624	4639	4654	4669	4683	4698	4713	4728	4742	4757	1	3	4	6	7	9	10	12	13
30	4771	4786	4800	4814	4829	4843	4857	4871	4886	4900	1	3	4	6	7	9	10	11	13
31	4914	4928	4942	4955	4969	4983	4997	5011	5024	5038	1	3	4	6	7	8	10	11	12
32	5051	5065	5079	5092	5105	5119	5132	5145	5159	5172	1	3	4	5	7	8	9	11	12
33	5185	5198	5211	5224	5237	5250	5263	5276	5289	5302	1	3	4	5	6	8	9	10	12
34	5315	5328	5340	5353	5366	5378	5391	5403	5416	5428	1	3	4	5	6	8	9	10	11
35	5441	5453	5465	5478	5490	5502	5514	5527	5539	5551	1	2	4	5	6	7	9	10	11
36	5563	5575	5587	5599	5611	5623	5635	5647	5658	5670	1	2	4	5	6	7	8	10	11
37	5682	5694	5705	5717	5729	5740	5752	5763	5775	5786	1	2	3	5	6	7	8	9	10
38	5798	5809	5821	5832	5843	5855	5866	5877	5888	5899	1	2	3	5	6	7	8	9	10
39	5911	5922	5933	5944	5955	5966	5977	5988	5999	6010	1	2	3	4	5	7	8	9	10
40	6021	6031	6042	6053	6064	6075	6085	6096	6107	6117	1	2	3	4	5	6	8	9	10
41	6128	6138	6149	6160	6170	6180	6191	6201	6212	6222	1	2	3	4	5	6	7	8	9
42	6232	6243	6253	6263	6274	6284	6294	6304	6314	6325	1	2	3	4	5	6	7	8	9
43	6335	6345	6355	6365	6375	6385	6395	6405	6415	6425	1	2	3	4	5	6	7	8	9
44	6435	6444	6454	6464	6474	6484	6493	6503	6513	6522	1	2	3	4	5	6	7	8	9
45	6532	6542	6551	6561	6571	6580	6590	6599	6609	6618	1	2	3	4	5	6	7	8	9
46	6628	6637	6646	6656	6665	6675	6684	6693	6702	6712	1	2	3	4	5	6	7	7	8
47	6721	6730	6739	6749	6758	6767	6776	6785	6794	6803	1	2	3	4	5	5	6	7	8
48	6812	6821	6830	6839	6848	6857	6866	6875	6884	6893	1	2	3	4	4	5	6	7	8
49	6902	6911	6920	6928	6937	6946	6955	6964	6972	6981	1	2	3	4	4	5	6	7	8
50	6990	6998	7007	7016	7024	7033	7042	7050	7059	7067	1	2	3	3	4	5	6	7	8
51	7076	7084	7093	7101	7110	7118	7126	7135	7143	7152	1	2	3	3	4	5	6	7	8
52	7160	7168	7177	7185	7193	7202	7210	7218	7226	7235	1	2	2	3	4	5	6	7	7
53	7243	7251	7259	7267	7275	7284	7292	7300	7308	7316	1	2	2	3	4	5	6	6	7
54	7324	7332	7340	7348	7356	7364	7372	7380	7388	7396	1	2	2	3	4	5	6	6	7
	0	1	2	3	4	5	6	7	8	9	1	2	3	4	5	6	7	8	9

LOGARITHMS

	0	1	2	3	4	5	6	7	8	9	1	2	3	4	5	6	7	8	9
55	7404	7412	7419	7427	7435	7443	7451	7459	7466	7474	1	2	2	3	4	5	5	6	7
56	7482	7490	7497	7505	7513	7520	7528	7536	7543	7551	1	2	2	3	4	5	5	6	7
57	7559	7566	7574	7582	7589	7597	7604	7612	7619	7627	1	2	2	3	4	5	5	6	7
58	7634	7642	7649	7657	7664	7672	7679	7686	7694	7701	1	1	2	3	4	4	5	6	7
59	7709	7716	7723	7731	7738	7745	7752	7760	7767	7774	1	1	2	3	4	4	5	6	7
60	7782	7789	7796	7803	7810	7818	7825	7832	7839	7846	1	1	2	3	4	4	5	6	6
61	7853	7860	7868	7875	7882	7889	7896	7903	7910	7917	1	1	2	3	4	4	5	6	6
62	7924	7931	7938	7945	7952	7959	7966	7973	7980	7987	1	1	2	3	3	4	5	6	6
63	7993	8000	8007	8014	8021	8028	8035	8041	8048	8055	1	1	2	3	3	4	5	5	6
64	8062	8069	8075	8082	8089	8096	8102	8109	8116	8122	1	1	2	3	3	4	5	5	6
65	8129	8136	8142	8149	8156	8162	8169	8176	8182	8189	1	1	2	3	3	4	5	5	6
66	8195	8202	8209	8215	8222	8228	8235	8241	8248	8254	1	1	2	3	3	4	5	5	6
67	8261	8267	8274	8280	8287	8293	8299	8306	8312	8319	1	1	2	3	3	4	5	5	6
68	8325	8331	8338	8344	8351	8357	8363	8370	8376	8382	1	1	2	3	3	4	4	5	6
69	8388	8395	8401	8407	8414	8420	8426	8432	8439	8445	1	1	2	2	3	4	4	5	6
70	8451	8457	8463	8470	8476	8482	8488	8494	8500	8506	1	1	2	2	3	4	4	5	6
71	8513	8519	8525	8531	8537	8543	8549	8555	8561	8567	1	1	2	2	3	4	4	5	5
72	8573	8579	8585	8591	8597	8603	8609	8615	8621	8627	1	1	2	2	3	4	4	5	5
73	8633	8639	8645	8651	8657	8663	8669	8675	8681	8686	1	1	2	2	3	4	4	5	5
74	8692	8698	8704	8710	8716	8722	8727	8733	8739	8745	1	1	2	2	3	4	4	5	5
75	8751	8756	8762	8768	8774	8779	8785	8791	8797	8802	1	1	2	2	3	3	4	5	5
76	8808	8814	8820	8825	8831	8837	8842	8848	8854	8859	1	1	2	2	3	3	4	5	5
77	8865	8871	8876	8882	8887	8893	8899	8904	8910	8915	1	1	2	2	3	3	4	4	5
78	8921	8927	8932	8938	8943	8949	8954	8960	8965	8971	1	1	2	2	3	3	4	4	5
79	8976	8982	8987	8993	8998	9004	9009	9015	9020	9025	1	1	2	2	3	3	4	4	5
80	9031	9036	9042	9047	9053	9058	9063	9069	9074	9079	1	1	2	2	3	3	4	4	5
81	9085	9090	9096	9101	9106	9112	9117	9122	9128	9133	1	1	2	2	3	3	4	4	5
82	9138	9143	9149	9154	9159	9165	9170	9175	9180	9186	1	1	2	2	3	3	4	4	5
83	9191	9196	9201	9206	9212	9217	9222	9227	9232	9238	1	1	2	2	3	3	4	4	5
84	9243	9248	9253	9258	9263	9269	9274	9279	9284	9289	1	1	2	2	3	3	4	4	5
85	9294	9299	9304	9309	9315	9320	9325	9330	9335	9340	1	1	2	2	3	3	4	4	5
86	9345	9350	9355	9360	9365	9370	9375	9380	9385	9390	1	1	2	2	3	3	4	4	5
87	9395	9400	9405	9410	9415	9420	9425	9430	9435	9440	0	1	1	2	2	3	3	4	4
88	9445	9450	9455	9460	9465	9469	9474	9479	9484	9489	0	1	1	2	2	3	3	4	4
89	9494	9499	9504	9509	9513	9518	9523	9528	9533	9538	0	1	1	2	2	3	3	4	4
90	9542	9547	9552	9557	9562	9566	9571	9576	9581	9586	0	1	1	2	2	3	3	4	4
91	9590	9595	9600	9605	9609	9614	9619	9624	9628	9633	0	1	1	2	2	3	3	4	4
92	9638	9643	9647	9652	9657	9661	9666	9671	9675	9680	0	1	1	2	2	3	3	4	4
93	9685	9689	9694	9699	9703	9708	9713	9717	9722	9727	0	1	1	2	2	3	3	4	4
94	9731	9736	9741	9745	9750	9754	9759	9763	9768	9773	0	1	1	2	2	3	3	4	4
95	9777	9782	9786	9791	9795	9800	9805	9809	9814	9818	0	1	1	2	2	3	3	4	4
96	9823	9827	9832	9836	9841	9845	9850	9854	9859	9863	0	1	1	2	2	3	3	4	4
97	9868	9872	9877	9881	9886	9890	9894	9899	9903	9908	0	1	1	2	2	3	3	4	4
98	9912	9917	9921	9926	9930	9934	9939	9943	9948	9952	0	1	1	2	2	3	3	4	4
99	9956	9961	9965	9969	9974	9978	9983	9987	9991	9996	0	1	1	2	2	3	3	3	4
	0	1	2	3	4	5	6	7	8	9	1	2	3	4	5	6	7	8	9

NAPIERIAN OR NATURAL LOGARITHMS (log$_e$ N)

	0	1	2	3	4	5	6	7	8	9	1 2 3	4 5 6	7 8 9
1·0	0·0000	0099	0198	0296	0392	0488	0583	0677	0770	0862	10 19 29	38 48 57	67 76 86
1·1	·0953	1044	1133	1222	1310	1398	1484	1570	1655	1740	9 17 26	35 44 52	61 70 78
1·2	·1823	1906	1989	2070	2151	2231	2311	2390	2469	2546	8 16 24	32 40 48	56 64 72
1·3	·2624	2700	2776	2852	2927	3001	3075	3148	3221	3293	7 15 22	30 37 44	52 59 67
1·4	·3365	3436	3507	3577	3646	3716	3784	3853	3920	3988	7 14 21	28 35 41	48 55 62
1·5	·4055	4121	4187	4253	4318	4383	4447	4511	4574	4637	6 13 19	26 32 39	45 52 58
1·6	·4700	4762	4824	4886	4947	5008	5068	5128	5188	5247	6 12 18	24 30 36	42 48 55
1·7	·5306	5365	5423	5481	5539	5596	5653	5710	5766	5822	6 11 17	24 29 34	40 46 51
1·8	·5878	5933	5988	6043	6098	6152	6206	6259	6313	6366	5 11 16	22 27 32	38 43 49
1·9	·6419	6471	6523	6575	6627	6678	6729	6780	6831	6881	5 10 15	20 26 31	36 41 46
2·0	·6931	6981	7031	7080	7129	7178	7227	7275	7324	7372	5 10 15	20 24 29	34 39 44
2·1	·7419	7467	7514	7561	7608	7655	7701	7747	7793	7839	5 9 14	19 23 28	33 37 42
2·2	·7885	7930	7975	8020	8065	8109	8154	8198	8242	8286	4 9 13	18 22 27	31 36 40
2·3	·8329	8372	8416	8459	8502	8544	8587	8629	8671	8713	4 9 13	17 21 26	30 34 38
2·4	·8755	8796	8838	8879	8920	8961	9002	9042	9083	9123	4 8 12	16 20 24	29 33 37
2·5	·9163	9203	9243	9282	9322	9361	9400	9439	9478	9517	4 8 12	16 20 24	27 31 35
2·6	·9555	9594	9632	9670	9708	9746	9783	9821	9858	9895	4 8 11	15 19 23	26 30 34
2·7	·9933	9969	1·0006	0043	0080	0116	0152	0188	0225	0260	4 7 11	15 18 22	25 29 33
2·8	1·0296	0332	0367	0403	0438	0473	0508	0543	0578	0613	4 7 11	14 18 21	25 28 32
2·9	1·0647	0682	0716	0750	0784	0818	0852	0886	0919	0953	3 7 10	14 17 20	24 27 31
3·0	1·0986	1019	1053	1086	1119	1151	1184	1217	1249	1282	3 7 10	13 16 20	23 26 30
3·1	1·1314	1346	1378	1410	1442	1474	1506	1537	1569	1600	3 6 10	13 16 19	22 25 29
3·2	1·1632	1663	1694	1725	1756	1787	1817	1848	1878	1909	3 6 9	12 15 18	22 25 28
3·3	1·1939	1969	2000	2030	2060	2090	2119	2149	2179	2208	3 6 9	12 15 18	21 24 27
3·4	1·2238	2267	2296	2326	2355	2384	2413	2442	2470	2499	3 6 9	12 15 17	20 23 26
3·5	1·2528	2556	2585	2613	2641	2669	2698	2726	2754	2782	3 6 8	11 14 17	20 23 25
3·6	1·2809	2837	2865	2892	2920	2947	2975	3002	3029	3056	3 5 8	11 14 16	19 22 25
3·7	1·3083	3110	3137	3164	3191	3218	3244	3271	3297	3324	3 5 8	11 13 16	19 21 24
3·8	1·3350	3376	3403	3429	3455	3481	3507	3533	3558	3584	3 5 8	10 13 16	18 21 23
3·9	1·3610	3635	3661	3686	3712	3737	3762	3788	3813	3838	3 5 8	10 13 15	18 20 23
4·0	1·3863	3888	3913	3938	3962	3987	4012	4036	4061	4085	2 5 7	10 12 15	17 20 22
4·1	1·4110	4134	4159	4183	4207	4231	4255	4279	4303	4327	2 5 7	10 12 14	17 19 22
4·2	1·4351	4375	4398	4422	4446	4469	4493	4516	4540	4563	2 5 7	9 12 14	16 19 21
4·3	1·4586	4609	4633	4656	4679	4702	4725	4748	4770	4793	2 5 7	9 12 14	16 18 21
4·4	1·4816	4839	4861	4884	4907	4929	4951	4974	4996	5019	2 5 7	9 11 14	16 18 20
4·5	1·5041	5063	5085	5107	5129	5151	5173	5195	5217	5239	2 4 7	9 11 13	15 18 20
4·6	1·5261	5282	5304	5326	5347	5369	5390	5412	5433	5454	2 4 6	9 11 13	15 17 19
4·7	1·5476	5497	5518	5539	5560	5581	5602	5623	5644	5665	2 4 6	8 11 13	15 17 19
4·8	1·5686	5707	5728	5748	5769	5790	5810	5831	5851	5872	2 4 6	8 10 12	14 16 19
4·9	1·5892	5913	5933	5953	5974	5994	6014	6034	6054	6074	2 4 6	8 10 12	14 16 18
5·0	1·6094	6114	6134	6154	6174	6194	6214	6233	6253	6273	2 4 6	8 10 12	14 16 18
5·1	1·6292	6312	6332	6351	6371	6390	6409	6429	6448	6467	2 4 6	8 10 12	14 16 18
5·2	1·6487	6506	6525	6544	6563	6582	6601	6620	6639	6658	2 4 6	8 10 11	13 15 17
5·3	1·6677	6696	6715	6734	6752	6771	6790	6808	6827	6845	2 4 6	7 9 11	13 15 17

Hyperbolic or Napierian Logarithms of 10^{-n}

n	1	2	3	4	5	6	7	8	9
log$_e$ 10^n	2·3026	4·6052	6·9078	9·2103	11·5129	13·8155	16·1181	18·4207	20·7233

NAPIERIAN OR NATURAL LOGARITHMS (log$_e$ N)

Differences (ADD)

	0	1	2	3	4	5	6	7	8	9	1 2 3	4 5 6	7 8 9
5·4	1·6864	6882	6901	6919	6938	6956	6974	6993	7011	7029	2 4 5	7 9 11	13 15 17
5·5	1·7047	7066	7084	7102	7120	7138	7156	7174	7192	7210	2 4 5	7 9 11	13 14 16
5·6	1·7228	7246	7263	7281	7299	7317	7334	7352	7370	7387	2 4 5	7 9 11	12 14 16
5·7	1·7405	7422	7440	7457	7475	7492	7509	7527	7544	7561	2 3 5	7 9 10	12 14 16
5·8	1·7579	7596	7613	7630	7647	7664	7681	7699	7716	7733	2 3 5	7 9 10	12 14 15
5·9	1·7750	7766	7783	7800	7817	7834	7851	7867	7884	7901	2 3 5	7 8 10	12 13 15
6·0	1·7918	7934	7951	7967	7984	8001	8017	8034	8050	8066	2 3 5	7 8 10	12 13 15
6·1	1·8083	8099	8116	8132	8148	8165	8181	8197	8213	8229	2 3 5	6 8 10	11 13 15
6·2	1·8245	8262	8278	8294	8310	8326	8342	8358	8374	8390	2 3 5	6 8 10	11 13 14
6·3	1·8405	8421	8437	8453	8469	8485	8500	8516	8532	8547	2 3 5	6 8 9	11 13 14
6·4	1·8563	8579	8594	8610	8625	8641	8656	8672	8687	8703	2 3 5	6 8 9	11 12 14
6·5	1·8718	8733	8749	8764	8779	8795	8810	8825	8840	8856	2 3 5	6 8 9	11 12 14
6·6	1·8871	8886	8901	8916	8931	8946	8961	8976	8991	9006	2 3 5	6 8 9	11 12 14
6·7	1·9021	9036	9051	9066	9081	9095	9110	9125	9140	9155	1 3 4	6 7 9	10 12 13
6·8	1·9169	9184	9199	9213	9228	9242	9257	9272	9286	9301	1 3 4	6 7 9	10 12 13
6·9	1·9315	9330	9344	9359	9373	9387	9402	9416	9430	9445	1 3 4	6 7 9	10 12 13
7·0	1·9459	9473	9488	9502	9516	9530	9544	9559	9573	9587	1 3 4	6 7 9	10 11 13
7·1	1·9601	9615	9629	9643	9657	9671	9685	9699	9713	9727	1 3 4	6 7 8	10 11 13
7·2	1·9741	9755	9769	9782	9796	9810	9824	9838	9851	9865	1 3 4	6 7 8	10 11 12
7·3	1·9879	9892	9906	9920	9933	9947	9961	9974	9988	2·0001	1 3 4	5 7 8	10 11 12
7·4	2·0015	0028	0042	0055	0069	0082	0096	0109	0122	0136	1 3 4	5 7 8	9 11 12
7·5	2·0149	0162	0176	0189	0202	0215	0229	0242	0255	0268	1 3 4	5 7 8	9 11 12
7·6	2·0281	0295	0308	0321	0334	0347	0360	0373	0386	0399	1 3 4	5 7 8	9 10 12
7·7	2·0412	0425	0438	0451	0464	0477	0490	0503	0516	0528	1 3 4	5 6 8	9 10 12
7·8	2·0541	0554	0567	0580	0592	0605	0618	0631	0643	0656	1 3 4	5 6 8	9 10 11
7·9	2·0669	0681	0694	0707	0719	0732	0744	0757	0769	0782	1 3 4	5 6 8	9 10 11
8·0	2·0794	0807	0819	0832	0844	0857	0869	0882	0894	0906	1 3 4	5 6 7	9 10 11
8·1	2·0919	0931	0943	0956	0968	0980	0992	1005	1017	1029	1 2 4	5 6 7	9 10 11
8·2	2·1041	1054	1066	1078	1090	1102	1114	1126	1138	1150	1 2 4	5 6 7	9 10 11
8·3	2·1163	1175	1187	1199	1211	1223	1235	1247	1258	1270	1 2 4	5 6 7	8 10 11
8·4	2·1282	1294	1306	1318	1330	1342	1353	1365	1377	1389	1 2 4	5 6 7	8 9 11
8·5	2·1401	1412	1424	1436	1448	1459	1471	1483	1494	1506	1 2 4	5 6 7	8 9 11
8·6	2·1518	1529	1541	1552	1564	1576	1587	1599	1610	1622	1 2 3	5 6 7	8 9 10
8·7	2·1633	1645	1656	1668	1679	1691	1702	1713	1725	1736	1 2 3	5 6 7	8 9 10
8·8	2·1748	1759	1770	1782	1793	1804	1815	1827	1838	1849	1 2 3	5 6 7	8 9 10
8·9	2·1861	1872	1883	1894	1905	1917	1928	1939	1950	1961	1 2 3	4 6 7	8 9 10
9·0	2·1972	1983	1994	2006	2017	2028	2039	2050	2061	2072	1 2 3	4 6 7	8 9 10
9·1	2·2083	2094	2105	2116	2127	2138	2148	2159	2170	2181	1 2 3	4 5 7	8 9 10
9·2	2·2192	2203	2214	2225	2235	2246	2257	2268	2279	2289	1 2 3	4 5 6	8 9 10
9·3	2·2300	2311	2322	2332	2343	2354	2364	2375	2386	2396	1 2 3	4 5 6	7 9 10
9·4	2·2407	2418	2428	2439	2450	2460	2471	2481	2492	2502	1 2 3	4 5 6	7 8 10
9·5	2·2513	2523	2534	2544	2555	2565	2576	2586	2597	2607	1 2 3	4 5 6	7 8 9
9·6	2·2618	2628	2638	2649	2659	2670	2680	2690	2701	2711	1 2 3	4 5 6	7 8 9
9·7	2·2721	2732	2742	2752	2762	2773	2783	2793	2803	2814	1 2 3	4 5 6	7 8 9
9·8	2·2824	2834	2844	2854	2865	2875	2885	2895	2905	2915	1 2 3	4 5 6	7 8 9
9·9	2·2925	2935	2946	2956	2966	2976	2986	2996	3006	3016	1 2 3	4 5 6	7 8 9
10·0	2·3026												

Hyperbolic or Napierian Logarithms of 10^n

n	1	2	3	4	5	6	7	8	9
log$_e$ 10^{-n}	$\overline{3}$·6974	$\overline{5}$·3948	$\overline{7}$·0922	$\overline{10}$·7897	$\overline{12}$·4871	$\overline{14}$·1845	$\overline{17}$·8819	$\overline{19}$·5793	$\overline{21}$·2767

NATURAL SINES (left-hand column and top)

	0'	6'	12'	18'	24'	30'	36'	42'	48'	54'	—	1'	2'	3'	4'	5'	
0°	·0000	0017	0035	0052	0070	0087	0105	0122	0140	0157	·0175	3	6	9	12	15	89°
1	·0175	0192	0209	0227	0244	0262	0279	0297	0314	0332	·0349	3	6	9	12	15	88
2	·0349	0366	0384	0401	0419	0436	0454	0471	0488	0506	·0523	3	6	9	12	15	87
3	·0523	0541	0558	0576	0593	0610	0628	0645	0663	0680	·0698	3	6	9	12	15	86
4	·0698	0715	0732	0750	0767	0785	0802	0819	0837	0854	·0872	3	6	9	12	14	85
5	·0872	0889	0906	0924	0941	0958	0976	0993	1011	1028	·1045	3	6	9	12	14	84
6	·1045	1063	1080	1097	1115	1132	1149	1167	1184	1201	·1219	3	6	9	12	14	83
7	·1219	1236	1253	1271	1288	1305	1323	1340	1357	1374	·1392	3	6	9	12	14	82
8	·1392	1409	1426	1444	1461	1478	1495	1513	1530	1547	·1564	3	6	9	12	14	81
9	·1564	1582	1599	1616	1633	1650	1668	1685	1702	1719	·1736	3	6	9	11	14	80
10	·1736	1754	1771	1788	1805	1822	1840	1857	1874	1891	·1908	3	6	9	11	14	79
11	·1908	1925	1942	1959	1977	1994	2011	2028	2045	2062	·2079	3	6	9	11	14	78
12	·2079	2096	2113	2130	2147	2164	2181	2198	2215	2233	·2250	3	6	9	11	14	77
13	·2250	2267	2284	2300	2317	2334	2351	2368	2385	2402	·2419	3	6	8	11	14	76
14	·2419	2436	2453	2470	2487	2504	2521	2538	2554	2571	·2588	3	6	8	11	14	75
15	·2588	2605	2622	2639	2656	2672	2689	2706	2723	2740	·2756	3	6	8	11	14	74
16	·2756	2773	2790	2807	2823	2840	2857	2874	2890	2907	·2924	3	6	8	11	14	73
17	·2924	2940	2957	2974	2990	3007	3024	3040	3057	3074	·3090	3	6	8	11	14	72
18	·3090	3107	3123	3140	3156	3173	3190	3206	3223	3239	·3256	3	6	8	11	14	71
19	·3256	3272	3289	3305	3322	3338	3355	3371	3387	3404	·3420	3	5	8	11	14	70
20	·3420	3437	3453	3469	3486	3502	3518	3535	3551	3567	·3584	3	5	8	11	14	69
21	·3584	3600	3616	3633	3649	3665	3681	3697	3714	3730	·3746	3	5	8	11	14	68
22	·3746	3762	3778	3795	3811	3827	3843	3859	3875	3891	·3907	3	5	8	11	13	67
23	·3907	3923	3939	3955	3971	3987	4003	4019	4035	4051	·4067	3	5	8	11	13	66
24	·4067	4083	4099	4115	4131	4147	4163	4179	4195	4210	·4226	3	5	8	11	13	65
25	·4226	4242	4258	4274	4289	4305	4321	4337	4352	4368	·4384	3	5	8	11	13	64
26	·4384	4399	4415	4431	4446	4462	4478	4493	4509	4524	·4540	3	5	8	10	13	63
27	·4540	4555	4571	4586	4602	4617	4633	4648	4664	4679	·4695	3	5	8	10	13	62
28	·4695	4710	4726	4741	4756	4772	4787	4802	4818	4833	·4848	3	5	8	10	13	61
29	·4848	4863	4879	4894	4909	4924	4939	4955	4970	4985	·5000	3	5	8	10	13	60
30	·5000	5015	5030	5045	5060	5075	5090	5105	5120	5135	·5150	3	5	8	10	13	59
31	·5150	5165	5180	5195	5210	5225	5240	5255	5270	5284	·5299	2	5	7	10	12	58
32	·5299	5314	5329	5344	5358	5373	5388	5402	5417	5432	·5446	2	5	7	10	12	57
33	·5446	5461	5476	5490	5505	5519	5534	5548	5563	5577	·5592	2	5	7	10	12	56
34	·5592	5606	5621	5635	5650	5664	5678	5693	5707	5721	·5736	2	5	7	10	12	55
35	·5736	5750	5764	5779	5793	5807	5821	5835	5850	5864	·5878	2	5	7	9	12	54
36	·5878	5892	5906	5920	5934	5948	5962	5976	5990	6004	·6018	2	5	7	9	12	53
37	·6018	6032	6046	6060	6074	6088	6101	6115	6129	6143	·6157	2	5	7	9	12	52
38	·6157	6170	6184	6198	6211	6225	6239	6252	6266	6280	·6293	2	5	7	9	11	51
39	·6293	6307	6320	6334	6347	6361	6374	6388	6401	6414	·6428	2	4	7	9	11	50
40	·6428	6441	6455	6468	6481	6494	6508	6521	6534	6547	·6561	2	4	7	9	11	49
41	·6561	6574	6587	6600	6613	6626	6639	6652	6665	6678	·6691	2	4	7	9	11	48
42	·6691	6704	6717	6730	6743	6756	6769	6782	6794	6807	·6820	2	4	6	9	11	47
43	·6820	6833	6845	6858	6871	6884	6896	6909	6921	6934	·6947	2	4	6	8	11	46
44	·6947	6959	6972	6984	6997	7009	7022	7034	7046	7059	·7071	2	4	6	8	10	45
—	—	54'	48'	42'	36'	30'	24'	18'	12'	6'	0'	1'	2'	3'	4'	5'	

NATURAL COSINES (right-hand column and bottom)

NATURAL SINES (left-hand column and top)

	0'	6'	12'	18'	24'	30'	36'	42'	48'	54'	—	1'	2'	3'	4'	5'	
45°	·7071	7083	7096	7108	7120	7133	7145	7157	7169	7181	·7193	2	4	6	8	10	**44**°
46	·7193	7206	7218	7230	7242	7254	7266	7278	7290	7302	·7314	2	4	6	8	10	**43**
47	·7314	7325	7337	7349	7361	7373	7385	7396	7408	7420	·7431	2	4	6	8	10	**42**
48	·7431	7443	7455	7466	7478	7490	7501	7513	7524	7536	·7547	2	4	6	8	10	**41**
49	·7547	7559	7570	7581	7593	7604	7615	7627	7638	7649	·7660	2	4	6	8	9	**40**
50	·7660	7672	7683	7694	7705	7716	7727	7738	7749	7760	·7771	2	4	6	7	9	**39**
51	·7771	7782	7793	7804	7815	7826	7837	7848	7859	7869	·7880	2	4	5	7	9	**38**
52	·7880	7891	7902	7912	7923	7934	7944	7955	7965	7976	·7986	2	4	5	7	9	**37**
53	·7986	7997	8007	8018	8028	8039	8049	8059	8070	8080	·8090	2	3	5	7	9	**36**
54	·8090	8100	8111	8121	8131	8141	8151	8161	8171	8181	·8192	2	3	5	7	8	**35**
55	·8192	8202	8211	8221	8231	8241	8251	8261	8271	8281	·8290	2	3	5	7	8	**34**
56	·8290	8300	8310	8320	8329	8339	8348	8358	8368	8377	·8387	2	3	5	6	8	**33**
57	·8387	8396	8406	8415	8425	8434	8443	8453	8462	8471	·8480	2	3	5	6	8	**32**
58	·8480	8490	8499	8508	8517	8526	8536	8545	8554	8563	·8572	2	3	5	6	8	**31**
59	·8572	8581	8590	8599	8607	8616	8625	8634	8643	8652	·8660	1	3	4	6	7	**30**
60	·8660	8669	8678	8686	8695	8704	8712	8721	8729	8738	·8746	1	3	4	6	7	**29**
61	·8746	8755	8763	8771	8780	8788	8796	8805	8813	8821	·8829	1	3	4	6	7	**28**
62	·8829	8838	8846	8854	8862	8870	8878	8886	8894	8902	·8910	1	3	4	5	7	**27**
63	·8910	8918	8926	8934	8942	8949	8957	8965	8973	8980	·8988	1	3	4	5	6	**26**
64	·8988	8996	9003	9011	9018	9026	9033	9041	9048	9056	·9063	1	3	4	5	6	**25**
65	·9063	9070	9078	9085	9092	9100	9107	9114	9121	9128	·9135	1	2	4	5	6	**24**
66	·9135	9143	9150	9157	9164	9171	9178	9184	9191	9198	·9205	1	2	3	5	6	**23**
67	·9205	9212	9219	9225	9232	9239	9245	9252	9259	9265	·9272	1	2	3	4	6	**22**
68	·9272	9278	9285	9291	9298	9304	9311	9317	9323	9330	·9336	1	2	3	4	5	**21**
69	·9336	9342	9348	9354	9361	9367	9373	9379	9385	9391	·9397	1	2	3	4	5	**20**
70	·9397	9403	9409	9415	9421	9426	9432	9438	9444	9449	·9455	1	2	3	4	5	**19**
71	·9455	9461	9466	9472	9478	9483	9489	9494	9500	9505	·9511	1	2	3	4	5	**18**
72	·9511	9516	9521	9527	9532	9537	9542	9548	9553	9558	·9563	1	2	3	4	4	**17**
73	·9563	9568	9573	9578	9583	9588	9593	9598	9603	9608	·9613	1	2	2	3	4	**16**
74	·9613	9617	9622	9627	9632	9636	9641	9646	9650	9655	·9659	1	2	2	3	4	**15**
75	·9659	9664	9668	9673	9677	9681	9686	9690	9694	9699	·9703	1	1	2	3	4	**14**
76	·9703	9707	9711	9715	9720	9724	9728	9732	9736	9740	·9744	1	1	2	3	3	**13**
77	·9744	9748	9751	9755	9759	9763	9767	9770	9774	9778	·9781	1	1	2	3	3	**12**
78	·9781	9785	9789	9792	9796	9799	9803	9806	9810	9813	·9816	1	1	2	2	3	**11**
79	·9816	9820	9823	9826	9829	9833	9836	9839	9842	9845	·9848	1	1	2	2	3	**10**
80	·9848	9851	9854	9857	9860	9863	9866	9869	9871	9874	·9877	0	1	1	2	2	**9**
81	·9877	9880	9882	9885	9888	9890	9893	9895	9898	9900	·9903	0	1	1	2	2	**8**
82	·9903	9905	9907	9910	9912	9914	9917	9919	9921	9923	·9925	0	1	1	2	2	**7**
83	·9925	9928	9930	9932	9934	9936	9938	9940	9942	9943	·9945	0	1	1	1	2	**6**
84	·9945	9947	9949	9951	9952	9954	9956	9957	9959	9960	·9962	0	1	1	1	1	**5**
85	·9962	9963	9965	9966	9968	9969	9971	9972	9973	9974	·9976	0	0	1	1	1	**4**
86	·9976	9977	9978	9979	9980	9981	9982	9983	9984	9985	·9986	0	0	1	1	1	**3**
87	·9986	9987	9988	9989	9990	9990	9991	9992	9993	9993	·9994						**2**
88	·9994	9995	9995	9996	9996	9997	9997	9997	9998	9998	·9998						**1**
89	·9998	9999	9999	9999	9999	1·000	1·000	1·000	1·000	1·000	1·0000						**0**
—		54'	48'	42'	36'	30'	24'	18'	12'	6'	0'	1'	2'	3'	4'	5'	

The bold type indicates that the integer changes.

NATURAL COSINES (right-hand column and bottom)

Index

356